THE SAFE AND EFFECTIVE USE OF PESTICIDES

PESTICIDE
APPLICATION
COMPENDIUM

1

The Safe and Effective Use of Pesticides

Patrick J. Marer, Author
Senior Writer, IPM Manual Group
University of California, Davis

Mary Louise Flint, Technical Editor
Director, IPM Manual Group
University of California, Davis

Michael W. Stimmann
Statewide Pesticide Coordinator, OPIC
University of California, Davis

University of California
Statewide Integrated Pest Management Project
Division of Agriculture and Natural Resources
Publication 3324

1988

ORDERING
For information about ordering this publication, write to:

Publications
Division of Agriculture and Natural Resources
University of California
6701 San Pablo Avenue
Oakland, California 94608-1239

or telephone (415) 642-2431

Publication #3324

ISBN 0-931876-83-4
Library of Congress Catalog Card No. 87-73550

Printed in the United States of America

10m-rep-10/90

To simplify information, trade names of products have been used. No endorsement of named or illustrated products is intended, nor is criticism implied of similar products that are not mentioned or illustrated.

Acknowledgments

This manual was produced under the auspices of the University of California Statewide Integrated Pest Management Project through a Memorandum of Understanding between the University of California and the California Department of Food and Agriculture. It was prepared under the direction of Mary Louise Flint, Director, IPM Manual Group, University of California, Davis, and Michael W. Stimmann, Co-ordinator, Office of Pesticide Information and Coordination, University of California.

Production

Design and Production Coordination: Naomi Schiff
Photographs: Jack Kelly Clark
Drawings: David Kidd
Editing: Louise Eubanks/Janine Hannel

Technical Committee and Principal Reviewers

The following people provided ideas, information, and suggestions and reviewed the many manuscript drafts. Some served on an advisory committee that was formed prior to writing the book; others were key resources and reviewers in specific subject matter areas.

H. Agamalian, University of California, Cooperative Extension
M. Barnes, University of California, Riverside
J. Coburn, Western Farm Service
S. Cohen, Monterey County Agricultural Commissioner's Office
C. Elmore, University of California, Davis
D. Farnham, University of California, Cooperative Extension
M. Flint, University of California, Davis
J. Ford, Pesticide Applicators' Professional Association
T. Gantenbein, John Taylor Fertilizer Company
B. Grafton-Cardwell, University of California, Davis
J. Grieshop, University of California, Davis
P. Kurtz, California Department of Food and Agriculture
T. Lanini, University of California, Davis
J. Munro, Pest Control Operators of California
R. Norris, University of California, Davis
A. Paulus, University of California, Riverside
M. Rucker, University of California, Davis

P. Scheuber, Modesto Junior College
S. Segal, U.S. Environmental Protection Agency
M. Stimmann, University of California
S. Strew, California Agricultural Production Consultants Association
M. Takeda, California Department of Food and Agriculture
M. Tanner, Merced County Agricultural Commissioner

Special Thanks

The following persons have generously provided information, offered suggestions, reviewed draft manuscripts, or assisted in setting up photographs:

R. Adamchak, University of California, Davis
H. Alford, Pesticide Impact Assessment Program
O. Bacon, University of California, Davis
W. Bagley, Stull Chemical Company
J. Bailey, University of California, Riverside
R. Barrett, Bugman Pest Control, Inc.
S. Besemer , University of California, Cooperative Extension
J. Borieko, Tingley Rubber Corporation
L. Breschini, John Pryor Company
S. Brown, California Highway Patrol
M. Brush, University of California, Davis
K. Burroughs, Harmony Farm Supply
M. Connelly, Yoder Brothers, Inc.
M. Corodemas, San Diego County Agricultural Commissioner's Office
L. Craft, Jr., Santa Cruz County Agricultural Commissioner
A. Craigmill, University of California, Davis
V. Crawford, Wilbur Ellis Company
D. Crutchfield, Pest Control Operators of California
D. Dahlvang, Racal Airstream, Inc.
J. Dibble, University of California, Kearney Agricultural Center
S. DiGangi, Zabala Vineyards
R. Donley, Los Angeles County Agricultural Commissioner's Office
B. Enos, Western Farm Service
M. Fike, Sacramento-Yolo Mosquito Abatement District
S. Fleming, Harbor Pest Control
N. Frost, U.S. Environmental Protection Agency
T. Fukuto, University of California, Riverside
J. Garcia, Soilserve, Inc.
G. Georghiou, University of California, Riverside
K. Giles, University of California, Davis
D. Harlow, U.S. Fish and Wildlife Service
J. Hinojos, Bugman Pest Control, Inc.
G. Hurley, Western Farm Service
K. Inouye, Disneyland
R. Jackson, California Department of Health Services
T. Jacobs, Tufts Ranch
M. Jordan, Wilbur Ellis Company
B. Kennedy, Hartnell College
A. Keshner, Bugman Pest Control, Inc.
C. Koehler, University of California, Berkeley
J. LaBoyteaux
E. Laird, University of California, Riverside.

M. Lawton, Western Exterminator Company
C. Lindeman, Lindeman Valley Farm
E. Littrell, California Department of Fish and Game
E. Loomis, University of California, Davis
K. Maddy, California Department of Food and Agriculture
W. McHenry, University of California, Davis
C. McNamara, Sierra Orchards
J. Meyer, University of California, Riverside
T. Miller, University of California, Riverside
L. Miner, Soilserve, Inc.
J. Mitchell, Tingley Rubber Corporation
L. Mitich, University of California, Davis
D. Morris, University of California, Davis
B. Mullens, University of California, Riverside
R. Nelson, Nelson Manufacturing Company
R. Neumann, California Department of Health Services
H. Ohr, University of California, Riverside
R. Oshima, California Department of Food and Agriculture
M. Parrella, University of California, Riverside
J. Parker, Cherlor Manufacturing Company
T. Perring, University of California, Riverside
M. Ravera, Chevron Chemical Company
W. Reil, University of California, Cooperative Extension
L. Riehl, University of California, Riverside
T. Salmon, University of California, Davis
E. Sanders, Bugman Pest Control, Inc.
G. Sayre
D. Schulteis, Wilbur Ellis Company
D. Sites, Soilserve, Inc.
A. Slater, University of California, Berkeley
T. Stamen, University of California, Cooperative Extension
J. Strand, University of California, Davis
L. Strand, University of California, Davis
C. Swanson, Airstream Helmet Company
R. Van Steenwyk, University of California, Berkeley
N. Vasquez, Home Acres Farm
R. Wagner, University of California, Riverside
G. Walker, University of California, Riverside
B. Westerdahl, University of California, Davis
R. Winn, Mine Safety Appliances Company
W. Yates, University of California, Davis
G. Young, Y and B Ag Services
M. Zavala, University of California, Davis

Contents

1 Introduction

Pesticide application is a highly skilled occupation requiring specialized training.

Applying pesticides requires many special skills and responsibilities. It is an important occupation on its own and an indispensable part of many other occupations. As a person who applies pesticides or supervises pesticide applicators, you must be sure pesticides are handled properly and safely. It is often necessary to identify pests and then select the best methods for their control. For your own safety, as well as that of co-workers and family, it is essential to understand the hazards of pesticides and how to avoid injury. Protecting the environment is also a major concern. Additionally, you need to be familiar with all state and federal laws regulating the use, storage, transportation, application, and disposal of pesticides.

This book is Volume I of the *Pesticide Application Compendium*. It is designed to help you learn safe and effective ways of using pesticides and to show you how to reduce accidents to avoid injury and environmental problems. Should problems occur, this book describes how to handle them. Volume I also includes general information on pesticides, chemical pest control, and other pest management methods. Use it as a study guide if you are preparing for the State of California Qualified Pesticide Applicator License (QL) or Qualified Pesticide Applicator Certificate (QC) examinations. The California Department of Food and Agriculture (CDFA) gives these tests to pest control operators and their authorized agents, commercial applicators, landscape maintenance personnel, researchers, pesticide dealers and their designated agents, and anyone else applying pesticides as part of their work. The *Pesticide Application Compendium* is also a useful reference for growers, structural pest control operators, pest control advisers, pest management students, homeowners, or anyone involved in pesticide use decisions. Finally, this book is a helpful instructional guide for training people in the use of pesticides.

Other volumes of the *Pesticide Application Compendium* cover the 11 occupational areas in which pesticides are used (Table 1-1). Applicants for a Qualified Applicator Certificate or Qualified Applicator License are required to take an examination in one or more of these specialized areas. If you plan to take some of these exams, thoroughly study the sections that are appropriate to your work, without neglecting areas in which you do not plan to be tested. Areas of pesticide use often overlap and knowledge gained in one section will help to understand the concepts in another.

There are many excellent books and publications dealing with pesticide use, how pesticides work, effects of pesticides on the environment, and other related topics. Some of these are listed in the References section at the end of this volume; use this as a reading list for additional information on the topics presented in each chapter. The Glossary includes definitions of technical terms and expressions that are used in pesticide application. An Index is also included to help locate information quickly.

In addition to this text and the references cited here, there are two important sources you should rely on for information regarding pesticides:

County agricultural commissioners are regulatory officials of the California Department of Food and Agriculture (CDFA). Their offices throughout the state have the responsibility, among other functions, for issuing permits for restricted pesticides, monitoring pesticide use, storage, and disposal, and enforcing pesticide worker safety regulations. Agricultural commissioners' offices provide local information on pesticide use, storage, transportation, disposal, and hazards. They should be contacted in the event of any pesticide emergency.

The University of California, through its Cooperative Extension Program, maintains offices in most counties of the state. Cooperative Extension offices are staffed by specialists who provide pest identification, pest management, and pesticide use information for home, structural, agricultural, livestock and poultry, rangeland, wildlife, turf and landscape, forest, and aquatic areas. Farm advisors work closely with other University of California researchers and specialists.

TABLE 1-1

Specialized Areas of Pesticide Application

SPECIALIZED AREA	DESCRIPTION OF PESTICIDE APPLICATOR'S WORK	TYPE OF PESTS
Residential, Industrial, and Institutional Pest Control	Performs pest control in apartments, restaurants, hospitals, offices, warehouses, grocery stores, and other similar buildings as part of employment by owner or operator of the building. Selects pest control methods and pesticides to use. Performs postharvest fumigation and insecticide and fungicide applications to agricultural products. Applies pesticides to stored agricultural products. Controls weeds around commercial and industrial structures. Work is often closely associated with people and their pets. A special subsection of this category, which requires a separate examination, relates to the application of pesticides for preservation of wood products such as lumber, posts, and other structural wood.	*Invertebrates:* cockroaches, bugs, stored product pests, flies, fleas, mosquitoes, termites, ants, other insects; spiders and mites. *Vertebrates:* rats, mice, bats, and birds. *Weeds.* *Microorganisms:* wood-decaying fungi.
Landscape Maintenance Pest Control	Controls pests on or around ornamental and fruit trees, shrubs, small fruits and berries, turf, and flowers; works around homes, businesses, cemeteries, theme parks, public parks, indoor malls, and house plants. Pesticide application is often part of a landscape maintenance business. Applicator makes decisions regarding pest control methods, irrigation, and plant nutrition. Work is closely associated with human activities.	*Invertebrates:* aphids, scales, flies, bees, wasps, earwigs, moths, beetles, and bugs; spiders, mites, and centipedes. *Vertebrates:* rats, mice, gophers, moles, squirrels, rabbits, birds, snakes, lizards. *Microorganisms:* fungi, bacteria, viruses. *Weeds:* various types of terrestrial weeds.

SPECIALIZED AREA	DESCRIPTION OF PESTICIDE APPLICATOR'S WORK	TYPE OF PESTS
Right-of-Way Pest Control	Performs pesticide applications along roads, rail lines, utility accesses, and drainage ditches to keep these areas free of undesirable weeds, to prevent fire hazards and obstruction of access or view. Applies pesticides for control of vertebrates and insects that interfere with desirable foliage or water drainage. A special subsection of this category, which requires a separate examination, relates to the application of pesticides for the preservation of wood products and utility poles.	*Weeds*: various types of terrestrial weeds. *Vertebrates*: Squirrels, mice, gophers, moles, rabbits, birds. *Invertebrates*: pests of foliage and wood products.
Forest Pest Control	Applies pesticides in forest locations. Is responsible for protecting wildlife, watershed, and lakes and streams.	*Invertebrates*: boring and defoliating insects of forest trees; mites. *Weeds*: mostly undesirable plant species competing with forest trees, parasitic plants. *Microorganisms*: plant disease agents affecting forest trees. *Vertebrates*: squirrels, voles, gophers, and others.
Aquatic Pest Control	Applies pesticides for control of aquatic weeds, pest fish, arthropods, and molluscs. Requires special skills to protect aquatic environments and nontarget organisms. Familiarity with aquatic ecosystems and the ultimate use of water is very important to protect people and crops.	*Aquatic weeds. Pest Fish. Invertebrates.*
Plant Agriculture Pest Control	Applies pesticides in and around agricultural crops. Often employed by a commercial applicator. Usually supervises equipment operators. Responsible for protecting field-workers, groundwater, and environment. May work with highly toxic materials.	*Invertebrates*: many different agricultural pest insects and mites, snails, nematodes. *Vertebrates*: squirrels, gophers, rabbits, birds. *Weeds*: many types of agricultural weeds and poisonous plants. *Microorganisms*: fungi, bacteria, and viruses that cause crop diseases.

SPECIALIZED AREA	DESCRIPTION OF PESTICIDE APPLICATOR'S WORK	TYPE OF PESTS
Animal Agriculture Pest Control	Applies pesticides for control of livestock and poultry pests. Requires familiarity with livestock and poultry and unique pest control techniques. Pesticide use is closely associated with animals.	*Invertebrates*: mosquitoes, lice, flies, and bugs; ticks and mites. *Vertebrates*: livestock and poultry predators. *Weeds*: poisonous plants and undesirable range weeds.
Seed Treatment	Performs or supervises the application of insecticides and fungicides to seeds used to produce agricultural crops. Requires familiarity with different methods of protecting seeds. Usually employed by a seed treatment company.	*Invertebrates*: seed feeding or damaging insects. *Microorganisms*: fungi and bacteria.
Regulatory Pest Control	Involved in the detection and eradication of imported pests that pose threats of economic harm to agriculture, livestock and poultry, or other segments of society. Must be familiar with suppression and eradication methods. Requires understanding of ways pests enter and disperse through an area. Usually works for a public agency.	*Invertebrates*: exotic insects and mites that threaten to cause economic or health damage. Nematodes and snails — damaging species that might be introduced from other areas. *Weeds*: aquatic, terrestrial, and exotic weeds. *Vertebrates*: reptiles, birds, and rodents and other mammals. *Microorganisms*: exotic plant disease organisms.
Demonstration and Research Pest Control	Evaluates pesticides for efficacy. Studies interactions between pests, nonpests, and environmental factors when pesticides are applied. Demonstrates proper and effective methods of using pesticides. May be pesticide chemical company field representative, farm advisor, university researcher, independent consultant, or contract researcher.	All types of agricultural and nonagricultural pests may be involved.
Public Health Pest Control	Involved in applying pesticides to control pests that transmit disease organisms to people. Usually employed by public agencies. Pesticide use is often closely associated with homes and workplaces.	*Invertebrates*: flies, fleas, cockroaches, mosquitoes, lice, bugs; ticks, mites, and spiders. *Vertebrates*: rats, mice, bats, birds,

Pest
Identification

Close examination may be necessary to identify pests which are causing damage.

Pests are organisms that compete with people for food or fiber, interfere with raising crops or livestock and poultry, damage property and personal belongings, disfigure ornamental plantings, transmit or cause plant or animal diseases, or are otherwise bothersome. Before trying to control a pest, it must be properly identified. Be certain that any injury or observed damage is actually due to the identified pest and not some other cause. Once you have identified the pest and confirmed that it is causing damage, become familiar with its life cycle, growth, and reproductive habits. Use this information to form your pest control plans. Misidentification and lack of information about a pest lead to use of improper control methods or incorrect control timing and are among the most frequent causes of pest control failure.

Four main groups of pests include: (1) weeds (undesired plants); (2) invertebrates (insects and their relatives, nematodes, snails, and slugs); (3) vertebrates (birds, reptiles, amphibians, fish, rodents, and other mammals); and (4) disease agents (bacteria, viruses, fungi, mycoplasmas, other microorganisms, and nonliving factors). This chapter reviews some of the fundamentals used to identify pests, including how to use identification aids.

HOW PLANTS AND ANIMALS ARE NAMED

Scientific Names

Classification systems are sets of rules used for organizing and naming living things. An elegant, standardized classification system used throughout the world is the basis for the scientific names given to plants and animals. This system reveals how different plants and animals are related. Scientific names are very useful when trying to locate information about an organism.

Living organisms are usually included in one of two major groups: the *Plant Kingdom* or the *Animal Kingdom*. Usually it is easy to distinguish between living organisms and nonliving objects and between plants and animals; however, microorganisms and algae are more difficult to classify because they may have characteristics that make them intermediate between plants and animals.

There are six subcategories within a kingdom in the typical classification system: *Phylum, Class, Order, Family, Genus,* and *Species.* Organisms are separated according to unique characteristics that set them apart from other organisms. For example, in the phylum Arthropoda all organisms have jointed appendages and an external skeleton, while animals in the phylum Chordata have a backbone, a spinal nerve cord, and an internal skeleton. Within a phylum there are several orders, each containing one or more families. A family is a group of related genera, and a genus is a collection of species. A species is unique from all other organisms, although there may be genetic variations among individuals in color and size differences, ability to attack a specific cultivar of a crop, or even in the ability to resist the effects of certain pesticides. The genus and species names form an organism's scientific name.

Common Names

Besides a scientific name, most plants and animals have one or more common names. Pesticide labels refer to most organisms by their common names. Common names are usually descriptive, such as "house fly," "American cockroach," "roof rat," "field bindweed," "giant foxtail," "apple scab," and "fireblight." The disadvantage of using common names is that they do not provide any information about the relationship of one organism to another. For example, johnsongrass (*Sorghum halepense*)—a weed, Sudangrass (*Sorghum sudanense*)—a forage plant, and grain sorghum (*Sorghum bicolor*)—a grain, belong to the genus *Sorghum*, but their common names do not indicate any relationship. Because common names vary with locality or host, an organism will often have more than one common name. The insect *Heliothis zea*, for example, is known as the corn earworm, bollworm, and tomato fruitworm, depending on which crop it is attacking (Figure 2-1). Organizations such as the Weed Science Society of America and the Entomological Society of America have developed lists of accepted common names that are being used to avoid confusion.

  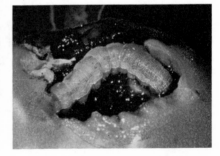

FIGURE 2-1.

Common names often are not a dependable way to identify pests, because some may have more than one common name. The Heliothis zea *shown here, for example, is known as the corn earworm, bollworm, and tomato fruitworm, depending on which crop it is infesting.*

WAYS TO IDENTIFY A PEST

You can identify a pest either by using the guidelines included in this chapter and then consulting identification books, or by having the pest examined and identified by a specialist. Always collect several specimens of the pest, because different species often look similar.

Certain insects and most mites, nematodes, and plant pathogens, are too small to be readily identified in the field. Their accurate identification requires the use of microscopes, special tests, or careful analysis of damage. Often the pest's host association and location are important to making positive identifications. The environmental conditions of the area where the pest was collected, as well as the time of year of collection, may also provide clues to the pest's identity.

Pest species may have different physical forms, depending on their life cycle or time of year. Weed seeds, for example, do not resemble a seedling or mature weed; many insect species undergo extreme changes in appearance as they develop from eggs through larval, pupal, and adult stages.

Identification Experts

Some pests, such as nematodes and most pathogens, can be positively identified only by trained experts using special techniques and equipment. Private laboratories are available to identify nematodes, mites, insects, plant pathogens, and often other pests. Farm advisors in each county have expertise in pest identification and are also in close contact with other University of California experts. County agricultural commissioners and their staff are helpful resources since they are associated with the California Department of Food and Agriculture, which maintains a pest identification laboratory. Many pest control companies have licensed pest control advisers and other experts on their staff that can identify some types of pests.

When sending samples for identification, be sure that the material is maintained fresh and undamaged. Provide complete information on where the pest was found and, if appropriate, include examples of pest damage. Weeds, pathogens, and arthropods all require different types of sampling and handling. Tables in the following sections provide instructions on how to collect and prepare specimens for shipment to an expert or identification laboratory.

Identification Keys

Identification keys provide descriptive clues to the identity of living organisms; use them when trying to identify weeds or arthropods. Many keys have been developed by experts for use by other experts, so they sometimes may be difficult to understand unless the user is familiar with the terms used to describe the pest's structures. Simple keys are usually available for common pests, however. *Dichotomous keys* consist of a series of sequentially paired statements. Select the statement that best fits the pest being identified and you will be directed to another pair of statements. Then continue working through the key in this manner until the pest's identity is revealed. Physical features are used in dichotomous keys for identification of plants or animals. Characteristics such as color or size may also be used, especially with weeds. Sometimes photographs or drawings are included to help illustrate features referred to in the key. Table 2-1 is an example of a dichotomous key.

Photographs and Drawings

Whenever possible, refer to photographs and drawings for identification because they provide a visual description of the pest (Figure 2-2) and of its damage. Unique or distinguishing features of a pest are easily illustrated. Photographs and drawings can be found in publications such as the California Department of Food and Agriculture *Vertebrate Pest Control Handbook*, the University of California *Growers' Weed Identifica-*

tion Handbook, the U.C. *Wildlife Pest Control Around Gardens and Homes*, the U.C. Integrated Pest Management Manuals, and other pest management publications, textbooks, and field guides.

Preserved Specimens

Plants, insects and other arthropods, reptiles, and mammals are commonly preserved for study and comparison (Figure 2-3); often these specimens have been identified by experts. Museums and herbaria at universities or other teaching and research institutions are the most common locations for large collections of preserved specimens, although you can purchase individual specimens and small collections of more common household and structural pests. Collections of weed seeds are

TABLE 2-1

Example of a Dichotomous Key.

KEY TO COMMON ADULT COCKROACHES*	
1a. Small, about ⅝" or shorter	2
1b. Medium to large, longer than ⅝"	4
2a. Pronotum without longitudinal black bars	3
2b. Pronotum with 2 longitudinal black bars	German Cockroach *Blatella germanica*
3a. Wings covering about half of abdomen. Pronotum about ¼" wide	Wood Roach *Parcoblatta* spp.
3b. Wings covering nearly all of abdomen or extending beyond. Pronotum narrower	Brownbanded Cockroach *Supella supellictilium*
4a. Wings covering abdomen, often extending beyond	5
4b. Wings absent or shorter than abdomen	Oriental Cockroach *Blatta orientalis*
5a. Pronotum more than ¼" wide	6
5b. Pronotum about ¼" wide with pale border	Wood Roach *Parcoblatta* spp.
6a. Front wing without pale streak. Pronotum solid color, or with pale design only moderately conspicuous	7
6b. Front wing with outer pale streak at base. Pronotum strikingly marked	Australian Cockroach *Periplaneta australasiae*
7a. Pronotum usually with some pale area. General color seldom darker than reddish chestnut	8
7b. Pronotum solid dark color. General color very dark brown to black	Smokybrown Cockroach *Periplaneta fuliginosa*
8a. Last segment of cercus not twice as long as wide	Brown Cockroach *Periplaneta brunnea*
8b. Last segment of cercus twice as long as wide	American Cockroach *Periplaneta americana*

*Adapted from: "Pictorial Key to Some Common Adult Cockroaches" by H. D. Platt, Department of Health, Education, and Welfare, Public Health Service Communicable Disease Center, Atlanta, GA, October, 1953.

available which are helpful identification aids. Most preserved material is fragile, so handle it carefully; store specimens in a safe place to prevent damage and deterioration.

Characteristic Signs

Pest animals may leave signs of their presence that can help you discover what they are. Birds and rodents build nests which are often characteristic to each species. Many insects can be identified by the type of feeding damage they cause. Rodents dig unique burrows in the ground, and often leave identifying gnaw marks on tree trunks or other objects; sometimes trails in grass or tracks in dirt are helpful clues to rodent identification. Rodent fecal pellets and insect frass are also distinctive and important identification aids. Weeds may have unique flowers, seeds, fruits, or unusual growth habits. Also look for remains of weed plants from the previous season. Fungi and other pathogens sometimes cause specific types of damage, deformation, or color changes to host tissues.

FIGURE 2-2.

Photographs such as these of a cabbage looper egg and larva can show unique physical characteristics or coloration patterns that are useful in the identification of pests.

Weeds

Weeds are plants that interfere with the growing of crops or ornamental plants, endanger livestock, affect the health of people, interfere with the safety or use of roads, utilities, and waterways, or are visual or physical nuisances. Grasses, broadleaved herbaceous plants, shrubs, and even trees are considered weeds if they interfere with the activities of people (Figure 2-4).

Weeds of major importance possess special characteristics that distinguish them from the occasional out-of-place plant. They adapt well to local climates, soils, and other external conditions and can compete successfully with cultivated plants for available resources. Most weeds produce large quantities of seeds, even under adverse conditions. Seeds

FIGURE 2-3.

Preserved specimens, such as these mounted insects, can aid in pest identification. These are the adults of (a) the variegated cutworm, (b) the beet armyworm, (c) the cabbage looper, (d) the tomato fruitworm, (e) the tobacco budworm, and (f) the western yellowstriped armyworm.

FIGURE 2-4.

Weeds are plants that interfere with the growing of agricultural or landscape plants or that endanger the health or safety of people or animals. This photo shows an infestation of field bindweed in a tomato field.

of some weeds often lie dormant in the soil for extended periods, sometimes 20 or more years, before germinating. Some weed seeds or fruits are specially adapted to promote dispersal (Figure 2-5). Because many weeds are capable of reproducing through vegetative structures, cultural activities, such as hoeing, mowing, or discing, may result in the production of new plants (Figure 2-6). Consequently, weeds can be persistent and difficult to eliminate.

Only about 3% of the identified plant species found in the world are considered weeds in agricultural, turf, or landscape settings. Usually, there are no more than 25 to 35 weed species at any one crop or landscape site. Groups of different weed species are commonly associated with specific types of crops or landscape, right-of-way, aquatic, or forest settings. Weeds associated with crops or other cultivated areas are usually the ones most difficult to control with herbicides because they are often so similar to the cultivated plants. Some of the major weed families are discussed and illustrated at the end of this section.

Although some weeds are native plants, most were brought inadvertently or intentionally from foreign countries. In their new location, imported weeds can become serious pests because natural enemies or diseases are not present to suppress them. Often local environments are ideal for their growth.

How Weeds Are Pests

Weeds compete with agricultural crops for water, nutrients, light, and space; they may also interfere with farming operations. Some weed species are toxic to livestock, while others release toxins into the soil that inhibit the growth of other plants. Uncontrolled weeds can contaminate products at harvest and harbor insects and pathogens.

Around homes, businesses, and industry, heavy populations of weeds often infest lawns and groundcovers and ornamental plantings. Weeds sometimes harbor undesirable insect or vertebrate pests and detract from the appearance of landscaping. Dry weeds can be fire hazards. Various weed pollens cause allergies in some people and certain weeds produce skin irritation. Some weeds are poisonous to people and animals.

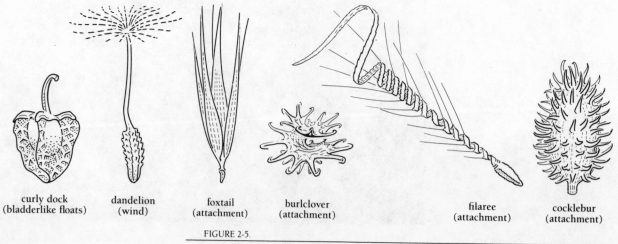

curly dock
(bladderlike floats)

dandelion
(wind)

foxtail
(attachment)

burclover
(attachment)

filaree
(attachment)

cocklebur
(attachment)

FIGURE 2-5.

Weed seeds such as those shown here have many adaptations enabling them to disperse. These special characteristics are among some of the ways weeds compete with other plants.

FIGURE 2-6.

Cultural activities such as discing can disperse some weeds because they break plants up into smaller segments that can reroot and form new plants.

Along roadsides and rights-of-way, large weeds interfere with travel and maintenance operations. Tall roadside weeds can obscure signs, while creeping species may clog road drains or erode pavement edges. Dense patches of weeds may provide shelter and food for rodents. In commercially operated forests, weeds may compete with tree seedlings, become fire hazards, interfere with normal cultural activities, and in some instances (like mistletoe) parasitize trees. Overgrown aquatic weeds clog irrigation canals and ditches, streams, rivers, and lakes, causing harm to native aquatic life and making the use of waterways difficult for people and animals. Large infestations of aquatic weeds also hinder fish growth and reproduction and promote mosquito problems by decreasing normal water flow and wave action.

Identifying Weeds

A simple way to begin to identify some common weeds is to compare specimens with color photographs and drawings. Use publications such as those listed in the "References." If you are unable to determine the species from these sources, it may be necessary to use identification keys or to compare the weeds with preserved specimens. Sometimes weed plants need to be sent to an expert for identification (Table 2-2).

Specialists identify plants, including weeds, by recognizing differences and similarities between flowers, leaves, stems, and roots; fruits, seeds, and special structures are also useful identification characteristics, as are the plant's growth habits. To identify weeds yourself, become familiar with the plant classification system and the plant's physical features, developmental stages, and life cycles.

Weed Classification. Most weeds fall within two major land plant groups, the *dicots* and the *monocots*. Dicots, often called broadleaves, are plants that produce two seedling leaves (cotyledons). Leaves of dicots usually have netlike veins. Dicots generally grow as *herbaceous* plants (leafy and herblike), or as *woody* plants (shrub- or treelike). Monocots are plants which produce only a single grasslike leaf in the seedling; leaves typically have veins which run parallel to their long axes. Grasses, sedges, and rushes are monocots.

Mosses (and liverworts) belong to a unique group of plants known as bryophytes. Mosses are different from the land plants because they lack a vascular system. They are occasional pests in aquatic settings and sometimes cause problems when growing in greenhouses or on buildings or ornamental plants.

Algae, nonflowering primitive aquatic plants that often clog streams, lakes, drainage ditches, and rice fields, are usually included among weed pests. Like more highly developed plants, algae carry out photosynthesis to convert light into energy. However, algae lack true stems, leaves, and flowers that are characteristic of higher plants. They reproduce through cell division or production of spores. Algae are discussed in more detail in Volume 4.

TABLE 2-2

Guidelines for Sampling and Sending Weeds for Weed Identification.

SAMPLING:
1. Choose several plants that represent the species.
2. Include stems, leaves, flowers (if present), and roots.
3. Dig up weeds to prevent damage to roots.
4. Shake plants lightly after digging to remove excess soil.

PREPARATION:
1. Keep plants in an ice chest while you are in the field. If they cannot be shipped immediately, store them in a refrigerator.
2. Place plants in plastic bags without moisture, or press them between sheets of absorbent paper and encase in heavy cardboard for protection.

LABELING:
Attach a label to the outside of each sample. Include the following information on labels:
1. Location where samples were taken, including names of nearby crossroads.
2. Description of specific characteristics of the site where the weeds were growing.
3. Whether plants are annuals or perennials.
4. Your name, address, and telephone number.
5. Date samples were taken.
6. Any other information that would help in the identification of the weeds.

SHIPPING:
1. Contact the person or laboratory who will receive samples to determine the best method of shipping and to inform them that samples will be arriving.
2. Pack samples in a sturdy, well-insulated container to prevent crushing or heat damage.
3. Mark package clearly and request shipper to keep it in a cool location.
4. Ship packages early in the week so they will arrive before a weekend.

In weed identification, it is usually easy to determine if the weed is a land plant, moss, or alga. With land weeds, you must distinguish between a woody or herbaceous dicot (broadleaf) or a monocot (grass, rush, or sedge).

Weed Development. Most weeds pass through several stages of development, beginning with the *seed*. Seeds are one of the ways weed plants disperse, and are the form in which many weed species can survive periods of stress such as cold, heat, drought, or competition. Sprouted seeds are known as *seedlings*; these are usually tender and vulnerable to environmental extremes and often quite susceptible to herbicides. Seedlings differ in appearance from mature plants and so may be difficult to recognize. Because they are the stage most readily controlled, it is necessary to become familiar with seedling identification.

From seedlings, weeds continue their *vegetative growth* stage marked by rapid foliage development and maximum growth. Plants then enter a *reproductive* period in which most energy is diverted for flowering and seed production. Weeds reach *maturity*, a post-reproductive period, once seeds have been formed. Perennial plants continue to repeat vegetative growth and reproductive cycles each year.

For identification, learn how to recognize the different growth stages of weeds. Understanding the growth stages is also important when selecting herbicides or other methods of weed control.

Weed Life Cycles. Weeds have either annual, biennial, or perennial life cycles. Occasionally, some weed species with one type of life cycle may behave as another type due to favorable weather conditions or abnormal or unusual changes in environmental influences; milder temperatures, for example, often promote longer life cycles.

Annuals: Annual weeds live one year or less. They sprout from a seed, mature, and produce seeds for the next generation during this period. Annuals are divided into two groups, *summer annuals* and *winter annuals* (Figure 2-7). Seeds of summer annuals sprout in the spring and plants produce seed and die during the summer or fall. Some common summer annual weeds include pigweed, puncturevine, barnyardgrass, Russian thistle, common purslane, and yellow foxtail. Seeds of winter annuals sprout in the fall; they grow over the winter, reproduce in the spring, but usually die before summer. Mustard, wild oats, annual bluegrass, burclover, and filaree are examples of winter annual weeds.

Biennials: Biennials are plants that live for two growing seasons. These sprout and undergo vegetative growth during the first season, then flower, produce seed, and die the following season. Bristly oxtongue, poison hemlock, wild carrot, common mullein, and scotch thistle are biennials.

Perennials: Perennial weeds live 3 or more years; some species live indefinitely. Many perennials lose their leaves or die back entirely during the winter (herbaceous perennials), but regrow each spring from roots or underground storage organs such as tubers, bulbs, or rhizomes. These storage organs provide the chief means of dispersal for a number of major perennial weeds as well. Examples of perennial weeds include curly dock, silverleaf nightshade, field bindweed, alkali sida, dandelion, yellow nutsedge, poison oak, johnsongrass, and bermudagrass. Woody plants such as trees and shrubs are perennials and are considered weeds under certain circumstances. Perennial weeds are the most difficult type to control with herbicides, cultural controls, or mechanical methods.

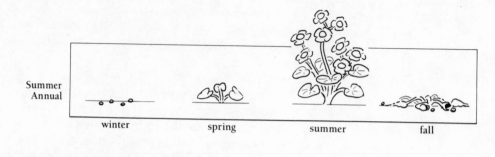

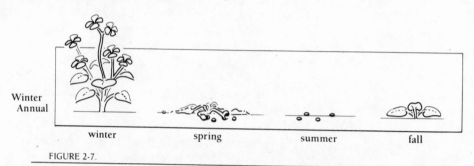

FIGURE 2-7.

This illustration shows the difference in the growth periods of winter and summer annual weeds.

Learn the life cycle of weeds to distinguish between annuals, biennials, and perennials. Once you know the life cycle, you will be able to select and time herbicide applications properly. For example, most perennial weeds are not susceptible to herbicides during early bloom stages.

Physical Features of Weeds. Weeds and most other plants have unique physical features which can be used for identification. These include flowers, stems, roots, and fruits and seeds.

Flowers. Flowers contain sexual reproductive organs and differ widely among species; these differences are helpful in weed identification. To use flowers as an identification aid, you must be familiar with the different flower parts (Figure 2-8). Flowers occur singly or as compound *inflorescences*. Inflorescences are groups of flowers arising from a common main stem. Different names, such as *panicle*, *raceme*, *spike*, *head*, *umbel*, and *catkin*, describe how flowers are arranged in an inflorescence (Figure 2-9).

Leaves. The arrangement and shape of leaves help in weed identification, as well as vein patterns and the presence of spines or hairs. When identifying seedlings, leaves are the only easily visible distinguishing structures. Cotyledons and occasionally the first true leaves of broad-leaved plants differ from the plant's mature foliage. Grasses are distinguished by the collar region of the plant where leaves separate from the main stem.

Stems. Stems form the weed's basic framework or skeleton and connect roots to other structures, providing support for leaves and flowers and channels for nutrient and water transport. Specialized stem modifications enable some weeds to reproduce vegetatively (Figure 2-10). *Rhizomes* are elongated underground stems that grow horizontally from the plant. *Tubers* also grow underground and are enlarged fleshy growths arising from stems or roots. A *stolon* is a stem that grows horizontally above the ground surface and roots at nodes.

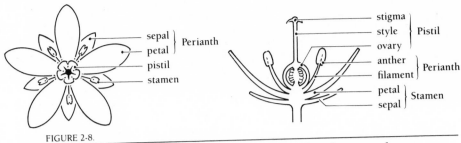

FIGURE 2-8.

Flowers have important parts that are useful in identifying weeds.

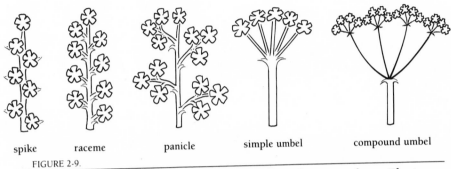

spike   raceme   panicle   simple umbel   compound umbel

FIGURE 2-9.

Flowers in infloresences may be arranged differently as seen here. The type of infloresence can often be used in weed identification.

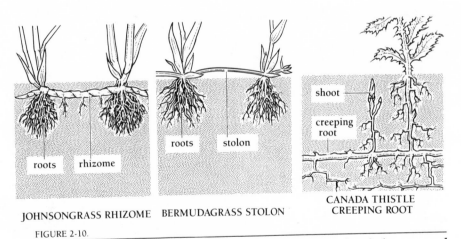

JOHNSONGRASS RHIZOME   BERMUDAGRASS STOLON

CANADA THISTLE CREEPING ROOT

FIGURE 2-10.

Many weeds can reproduce by several types of underground and aboveground rooting structures such as those illustrated here.

Roots. Roots absorb water and nutrients from the soil and store food. The creeping roots of some plants give rise to stems. Some weeds have a thick, elongated *taproot* from which short lateral rootlets arise (dandelions, for example). Other weeds have a network of fine branching *fibrous* roots as seen in grasses. Roots provide helpful clues in weed identification.

Fruits and Seeds. A fruit is the ripened ovary of a plant's flower. Seeds are the primary way weeds reproduce. Fruits and seeds are useful structures for identifying weeds, being unique in their shape, size, markings, and color.

Algae

Important Characteristics. Algae are primitive plants closely related to some fungi and protozoans. They reproduce by means of spores, cell division, or fragmentation. There are more than 17,000 identified species of algae. Pest algae fall within three general groups: planktonic or microscopic, filamentous, and attached-erect (Figure 2-11). Microscopic algae impart greenish or reddish colors to water. They may be seen as scums floating on the water surface. Filamentous algae form dense free-floating or attached mats. Attached erect algae, known as *Chara* and *Nitella*, resemble flowering plants with leaflike and stemlike structures. Algae clog irrigation channels, waterways, and ponds, render swimming pools unsightly, and plug irrigation equipment. Large algal buildup may cause oxygen depletion within a body of water and be responsible for fish death. Some forms release toxins into water as they decompose, which may cause poisoning of people or livestock.

Where Found. Algae occur in swimming pools, ponds, lakes, streams, rivers, and other bodies of water. They also become pests in irrigation canals. Some forms of algae are problems in flooded rice fields.

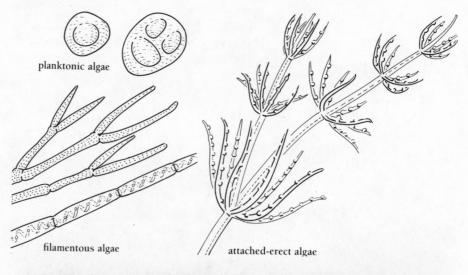

planktonic algae

filamentous algae

attached-erect algae

FIGURE 2-11.

There are three types of algae, planktonic, filamentous, and attached-erect. (Drawings are greatly enlarged.) Algae often clog waterways, as seen in the photograph.

Sedge Family

Important Characteristics. Sedges (Figure 2-12) are often perennial plants. They are grasslike, have fibrous root systems, and perennial species produce rhizomes or tubers. Elongated leaves are "V" shaped and arise from 3-sided stems. The shape of the stems of sedges distinguish them from grasses.

Where Found. Sedges are pests in orchards, vineyards, irrigated crops, and home gardens. They cause severe problems in rice fields. Sedges usually occur in marshy or poorly drained areas and along ditch and pond edges.

Examples. Yellow nutsedge or "nutgrass" (*Cyperus esculentus*), purple nutsedge (*C. rotundus*), blunt spikerush (*Eleocharis obtusa*), hardstem bulrush (*Scirpus acutus*), and river bulrush (*S. fluviatilis*) are examples of sedges.

FIGURE 2-12.

Sedges are grasslike plants with a fibrous root system, usually found in marshy or poorly drained areas.

Grass Family

Important Characteristics. Grasses (Figure 2-13) are a large family of annual or perennial plants that include many notable weeds as well as important cultivated crops such as grains. Some species provide substantial food sources for grazing livestock. Roots of grasses are dense and fibrous; several species reproduce by underground rhizomes. A major feature used in grass identification is the collar region (Figure 2-14).

FIGURE 2-13.

Grasses, like the wild oats shown here, are one of the largest families of important weeds.

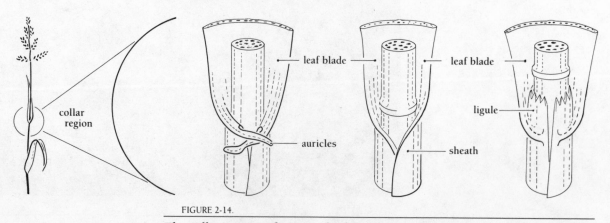

FIGURE 2-14.

The collar region of a grass leaf contains unique structures that are very important in identifying grass species.

A unique structure known as the *ligule* occurs at the collar region in many species of grasses. Several important grassy weeds are winter annuals; wild oats, for example, is one of the most widely distributed, troublesome winter annual weeds in California.

Where Found. Grassy weeds can be found in most cultivated and natural areas. They are often pests in fields, pastures, rangelands, orchards, vineyards, landscaped areas, turf, along roadsides and ditchbanks, and other locations.

Examples. Grass weeds include wild oats (*Avena fatua*), red brome (*Bromus rubens*), bermudagrass (*Cynodon dactylon*), smooth crabgrass (*Digitaria ischaemum*), barnyardgrass (*Echinochloa crus-galli*), foxtail fescue (*Festuca megalura*), deergrass (*Muhlenbergia rigens*), dallisgrass (*Paspalum dilatatum*), annual bluegrass (*Poa annua*), yellow foxtail (*Setaria glauca*), and johnsongrass (*Sorghum halepense*).

Mallow Family

Important Characteristics. Mallows are annual or perennial broad-leaved weeds, depending on the species (Figure 2-15). Most are resistant to many of the widely used herbicides, making them persistent pests. Most species are herbaceous plants that grow from 1/2 to 7 feet tall. They produce capsule or disclike fruits that enclose several seeds. Cotyledons are roundish to heart-shaped or pear-shaped. Leaves are usually round with serrated edges and have a characteristic palmate venation.

Where Found. Weeds in the mallow family are pests in annual crops, orchards, vineyards, and along roadsides, ditch banks, and in waste areas.

Examples. Venice mallow (*Hibiscus trionum*), velvetleaf (*Abutilon theophrasti*), little mallow (*Malva parviflora*), and alkali sida (*Sida hederacea*) are representative weeds of this family.

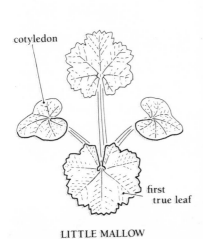

cotyledon

first true leaf

LITTLE MALLOW

FIGURE 2-15.

Malva parviflora – mallow family.

Mustard Family

Important Characteristics. Mustards are usually upright broadleaved weeds that grow from 1 to 5 feet tall, depending on the species (Figure 2-16). Many have yellow flowers. Seedlings usually have broad cotyledons which, in some species, are kidney shaped or indented at the tip. They have annual or biennial life cycles.

Where Found. Mustards occur in fields, orchards, vineyards, pastures, along roadsides and ditchbanks, and in vacant lots and waste areas.

Examples. Common weedy mustards include black mustard (*Brassica nigra*), birdsrape mustard (*Brassica rapa*), shepherdspurse (*Capsella bursa-pastoris*), hoary cress (*Cardaria draba*), wild radish (*Raphanus sativus*), and London rocket (*Sisymbrium irio*).

cotyledon

first
true
leaf

BIRDSRAPE MUSTARD
FIGURE 2-16.

Brassica rapa – *mustard family*

Goosefoot Family

Important Characteristics. Weeds of the goosefoot family range from annuals to perennials (Figure 2-17). Plants are variable in size depending on the species and where they are growing. Some reach heights of 6 feet; other species range in height between 8 and 24 inches. Mature leaves are often notched. Leaves are commonly tinged with purple. Cotyledon leaves of many species are elongated, being 4 to 6 times longer than their width.

Where Found. Weeds in the goosefoot family are widespread and found in agronomic, horticultural, and vegetable crops. They grow abundantly in waste places and along roadsides, irrigation, and drainage ditches.

Examples. Weeds in the goosefoot family include Australian saltbush (*Atriplex semibaccata*), common lambsquarters (*Chenopodium album*), nettleleaf goosefoot (*C. murale*), and Russian thistle (*Salsola iberica*).

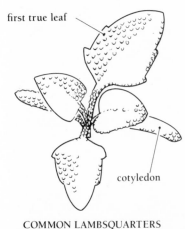

first true leaf

cotyledon

COMMON LAMBSQUARTERS
FIGURE 2-17.

Chenopodium album –
goosefoot family.

Amaranth Family

Important Characteristics. Amaranths are mostly upright herbaceous plants; some species reach 8 feet in height (Figure 2-18). They also include a low-growing prostrate species. The amaranths have small, inconspicuous greenish flowers. Cotyledons are narrow and elongate, and 4 to 5 times as long as they are wide. Seedling leaves are dull green to reddish on upper surfaces and magenta to bright red underneath. Some amaranths can cause nitrate poisoning in livestock under certain environmental conditions.

Where Found. Amaranths are pests in most cultivated crops, orchards, and vineyards. They are frequently found along ditch banks and roadsides and occur in waste areas.

Examples. Amaranths include tumble pigweed (*Amaranthus albus*), prostrate pigweed (*A. blitoides*), green amaranth (*A. hybridus*), and redroot pigweed (*A. retroflexus*).

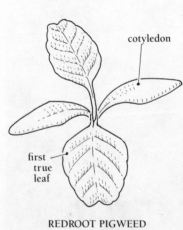

cotyledon

first true leaf

REDROOT PIGWEED

FIGURE 2-18.

Amaranthus retroflexus –
amaranth family.

Morningglory Family

Important Characteristics. Weeds in the morningglory family include annuals and perennials. These are low-growing vines (Figure 2-19). One species, field bindweed, is considered the most difficult to control perennial broadleaf weed in California. Cotyledons of plants in this family are large and roundish and are notched on the end. First true leaves are triangular.

Where Found. Weeds of the morningglory family occur in all agronomic and vegetable crops as well as along roadsides and ditchbanks.

Examples. Field bindweed (*Convolvulus arvensis*), ivyleaf morningglory (*Ipomoea hederacea*), and tall morningglory (*I. purpurea*) are examples of important weeds in the morningglory family.

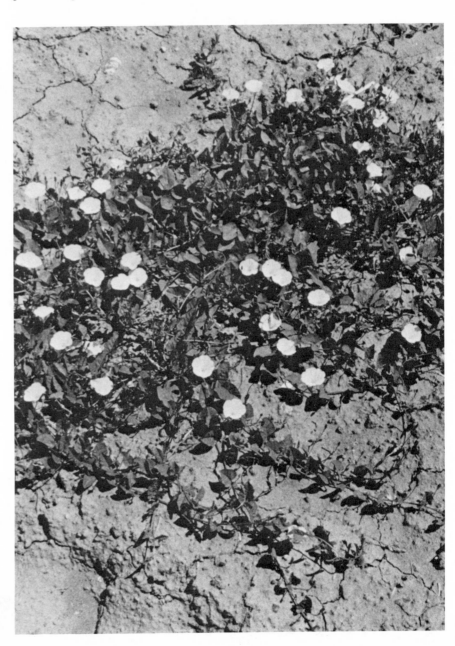

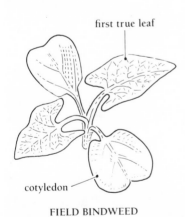

first true leaf

cotyledon

FIELD BINDWEED

FIGURE 2-19.

Convolvulus arvensis –
morningglory family.

Potato Family

Important Characteristics. Weeds in the potato family consist of a large group of pests. Most are low-growing, bushy plants reaching heights of about 3 feet (Figure 2-20). Tree tobacco reaches 12 feet. Cotyledons are longer than they are wide, sometimes by as much as 8 to 10 times. They are gently tapered to a point at the end. Some species produce thorny seedpods, while others produce rounded berries with many seeds. Several species are poisonous to people and livestock; seeds and young leaves contain high levels of alkaloids.

Where Found. Weeds in the potato family are found in agronomic crops and along roadsides, ditchbanks, and fence rows. They are also pests in vineyards and orchards.

Examples. This large family includes Chinese thornapple (*Datura ferox*), jimsonweed (*D. stramonium*), indian tobacco (*Nicotiana bigelovii*), tree tobacco (*N. glauca*), Wright groundcherry, lanceleaved groundcherry, and tomatillo groundcherry (*Physalis* spp.), and silverleaf nightshade, black nightshade, and hairy nightshade (*Solanum* spp.).

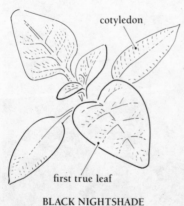

cotyledon

first true leaf

BLACK NIGHTSHADE

FIGURE 2-20.

Solanum nigrum –
potato family.

Aster Family

Important Characteristics. The very large aster family contains many important annual and perennial weeds; many of them are "thistles" (Figure 2-21), characterized by spines on their leaves, stems, or flowers. Weeds in the aster family are usually erect plants; many grow to heights of 1 to 3 feet while others are much taller. Most have showy flowers.

Where Found. Weed species of this family grow in almost all open or cultivated areas such as rangeland, along roadsides and ditchbanks, along fence rows, in vacant lots, cultivated crops, orchards, vineyards, and lawns.

Examples. Weeds in the aster family include common yarrow (*Achillea millefolium*), California mugwort (*Artemisia douglasiana*), Russian knapweed (*Centaurea repens*), yellow starthistle (*C. solstitialis*), Canada thistle (*Cirsium arvense*), bull thistle (*C. vulgare*), common sunflower (*Helianthus annuus*), Jerusalem artichoke (*H. tuberosus*), telegraphplant (*Heterotheca grandiflora*), prickly lettuce (*Lactuca scariola*), pineapple weed (*Matricaria matricarioides*), Scotch thistle (*Onopordum acanthium*), bristly oxtongue (*Picris echioides*), common groundsel (*Senecio vulgaris*), California goldenrod (*Solidago californica*), annual sowthistle (*Sonchus oleraceus*), dandelion (*Taraxacum officinale*), and cocklebur (*Xanthium strumarium*).

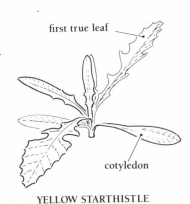

first true leaf

cotyledon

YELLOW STARTHISTLE

FIGURE 2-21.

Centaurea solstitialis –
aster family.

Invertebrates

Invertebrates are animals without backbones (vertebrae). These include nematodes and all other worms, snails and slugs, and the arthropods (insects, spiders, mites, and their relatives). Pest invertebrates affect people in many different ways. Some are parasites of livestock, poultry, or human beings; they feed on skin and hair, blood, or invade internal tissues. Many invertebrates transmit disease organisms to people, pets, livestock, poultry, or plants. A large number of invertebrates feed on growing plants. Invertebrates also consume or contaminate stored food products. Some invertebrates damage buildings and other structures, as well as books, fabrics, furniture, equipment, and many other items used by people.

TABLE 2-3

Ways in Which Arthropods Are Pests.

TYPE OF DAMAGE	EXAMPLE OF PEST
PLANT PESTS	
Chewing on leaves	Caterpillars, beetles, grasshoppers
Boring or tunneling into leaves, stems or fruit	Twig borers, leafminers, beetles
Sucking plant juices	Aphids, mites, scale, thrips, plant bugs
Feeding on roots	Beetles, aphids, flies
Feeding on fruit, nuts, berries	Moth larvae, beetles, earwigs
Causing malformations such as galls	Flies, wasps, mites
Transmitting diseases	Aphids, mites, leafhoppers
PESTS OF ANIMALS AND PEOPLE	
Have venomous bite or sting	Bees, wasps, ants, spiders, scorpions
Feed on flesh or blood	Flies, mosquitoes, bugs, ticks, fleas, lice, mites
Transmit diseases	Mosquitoes, bugs, flies, fleas, ticks
Cause allergic reactions	Bugs, bees, wasps, mites
Have offensive odors	Beetles, lacewings, bugs
Cause fear or are nuisances	Insects, spiders, scorpions, centipedes
Cause loss in livestock weight gain and reduction of milk or egg production	Fleas, ticks, mites
Damage and devalue hides and pelts. Loss of carcasses used for meat	Sheep ked, lice, ticks, mites, cattle grubs
Cause reduction in livestock's work and reproduction efficiency	Flies, mites
STORED PRODUCTS PESTS	
Eat or damage grains and other stored foodstuff	Beetles, cockroaches, moths, crickets
STRUCTURAL PESTS	
Damage buildings and other wood structures	Termites, beetles, ants, bees
PESTS OF HUMAN BELONGINGS	
Feed on clothing, carpeting, paper products	Moths, beetles, cockroaches, crickets
Feed on furniture, other wood products	Termites, beetles

Arthropods

Arthropods are one of the largest groups in the animal kingdom. The word *arthropod* means "jointed foot" and refers to organisms with an external skeleton and jointed body parts. Insects, spiders, ticks, and mites are part of this group, as are crabs, crayfish, shrimp, and lobsters. Centipedes, millipedes, scorpions, and sowbugs are also arthropods. Only about 3% of all the arthropod species that occur in the United States are considered pests, however. Table 2-3 lists some of the ways that arthropods interfere with human activities. Many arthropod species are beneficial to people, plants, or livestock (Table 2-4).

Sometimes an arthropod may be a pest in one situation but beneficial in another. For example, poisonous spiders are pests when they disturb or endanger people or animals, but in gardens and agricultural situations they help control harmful insects and other arthropod pests.

How to Identify Arthropods. Arthropods may range in size from several inches in length to microscopic. Important pests usually can be identified by comparing them to photographs or drawings found in pest management manuals, field guides, or other publications. Use a micro-

TABLE 2-4

Beneficial Arthropods.

TYPE OF BENEFIT	EXAMPLE OF ARTHROPOD
USEFUL PRODUCTS	
Silk	Silkworm moth
Beeswax, honey	Honey bees
Shellac	Scale insect
Pigments and dyes	Scale and gall insects
POLLINATION	
Figs	Wasp
Many fruits and vegetables	Honey bees, wild bees, bumble bees, flies, other insects
FOOD	
For fish	Mosquitoes, flies, many others
For birds	Butterflies, moths, beetles, many more
PARASITES AND PREDATORS OF HARMFUL INSECTS AND MITES	
Parasites	Wasps, flies
Predators	Beetles, lacewings, flies, bugs, spiders, wasps, mites
NATURAL CONTROL OF WEEDS	
Feeding damage	Moths, beetles, others
Introducing disease agents	Beetles
IMPROVED SOIL CONDITIONS	Beetles, other soil inhabiting insects
SCAVENGERS OF DEAD PLANT AND ANIMAL MATTER	Beetles, flies, many others
USE IN SCIENTIFIC STUDIES	Fruit flies, cockroaches, others
MEDICINAL USES	Bees, wasps, flies

scope or hand lens to examine small specimens. Identification keys are useful in determining the identity of some arthropods, but since they usually require a knowledge of body parts and structures, these are more difficult to use. Because it is not always easy to identify arthropods, you may need the help of a trained specialist. Table 2-5 explains how to prepare arthropods for shipment to specialists for identification.

Pest arthropods can be identified by distinguishing between the various types of body structures that are unique to different groups. Some insects undergo changes in their body form during their life (for example, caterpillars turn into moths or butterflies). It also is important to recognize developmental stages because the immature forms are often the ones that cause damage to plants or products. Adults of many pest insects possess wings which can be used for identification. Insect mouthpart

TABLE 2-5

Guidelines for Sampling and Sending Arthropods for Identification.

SAMPLING:

1. Collect plant-feeding insects and mites by snipping off portions of foliage or stems containing the pest and placing these into a plastic bag.
2. Use an insect net to collect flying insects.
3. For other insects, shake plant foliage onto a light-colored cloth sheet and funnel arthropods into a plastic bag or glass or plastic jar.
4. Keep all collected specimens cool by placing them in an ice chest or refrigerator.

PREPARATION:

1. Place insects, mites, and other arthropods into a glass vial with 70% isopropyl alcohol ("rubbing" alcohol also may be used). Vials must be sealed so that alcohol cannot leak out.
2. Include more than one individual of each species whenever possible. If other life stages are present (eggs, larvae, pupae, adults), include representative samples of these.
3. If the pest is associated with plants, send samples which show pest damage. Do not put plant material in the alcohol. Keep plants as fresh as possible by keeping them cool.

LABELING:

Attach a label to the outside of each vial and plant sample bag. Include the following information on labels:

1. Your name, address, and telephone number.
2. Name and variety of host plant, if applicable.
3. Date specimens were collected.
4. Whether pest is in commercial, agricultural, residential, nursery, or other setting. If the pest is not associated with plants, describe the site where it was collected.
5. Location of area where specimens were collected, including names of nearest crossroads.

SHIPPING:

1. Contact the person or laboratory who will receive samples to determine the best method of shipping and to inform them that samples will be arriving.
2. Pack samples in a sturdy, well-insulated container to prevent crushing or heat damage.
3. Mark package clearly and request shipper to keep it in a cool location.
4. If plant material is included, ship packages early in the week so they will arrive before a weekend.

types also are useful for identification; mouthparts may be modified for sucking, biting and chewing, or a combination of these.

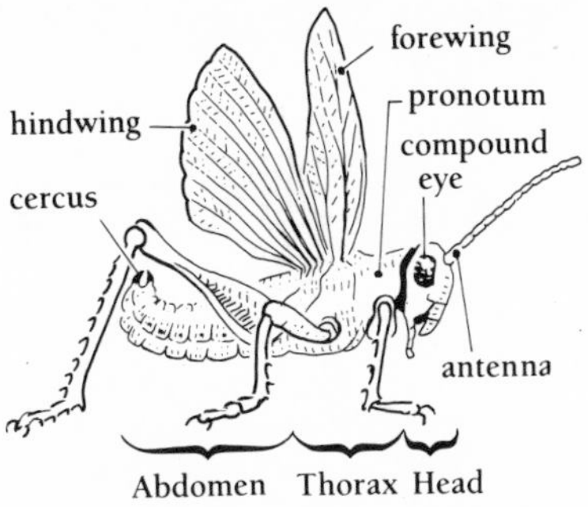

Arthropod Body Structure

Insects. The body of an adult insect is divided into three distinct parts—the head, thorax, and abdomen (Figure 2-22). Eyes and one pair of antennae arise from the head. There are usually several pairs of appendages that make up the mouthparts which are attached to the head. The thorax gives rise to three pairs of legs and often one or two pairs of wings. Insects have a segmented abdomen that is sometimes partially or entirely covered by the folded wings. Some insects have appendages extending from the tip of the abdomen, such as pinchers, a sting, or other structures.

Spiders. Spiders have two major body parts (Figure 2-23). The head and thorax are combined into one section called a *cephalothorax*. The cephalothorax gives rise to four pairs of legs, eyes, and mouthparts—including a pair of chelicerae (fangs). The remaining body part is a non-segmented abdomen that terminates with several pairs of spinnerets, the spider's web-spinning organs.

Ticks and Mites. Ticks and mites have only two body segments, but these are different from spiders (Figure 2-24). A small head (called a *gnathosoma*) with a few mouth appendages is attached to a large, combined thorax and abdomen called an *idiosoma*. Adult mites and ticks have four pairs of legs arising from the idiosoma, while immature forms usually have only three pairs.

Other Arthropods. Other arthropod groups have varying arrangements of body parts and legs. For example, centipedes and millipedes have many body segments with large numbers of legs (Figure 2-25). Scorpions have a long, segmented tail.

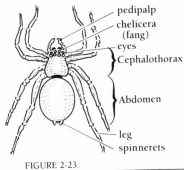

FIGURE 2-22.

The body parts of this grasshopper represent the general structures which usually can be seen on most adult insects. Some adult insects are wingless, however.

FIGURE 2-23.

Spiders have two main body regions rather than the three found in insects; they have four pairs of legs, a pair of pedipalps, and a pair of chelicerae. The chelicerae terminate in fangs which are used to inject venom. At the end of the abdomen are a cluster of spinnerets, part of the spider's web-producing mechanism.

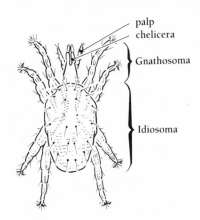

FIGURE 2-24.

The body arrangement of mites and ticks is different from insects or spiders. Adults generally have four pairs of legs; immature forms usually have three pairs.

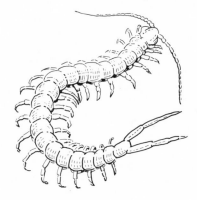

FIGURE 2-25.

Centipedes have many body segments, most of which give rise to a pair of legs. Centipedes have a pair of poisonous fangs which arise from the head segment.

Arthropod Developmental Stages

Arthropods hatch from eggs and increase in size by shedding their outer body covering and growing a new, larger one. This process is called *molting*. The period between one molt and the next is known as an *instar*. Immature arthropods pass through several instars before becoming adults.

Development from egg to adult in most insects is unique among the arthropods. In many insect orders, young may be different than the adults. Immature insects in these orders undergo a change in form known as *metamorphosis*. In a few orders, the young resemble adults, but are wingless; as they grow, they pass through a gradual metamorphosis and slowly develop wings. Insects in other orders go through an incomplete metamorphosis in which wingless young (resembling adults) change into winged adults. Insects that undergo complete metamorphosis pass through a larval (usually wormlike) stage, a pupal stage, and an adult stage.

SPIDERS: Class Arachnida – Order Araneae

Important Characteristics. Spiders have four pairs of legs and an unsegmented abdomen which is attached to the other main body part (the cephalothorax) by a narrow waist. At the tip of the abdomen are spinnerets, special organs used for producing different types of webbing.

Life Cycle. Immature spiders hatch from eggs and pass through several instars before becoming adults. Each instar is preceded by molting, a process in which the spider sheds its outer body covering, enabling it to grow larger. Most spiders live for 2 or 3 years, although some species may live as long as 20 to 30 years. Females generally live longer than males. Spiders feed on insects and other small arthropods.

Where Found. In buildings, spiders are found in corners of ceilings, behind and underneath furniture, and in basements, attics, and crawl spaces. Outdoors, spiders occur in most types of environments. They are commonly found on agricultural crops and landscape plants.

Damage. In California, black widow and brown recluse spiders can inflict painful bites that may require prompt medical attention. Other large spiders may bite occasionally. Spiders leave unattractive webbing on inside and outside surfaces of buildings.

Beneficial Aspects. Spiders are general predators of arthropods, including insects, and contribute to the natural control of many pest species.

TICKS AND MITES: Class Arachnida – Order Acarina

Important Characteristics. Mites and ticks have their abdomen broadly joined to the head and thorax (Figure 2-26). Adults usually have four pairs of legs, while immature mites and ticks most often have three or fewer pairs. Some species of mites produce fine webbing from silk glands located near their mouth. Most mites are very small and difficult to see without the aid of a hand lens or microscope.

Life Cycle. Ticks and mites hatch from eggs and pass through several immature stages before becoming adults. Immature ticks and mites re-

SIZE RELATIONSHIPS

The actual body lengths (not including legs, antennae, or wings) of arthropods in drawings such as the tick at the right are indicated by a size bar ▮━━━━━▮ *. Arthropods whose body length is 0.1 inch or smaller, such as the mite shown at the right, are indicated by a magnifying glass symbol* ○

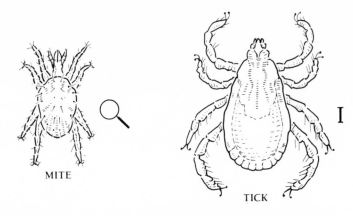

MITE

TICK

FIGURE 2-26.

Ticks and mites are closely related, although ticks are much larger. Ticks are parasites of vertebrates and feed on blood. Some mites are parasites, although many are serious plant pests.

semble adults. Mites usually develop from eggs to adults in a few days to a few weeks; some overwinter as adults, while other species overwinter as eggs. Ticks generally live much longer; some require 1 to 2 years to reach maturity and may live an additional 2 or 3 years as adults.

Where Found. Depending on the species, mites are found as parasites on plants or animals. Certain species are predatory on other mites. Ticks are blood-feeding parasites of vertebrates and require blood meals to develop and reproduce. They commonly are found on animal hosts or in or near their nests. They may also be found in cracks and crevices of buildings.

Damage. Plant-feeding mites often produce serious economic or visual damage, including leaf discoloration and defoliation. Some plant-feeding mites transmit disease-causing microorganisms. Feeding injuries produced by mites that infest animals and people may itch severely. Toxins injected by ticks during feeding sometimes cause paralysis of hosts; some tick species can transmit disease-causing microorganisms.

Beneficial Aspects. Several species of mites are predatory on pest mites and serve as an important component of natural and biological control programs.

BRISTLETAILS: Class Insecta – Order Thysanura (Silverfish and Firebrats)

Important Characteristics. Silverfish and firebrats have no wings. They usually have two or three long taillike structures (Figure 2-27). Mouthparts are of the chewing type. Most are brownish or silver in color. They are often nocturnal.

Life Cycle. Thysanurans undergo simple metamorphosis, therefore the young resemble adults, but are smaller. Individuals of some species live for 2 to 3 years.

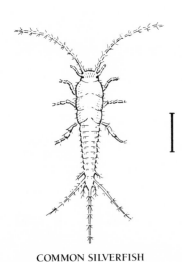

COMMON SILVERFISH

FIGURE 2-27.

Silverfish – Order Thysanura.

Where Found. Silverfish and firebrats are commonly found in homes, businesses, and libraries. They hide in cracks and crevices and other dark places. Silverfish prefer areas of high humidity and often may become trapped in sinks and tubs. Firebrats may be found in drier habitats.

Damage. These insects infest paper, cereals, and fabrics (including synthetics). They also feed on resins and glue used in books and picture mountings.

ORTHOPTERANS: Class Insecta – Order Orthoptera (Cockroaches, Crickets, Grasshoppers, Locusts, Katydids, Mantids, and Walkingsticks)

Important Characteristics. The order Orthoptera takes on many different forms (Figure 2-28), so it does not have any single identifying characteristic; all forms, however, have chewing mouthparts. Immature orthopterans resemble adults. Many insects in this order are winged, although most are not good fliers; numerous wingless species also exist.

Life Cycle. Orthopterans hatch from eggs which may have been glued to plant surfaces, inserted into plant tissues, laid in soil, or laid freely on the ground. Some species attach their eggs to other kinds of surfaces, usually in some type of egg capsule. Females of a few species carry their eggs about with them in an egg capsule. Immature forms do not have wings; they pass through a series of instars before reaching the adult stage. Most orthopterans may live 1 or more years.

Where Found. Orthopterans are found in most agricultural crops, landscaped areas, gardens, and in and around buildings. Many live in the soil, while others are found in trees or shrubs or in parts of buildings or other structures.

Damage. Some orthopterans feed on plants and may cause serious injury or defoliation. Cockroaches are troublesome pests in homes, restaurants, and other buildings; they infest stored food and can damage fabrics and paper products. A few species are known to carry disease

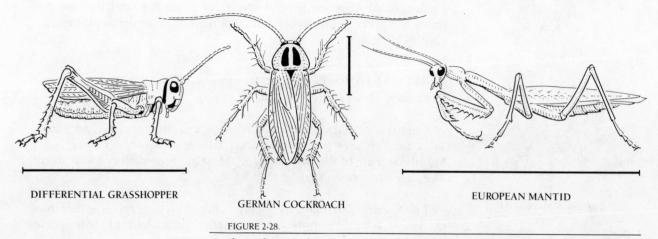

DIFFERENTIAL GRASSHOPPER   GERMAN COCKROACH   EUROPEAN MANTID

FIGURE 2-28.

Cockroach, grasshopper, and mantid – a few representatives of the Order Orthoptera.

organisms on their body and are capable of transmitting these to food or eating utensils.

Beneficial Aspects. Mantids are general predators of other arthropods and contribute to the natural control of some pests.

EARWIGS: Class Insecta – Order Dermaptera

Important Characteristics. Earwigs are elongated, sometimes wingless insects having characteristic forceplike cerci extending from the end of the abdomen (Figure 2-29). These "pinchers" are used for defense and for catching prey. Cerci of females of the European earwig are straight, while those of males are curved. Earwigs have chewing mouthparts.

Life Cycle. Earwigs hatch from eggs which have been deposited in a nest in the soil. Adult females protect their eggs and newly hatched young. Earwigs pass through gradual metamorphosis. They will develop from egg to adult in 2 to 3 months. They hibernate as adults.

Where Found. Earwigs usually nest in the ground and are found under boards, stones, and ground litter. Earwigs are very common insects in some parts of California.

Damage. Earwigs feed on vegetables, ripe fruit, garden flowers, and garbage; some are predaceous. They are occasional nuisances in buildings when wandering in from outdoors.

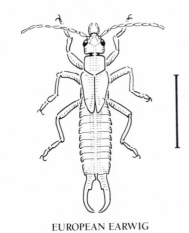

EUROPEAN EARWIG

FIGURE 2-29.

Earwig – Order Dermaptera.

TERMITES: Class Insecta – Order Isoptera

Important Characteristics. Termites superficially resemble ants, but have an abdomen broadly joined to the thorax. When compared to ants, there also are differences in body coloration and the shape and size of antennae (Figure 2-30). Some termites are winged, but lose their wings after dispersal; wings are much longer than the termite's body and both pair are of equal length (wings of ants are shorter—no more than the length of their body, with the front pair of wings being longest). Termites live in colonies having different types of individuals, called castes. Each caste performs specific functions in the colony. Many soldiers have enlarged heads with powerful pincerlike jaws. Queens have greatly enlarged abdomens. Termites are equipped with chewing type mouthparts.

Life Cycle. Upon hatching, young termites are often tended and fed by adults. The nymphal stage lasts 3 to 4 months or longer. Adult workers may live 3 to 5 years, while queens live much longer. Termite colonies may survive for many years, as aged individuals are continuously replaced by younger ones.

Where Found. Some species of termites live in the soil, but others construct nests in trees and wooden structures. Soil nesting species usually construct tunnels or tubes to wood sources which they use as food.

Damage. Termites cause serious damage to wood structures by feeding and constructing tunnels or galleries.

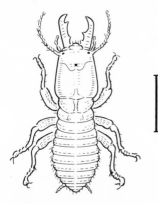

PACIFIC DAMPWOOD TERMITE
(SOLDIER)

FIGURE 2-30.

Termite – Order Isoptera.

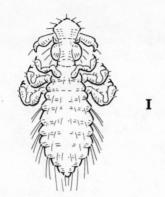

CHICKEN HEAD LOUSE
FIGURE 2-31.

Chewing lice – Order Mallophaga.

CHEWING LICE: Class Insecta – Order Mallophaga

Important Characteristics. Chewing lice (Figure 2-31) are very small oval or elongated wingless insects with chewing mouthparts. They have flattened bodies, sometimes with dark brown or black spots or bands. Chewing lice have a head that is wider than their thorax. A hand lens or microscope is needed to examine these tiny insects.

Life Cycle. Eggs of chewing lice are laid on hosts, usually attached to hair or feathers. They may pass through three or more nymphal stages before developing into adults. Most chewing lice develop into adults within 2 or 3 weeks after hatching.

Where Found. Chewing lice are parasites of birds, fowl, and a few mammals. Species are host specific—each species is only found on one type of animal.

Damage. These parasites feed on feathers and the outer skin and skin debris of birds, and on hair, blood, and skin of mammals. Poultry infested with chewing lice usually become restless and uncomfortable, have decreased weight gain, and have lowered egg production.

SUCKING LICE: Class Insecta – Order Anoplura

Important Characteristics. Sucking lice (Figure 2-32) are flat-bodied, wingless insects with piercing-sucking mouthparts. The width of the head is less than the thorax. A hand lens or microscope is necessary to examine these insects.

Life Cycle. Eggs are cemented to hairs of the host. After hatching, sucking lice pass through several instars and become adults within 1 to 2 weeks. Sucking lice pierce the skin of their hosts to feed on blood.

Where Found. Sucking lice are all host specific parasites of mammals, including people. The human body louse remains on clothing when removed and can survive off its host for short periods of time. Other species must remain on their hosts at all times.

Damage. Feeding by sucking lice causes irritation and itching. Some lice are capable of transmitting disease-causing organisms.

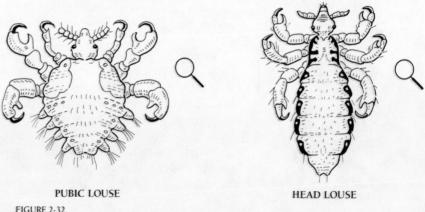

PUBIC LOUSE   **HEAD LOUSE**
FIGURE 2-32.

Sucking lice – Order Anoplura.

THRIPS: Class Insecta – Order Thysanoptera

Important Characteristics. Thrips are tiny, elongated insects with two pairs of wings (Figure 2-33). Their wings have a fringelike appearance. Thrips have modified sucking-rasping mouthparts.

Life Cycle. Thrips hatch from eggs and most species pass through four instars. During the first two instars the thrips actively feed. The final instars are more of a resting stage, which often take place in the soil. Wings are present after the thrips' final molt.

Where Found. Thrips are commonly found on plants, often in flowers and on tender, developing parts of leaves and fruits.

Damage. Thrips puncture plant cells and then suck up the fluid that escapes. This type of feeding damages cells, giving rise to deformed fruit and other plant structures. Some thrips are serious pests in greenhouses, gardens, and agricultural areas.

Beneficial Aspects. A few species of thrips are predatory. Predatory thrips play an important role in the natural control of several plant pests, including aphids and mites.

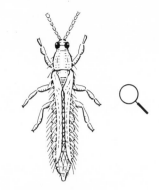

WESTERN FLOWER THRIPS

FIGURE 2-33.

Thrips – Order Thysanoptera.

TRUE BUGS: Class Insecta – Order Hemiptera
(Bed Bugs, Plant Bugs, Damsel Bugs, Assassin Bugs)

Important Characteristics. True bugs (Figure 2-34) can usually be recognized by the triangular shaped *scutellum* seen from above and the long, needlelike beak that folds under their body (piercing-sucking mouthparts). True bugs are various sizes depending on the species; some are nearly 2 inches long. They have two pairs of wings and most fly well; the second pair of wings are not visible while the insect is at rest. The wings are sometimes brightly colored.

Life Cycle. True bugs undergo gradual metamorphosis after hatching from eggs. Young resemble adults, but without wings. Life cycles vary among the many species of true bugs.

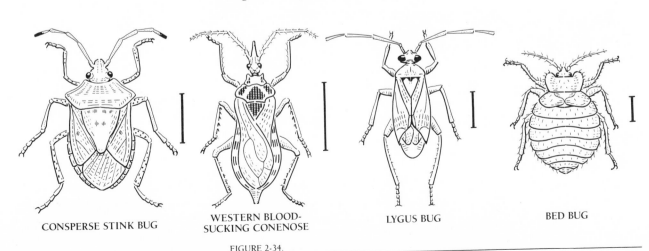

CONSPERSE STINK BUG

WESTERN BLOOD-
SUCKING CONENOSE

LYGUS BUG

BED BUG

FIGURE 2-34.

True bug – Order Hemiptera.

Where Found. True bugs feed on plants and animals, depending on the species. Most are free living, searching out appropriate hosts for food. There are several aquatic species.

Damage. Some species feed on blood of livestock, birds, rodents, and people. Feeding sites may become inflamed or infected and usually are very tender. A few species of true bugs transmit disease organisms. Plant-feeding bugs damage plant cells, causing deformities of fruits and other plant parts; some also inject chemicals into the plant that prevent or alter the plant's normal growth. Bugs are often serious plant pests in agricultural, home garden, and landscaped settings.

Beneficial Aspects. Some species of hemipterans are predators of other insects, including many insect pests. Examples of these are assassin bugs, bigeyed bugs, and minute pirate bugs.

HOMOPTERANS: Class Insecta – Order Homoptera
(Aphids, Leafhoppers, Cicadas, Psyllids, Whiteflies, Mealybugs, Scales, Phylloxeras, Spittlebugs, and Treehoppers)

Important Characteristics. Homopterans are a diverse group of somewhat soft-bodied insects. Most are winged, or have some winged forms. They all have piercing-sucking mouthparts. Some of the different homopterans are illustrated in Figure 2-35.

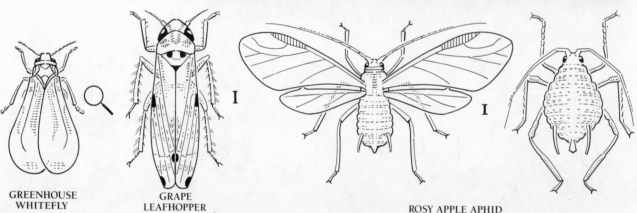

GREENHOUSE
WHITEFLY

GRAPE
LEAFHOPPER

ROSY APPLE APHID

FIGURE 2-35.

Leafhopper, whitefly, scale, and aphid – Order Homoptera. The photograph shows frosted scale insects on an English walnut tree. The round holes on some of the scale's outer body coverings are exit holes made by parasitic wasps.

Life Cycle. Because this group is so diverse, it is not possible to describe a typical life cycle. Some, such as the periodical cicada, have life cycles as long as 17 years, although many homopterans live less than 1 year.

Where Found. Homopterans are plant feeders, so are usually found on or near plants. Some occur in greenhouses and indoors on houseplants. Phylloxera often occur beneath the soil surface where they feed on plant roots.

Damage. Homopterans pierce plant tissues and suck out liquids. Feeding can cause deformity of leaves and fruit, loss of plant vigor, stunted growth, and dieback of plant parts. Most homopterans excrete a sticky substance called honeydew which supports the growth of black sooty mold fungi. Many species transmit disease-causing pathogens to host plants.

COLEOPTERANS: Class Insecta – Order Coleoptera (Beetles and Weevils)

Important Characteristics. Adult beetles and weevils range from "pinhead" size to several inches in length, depending on the species (Figure 2-36). Adults are usually recognized by their *elytra*, the hardened and often black first pair of wings that completely cover the abdomen and hide the insect's second pair of wings when it is not flying. Larvae are very diverse in appearance. Both larvae and adults have chewing mouthparts.

Life Cycle. Coleopterans undergo complete metamorphosis. Larvae pass through several instars before pupating; pupation often takes place in the soil. Some adults survive adverse weather conditions by going into a resting or dormant phase—hibernation during cold winter periods or aestivation during hot or dry periods.

Where Found. Beetle and weevil pests are commonly associated with plants, stored food products, wood, furniture, fabrics, and animal prod-

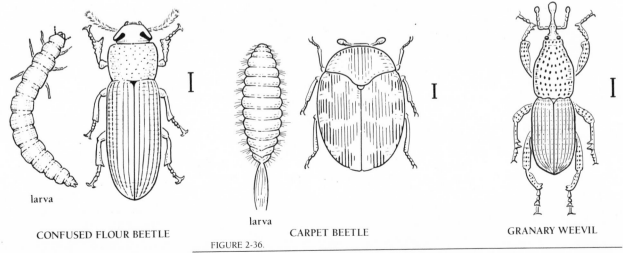

larva

CONFUSED FLOUR BEETLE

larva

CARPET BEETLE

GRANARY WEEVIL

FIGURE 2-36.

Beetles (left) and weevil (right) – Order Coleoptera.

ucts; some even bore into metal. Many species actively bore into items they use for food.

Damage. Beetles and weevils damage or destroy agricultural products, forest trees, stored foods, fabrics, furs, carpets, wood items, and landscape plants.

Beneficial Aspects. Several species of beetles are predatory on other insects, and contribute to the control of plant-feeding pests such as aphids and mites.

LEPIDOPTERANS: Class Insecta – Order Lepidoptera (Butterflies, Moths, and Skippers)

Important Characteristics. Adult butterflies, moths, and skippers (Figure 2-37) are distinguishable from other insects by their large, scale-covered and often brightly colored wings and coiled mouthparts. Larvae are wormlike with chewing mouthparts; adults have modified mouthparts in the form of a coiled tube which they can extend to suck up liquids. Butterflies and skippers can be separated from moths by their antennae; moths have tapering hairlike or feathery antennae, while those of butterflies and skippers are clubbed. (Some exceptions exist when moth antennae resemble butterflies, however these are rare and will generally not be encountered in pest management). Skippers have shorter, thicker bodies and smaller wings, setting them apart from butterflies. Moths are primarily nocturnal, while butterflies and skippers fly mostly during the day.

Life Cycle. Lepidopterans undergo complete metamorphosis. After hatching from eggs, larvae pass through several instars then enter the pupal stage and change into winged adults. Pupation sometimes takes place in the soil. Many pest lepidopterans overwinter as pupae. Life cycles from egg through adult vary according to the species, although many of the pest species can produce three or four generations per year.

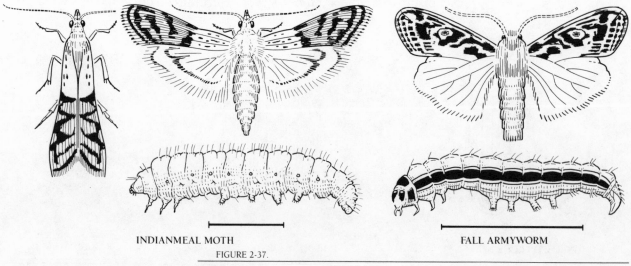

INDIANMEAL MOTH   FALL ARMYWORM

FIGURE 2-37.

Moths – Order Lepidoptera.

Where Found. Lepidopteran pests are found on or in plant parts, including fruits, in stored food products, and in fabrics. Adults are commonly attracted to lights.

Damage. Moth larvae are one of the most serious agricultural pests. They cause considerable damage to fruits and vegetables, nuts, grains, cotton, and forage crops. Moths are also prominent pests in stored foods and cause extensive damage to fabrics.

DIPTERANS: Class Insecta – Order Diptera (Flies, Mosquitoes, Gnats, and Midges)

Important Characteristics. Adult dipterans (Figure 2-38) are easily distinguished from other winged insects because they only have one pair of wings; in place of the second pair are *halteres*, small clublike organs believed to be involved in balance. Larvae are usually wormlike and are often called maggots. Most adults have modified structures for sucking, lapping, or piercing. Some adults have biting mouthparts.

Life Cycle. Dipterans undergo complete metamorphosis. Most species deposit eggs onto surfaces or into tissues of hosts, although in a few species eggs hatch inside the mother's body and larvae, rather than eggs, are deposited. Many species are noted for their rapid development from an egg through the adult stage, which may take as little as 3 or 4 days. Others have extended life cycles, taking 2 or more years to complete; often periods of adverse environmental conditions can be survived as resting pupae in the soil.

Where Found. Flies, mosquitoes, gnats, and midges may be found in most outdoor areas and inside buildings. Some larvae are internal parasites of animals; others invade plant tissues. Larvae of mosquitoes are found in aquatic habitats. Housefly larvae are commonly seen in garbage and animal feces.

Damage. Many dipteran species are serious pests. Larvae of some invade living animal tissues. Adult mosquitoes and the adults of some spe-

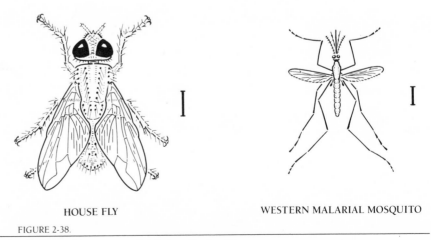

HOUSE FLY   WESTERN MALARIAL MOSQUITO

FIGURE 2-38.

Fly (left) and mosquito (right) – Order Diptera.

cies of flies, midges, and gnats are blood-feeding and are capable of transmitting several serious disease pathogens. Some species of flies are pests in agricultural crops and landscape settings, while others are nuisances in and around buildings and livestock or poultry areas.

Beneficial Aspects. Some species of flies are parasitic on pest insects and others are predators of certain insect orders. These are often important natural and biological control agents.

FLEAS: Class Insecta – Order Siphonaptera

Important Characteristics. Adult fleas are tiny dark brown or black wingless insects with laterally compressed bodies (Figure 2-39). Fleas are capable of hopping great distances. Larvae are wormlike. Fleas have piercing-sucking mouthparts.

Life Cycle. Fleas undergo complete metamorphosis. They usually pass through one complete cycle, from egg to adult, in 30 to 40 days.

Where Found. Eggs may be laid on the host, although they usually drop off before hatching. Larvae are free living, usually on the ground; they feed on skin debris and on the hair of their host. Adults must feed on the blood of warm-blooded animals and so are often found on their hosts or in their host's nest. They are usually host specific.

Damage. Flea bites are uncomfortable and may cause allergic reaction in some people and animals. Some species of fleas can transmit pathogens of diseases such as bubonic plague and murine typhus. Certain fleas serve as the intermediate hosts of tapeworms.

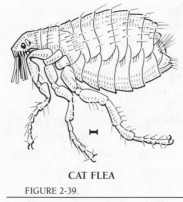

CAT FLEA

FIGURE 2-39.

Flea – Order Siphonaptera.

HYMENOPTERANS: Class Insecta – Order Hymenoptera
(Bees, Wasps, Ants, Sawflies, and Horntails)

Important Characteristics. Adult hymenopterans differ from other orders of insects by usually having their abdomen joined to the thorax by a very narrow waist (Figure 2-40); winged forms have two pairs of transparent, sometimes colored wings. Larvae are grublike. Certain spe-

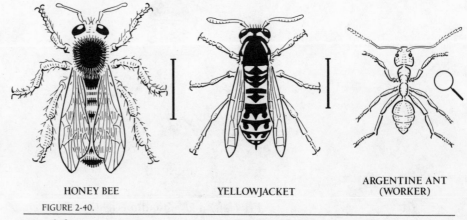

HONEY BEE   YELLOWJACKET   ARGENTINE ANT (WORKER)

FIGURE 2-40.

Bee (left), wasp (center), and ant (right) – Order Hymenoptera.

cies of bees, wasps, and ants are venomous; they have a sting at the tip of their abdomen. Several species are social rather than solitary; large numbers of individuals live together in a colony and members of the colony perform different tasks, depending on their caste. Larvae and some adult hymenopterans have chewing mouthparts.

Life Cycle. Hymenopterans undergo complete metamorphosis. Their larvae may be internal or external parasites of insects, spiders, and other arthropods; some species are plant parasites. Species that live in colonies forage for food for their larvae. Colonies of social hymenopterans may live for many years, although individuals will survive for only a few months to 1 or 2 years. Some species have several generations per year.

Where Found. Adult hymenopterans are commonly seen foraging for food on flowers and other plant parts. They are often attracted to sweets and meat. Social species live in nests in the ground or in buildings, trees, and other structures.

Damage. Many species of hymenoptera are venomous and can inflict painful stings which, in some people, may be fatal. Certain species of sawflies are serious agricultural and forest pests. Many bees and wasps are nuisances around homes and outdoor areas. Ants are persistent pests in food preparation areas and in agricultural crops. Some bees and ants damage wood structures.

Beneficial Aspects. Bees and other species of hymenoptera are important pollinators of agricultural and horticultural plants. Parasitic wasps play an extremely important role in the control of pest insects, including many pest homopterans, lepidopterans, dipterans, and coleopterans. Many are egg parasites and others attack the larval stages of their hosts.

Nematodes

Nematodes are a large group of nonsegmented worms (Figure 2-41). Most stages are eellike in shape, pointed at both ends, and quite distinct from segmented worms such as earthworms. Mature females of some

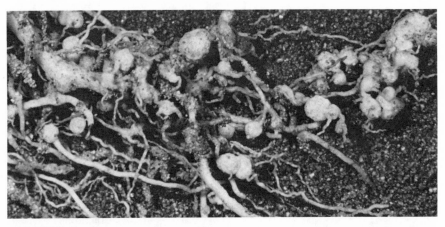

FIGURE 2-41.

Nematode – Phylum Nemathelminthes, Class Nematoda. The photograph shows nematode galls on the roots of a plant.

species have various rounded shapes. Most species are microscopic. Some species are plant parasites and others are animal parasites, but most are nonparasitic. Beneficial nematodes attack insects, plants, and other nematodes.

Most plant parasitic nematodes are internal or external parasites of plant roots. Some species inject tissues with salivary secretions that can cause root cells to enlarge and form knots or galls in which they live and feed. Others feed on root tips or in inconspicuous lesions on the root surface. Root damage caused by nematode feeding or migration prevents the plant from getting water and nutrients required for normal growth. Symptoms include stunting, loss of vigor, yellowing, and/or general decline. Although nematodes seldom cause the death of plants, nematode-infested plants may be highly susceptible to injury from other pests or environmental stresses. Some nematodes transmit viruses to plants, including grapevine fanleaf virus, tobacco rattle virus, arabis mosaic virus, and tomato black ring virus. A few nematode species attack aboveground parts of plants, causing damage such as stunting or production of galls in the place of seeds.

Nematodes are common parasites of many animals such as insects, snails, toads, frogs, birds, and most vertebrates (including human beings, poultry, and livestock). Hookworms and intestinal pinworms, for example, are nematodes. The canine heartworm, transmitted to dogs by a mosquito, is a nematode. Plant-damaging nematodes may be controlled through soil application of pesticides, although control is difficult. Control of human and animal internal parasites is performed by physicians or veterinarians, rather than certified pesticide applicators.

TABLE 2-6

Important Genera and Species of Nematodes.

PLANT PESTS:

citrus nematode	*Tylenchulus semipenetrans*
cyst nematode	*Heterodera* spp.
dagger nematode	*Xiphinema* spp.
foliar nematode	*Aphelenchoides* spp.
lesion nematode	*Pratylenchus* spp.
needle nematode	*Longidorus africanus*
pin nematode	*Paratylenchus* spp.
potato rot nematode	*Ditylenchus destructor*
rice root nematode	*Hirschmanniella* spp.
ring nematode	*Criconemoides* spp.
root-knot nematode	*Meloidogyne* spp.
seed-gall nematode	*Anguina* spp.
spiral nematode	*Rotylenchus* spp., *Helicotylenchus* spp.
stem nematode	*Ditylenchus dipsaci*
stubby root nematode	*Trichodorus* spp., *Paratrichodorus* spp.
stunt nematode	*Tylenchorhynchus* spp.

ANIMAL PESTS:

canine heartworm	*Dirofilaria immitis*
filariasis nematode	*Wucheria bancrofti*
hookworm	*Ancylostoma duodenale*
pig lungworm	*Metastrongylus apri*
pinworm	*Enterobius vermicularis*

Nematodes are difficult to identify because of the large number of species, their morphological complexity, and because most cannot be seen without the use of a microscope. Table 2-6 lists important genera and species. Many species of nematodes feed on fungi, insects, bacteria, and other nematodes; some of these nematodes are used as biological control agents for insect pests.

Identification of plant pest nematodes requires samples of the soil, roots, and/or other affected plant parts; send samples to a laboratory for identification (Table 2-7). In agricultural situations, samples should be taken randomly from several locations to provide information on distribution; take samples from infected areas as well as those that appear normal. Roots should be placed in the same bag with the soil to prevent them from drying out. Aboveground plant parts must be kept separate from roots and soil. When working in areas infested with nematodes, always clean boots, tools, and equipment thoroughly with hot water to avoid spreading the organisms to other areas.

TABLE 2-7

Guidelines for Sampling and Sending Plants for Identification of Plant Infesting Nematodes.

SAMPLING:

1. Take random samples, but sample in a consistent manner. Mark off area to be sampled into grids. Randomly select areas to be sampled.

2. Keep samples from suspected diseased areas separated from samples taken in healthy areas.

3. Dig up and include roots and soil of diseased and healthy plants. Take root samples from below the level of surface feeder roots because temperature and moisture fluctuations at the surface will affect the nematode species living there.

4. Include partially rotted or decayed roots.

5. Include aboveground plant parts if it is suspected that nematodes are infesting these areas. Keep aboveground parts separate from root and soil samples.

PREPARATION:

1. Place samples in clean plastic bags and keep them out of the sunlight. Keep samples in an insulated ice chest or refrigerator until they can be shipped. Do not freeze.

2. Pack samples in a well-insulated carton or disposable foam ice chest. Be sure container is sturdy to prevent damage to its contents.

LABELING:

Attach a label to the outside of each sample bag. Include the following information on labels:

1. Name and address of grower or property owner.

2. Crop or plant type, including variety.

3. Location of field or property (names of nearby cross roads).

4. Portion of planted area that sample represents. Include a map if necessary.

5. Brief description of the crop history (previous crops).

6. Observations by you and the owner or operator of previous problems and of when present problem was first detected.

7. Date samples were taken.

SHIPPING:

1. Contact the person or laboratory who will receive samples to determine the best method of shipping and to inform them that samples will be arriving.

2. Mark package clearly and request shipper to keep it in a cool location.

3. Ship packages early in the week so they will arrive before a weekend.

Plant Parasites. Most of the important nematode plant pests damage roots or other underground parts such as tubers or rhizomes; however a few feed above ground and damage stems, leaves, flowers, or seeds. Some nematodes are capable of transmitting virus diseases to plants.

Root feeders impair the root system and prevent the plant from obtaining water and nutrients. Infected plants wilt under temperature and water stress, may become stunted or yellow, and are also susceptible to infections by secondary pathogens such as fungi. The root-knot nematodes, *Meloidogyne* species, are pests of many vegetable, field, and fruit and nut tree crops. These usually cause the formation of characteristic galls on roots of infested plants. Eggs and larvae of cyst nematodes (*Heterodera* species) develop within cysts that are the bodies of female parents; they survive inside these cysts for several years. Other root-feeding species do not cause characteristic growths on roots and cannot be detected without careful examination under a microscope.

Only a few California crops are subject to attack by nematodes that infest aboveground plant parts. Stem- and leaf-feeding species can cause leaf spots, deformities, and decay of plant parts, and may kill growing buds. Some cause the formation of galls on stems, leaves, seeds, or in place of flowers; their larvae may live many years inside galls.

Animal Parasites. Filaria are nematodes that are internal parasites of people and other vertebrates; certain mosquito, tick, mite, fly, and flea species spread filaria from one infected animal to another. Elephantiasis and dog heartworm are nematode-caused diseases. Trichinosis, often transmitted to people through undercooked pork, is caused by the nematode *Trichinella spiralis*. The larvae of this nematode live in cysts in the muscle tissues of hosts and can survive for long periods of time.

Snails and Slugs: Phylum Mollusca

Snails and slugs belong to the phylum Mollusca; they are related to clams, scallops, abalone, octopi, and squid (Figure 2-42). Other pests that belong to this group are the shipworms (shipworms bore into wood that is in contact with salt or brackish water) and the pholads, clamlike marine organisms that attack and destroy submerged wood. Snails have shells that distinguish them from slugs. Generally, identification of a snail or a slug species is not required for successful control, although it is important to confirm that these pests are the actual cause of observed damage before control measures are started. In southern California, a predatory snail is used to control pest snails in citrus, so these must be distinguished from pest species.

Important Characteristics. Snails prefer cool moist surroundings, but their shells afford them protection from heat and dryness, enabling them to exploit many different environments. They seal themselves into their shells and become dormant for up to 4 years during dry periods. Slugs are shell-less and so are vulnerable to high temperatures and dry weather; they are more likely found in cool, damp locations.

FIGURE 2-42.

Snail – Phylum Mollusca.

Life Cycle. Snails and slugs lay between 10 and 200 eggs beneath the surface of the soil. Egg masses are laid several times each year from spring through fall. In cooler areas, egg laying activity stops during the winter. It takes from 1 to 3 years for newly hatched snails or slugs to reach maturity.

Where Found. Snails and slugs are commonly found in damp areas, soil litter, and foliage of plants. They are usually nocturnal and hide under boards, stones, among ivy, dense shrubbery, or damp refuse during the day. At night and early mornings, or during cool, damp periods, they forage for food. They will often return to the same resting area each day unless the area becomes too dry or has been disturbed.

Damage. Snails and slugs feed on landscape plants, fruits, berries, and vegetables. Snails can be serious pests in citrus where they feed on developing fruit. They are also pests in greenhouses. In addition to feeding damage, both snails and slugs leave a slime trail that detracts from the appearance of produce and foliage. Some aquatic snails vector disease organisms which can be transmitted to people.

Beneficial Aspects. Some species of snails are predaceous and may be established in programs of biological control of pest snails.

Vertebrates

Vertebrates belong to the phylum Chordata and are animals with internal skeletons and backbones. They include fish, amphibians (frogs, toads, and salamanders), reptiles (turtles, lizards, and snakes), birds, and mammals. They become pests if they are reservoirs of pathogens that cause disease (such as plague and rabies), damage crops or stored products, prey on livestock, or interfere with the activities or needs of people.

FIGURE 2-43.

Birds can be pests when they damage agricultural crops, nest on buildings, or interfere with aircraft operation.

FIGURE 2-44.

Ground squirrels are often pests because they compete with people for agricultural products. Squirrels also may damage levees and bridge foundations through their burrowing activities. They may vector diseases that can be transmitted to people or livestock by fleas or other insects.

Some of the most important vertebrate pests are rats and mice because they inhabit buildings, damage or soil furnishings and other items, consume and destroy stored food, and sometimes harbor disease organisms.

Identification of vertebrates is often easier than microorganisms, invertebrates, and weeds because there are fewer species involved. Many vertebrates are protected under endangered species, migrant species, or other wildlife laws, so knowing the species' identity is essential before beginning any control program. Vertebrates are commonly identified by comparing them to photographs and drawings, such as those in the CDFA *Vertebrate Pest Control Handbook* and other vertebrate publications. For help in identifying vertebrates, contact the U.S. Fish and Wildlife Service or the California Department of Fish and Game. Permits may be required from these agencies for control of some vertebrate species.

Besides observing the animals, another useful way to identify vertebrates is by recognizing their characteristic burrows, nests, tracks, and fecal droppings.

Fish

Some fish are pests because they eat other fish that are more important to people, compete with important fish for limited food and space, or occur in places where they are unwanted. Certain fish species may be undesirable because they serve as intermediate hosts for parasites, such as nematodes, that infect animals and people.

Amphibians and Reptiles

A few species of toads, frogs, snakes, lizards, and their relatives are poisonous to people and livestock. These species are pests if they occur in areas where people or livestock live. In addition, nonpoisonous species may require control when their presence cannot be tolerated by people. Amphibians will occasionally clog water outlets, filters, pipes, hoses, and other equipment associated with irrigation systems and drains.

Birds

Birds are generally appreciated because of their song, beauty, or predaceous habits. Many birds catch insects and some kill snakes, rats, mice, and other pests. Pest birds (Figure 2-43), however, harbor pathogens that can be transmitted to commercial poultry, eat or damage crops, roost in massive numbers in trees or on window ledges, interfere with aircraft, cause damage to buildings, or make too much noise. Bird lice and mites associated with nesting areas infest homes, hospitals, and offices.

Mammals

Mammals that interfere with the activities of people or cause harm to their crops, livestock, or possessions are usually considered pests (Figure 2-44). Rodents, such as rats, mice, and squirrels, compete with people and livestock for food and often carry certain fleas and mites that harbor and transmit disease organisms. Infected skunks and bats transmit rabies. Larger mammals, like foxes, coyotes, mountain lions, and bears, are occasional pests when they endanger people and their possessions or prey on livestock. Foraging deer can be serious pests of agricultural crops and residential landscaping. Sea otters, seals, and sea lions may be considered pests when they interfere with human activities or compete with people for food.

Disease Agents

Another important group of pests are plant and animal disease agents. Diseases in plants and animals may be caused by nonliving (abiotic) factors or by living (biotic) pathogenic microorganisms. Biotic and abiotic factors alter or interfere with the chemical processes that take place within an organism's cells, producing changes which are often identified as disease symptoms or disorders. To avoid applying unnecessary pest management treatments, it is very important to recognize abiotic factors that produce diseaselike symptoms in plants and animals.

Important Characteristics. Familiar signs of abiotic disorders in plants include poor growth, yellow foliage, deformed leaves, poor fruit or seed production, and deformed or rotten fruit. Different plant or animal organisms have different susceptibilities to the effects of abiotic factors; an organism's susceptibility often depends on a combination of factors including its genetic makeup, nutritional state, and certain climatic conditions.

Plants and animals may exhibit symptoms resembling diseases as a result of physical factors such as water stress or excess, nutrient deficiency or toxicity, salt buildup, prolonged wind and other weather extremes, or exposure to toxic materials such as air pollutants or certain pesticides (Table 2-8). Disorders caused by abiotic factors are a result of an organism's response to the conditions in its environment and cannot be transmitted from one plant or animal to another. Pesticides cannot be used to correct or prevent abiotic diseases.

Biotic disease factors are pathogens capable of spreading from one host to another and producing disease symptoms. The ability of the organism to spread and infect may depend on climatic factors and the host's genetic makeup or nutritional state. Several different types of microorganisms can produce diseases in plants and animals; the most important are bacteria, fungi, viruses, viroids, and mycoplasmas.

Identifying Disease-Causing Pathogens. Many pathogens are submicroscopic, making identification difficult. Often the use of electron microscopes or complex biochemical tests is required to confirm the identity of a pathogen. It may even be necessary to grow the organism on a nutritive substrate in the laboratory or to inject plants or laboratory animals with the pathogen to make a positive identification. Trained technicians and specialized laboratory equipment are required for these types of tests. Table 2-9 contains information on using laboratory identification services and how to send material for analysis.

Another way to identify disease-causing microorganisms in plants and animals is by studying the observed symptoms. Each host plant or animal usually has specific diseases to which it is susceptible and will produce specific symptoms or groups of symptoms. Learning to recognize disease symptoms of the plants or animals you work with will help you sort

TABLE 2-8

Soil, Climatic, and Other Factors Contributing to Plant Growth Problems.

EFFECTS OF TEMPERATURE:

Freezing	May cause dieback, leaf scorch, frost nip, bark split. Often causes separation of the epidermis from underlying tissues; use hand lens to see this.
Cool temperatures	May reduce pollination or fertilization and lower production of produce.
Cool soil temperatures	Slows or shuts down water and nutrient uptake, resulting in wilting or nutrient deficiency symptoms even when soil contains adequate amounts.
Inadequate chilling	Delays or interferes with normal breaking of dormancy; produces irregular bloom resulting in poor crop production.
High temperatures	Kills pollen, resulting in poor crop production. Causes sunburn of fruit and foliage. Results in sun scald, heat cankers, sunken lesions, pit burn, blossom end rot, and water core. May delay maturity of crops or reduce quality.

EFFECTS OF WATER:

Too little water	Results in water stress, wilting, poor plant growth, leaf or fruit drop, leaf scorch, and nutrient deficiencies. Leaves may have dull coloration and may roll at edges.
Too much water	Lowers oxygen available to roots; may cause flower, fruit, or leaf drop; lowers soil temperature; may kill roots. Plants may be more susceptible to soil fungi.

EFFECTS OF LIGHT:

Too little light	Causes yellowing of leaves and elongation of stems. Plants become spindly. May cause twig or shoot dieback. Reduces fruit size and slows maturity.
Too much light	Causes scalding or russeting of fruit or leaves; sunburn.

EFFECTS OF NUTRIENTS:

Nutrient deficiencies	Causes leaf discoloration, leaf spotting, deformities of leaves and fruit; reduces growth, production, and shortens life of plant; plant unable to tolerate other stresses; causes leaf and fruit drop.
Nutrient excesses	Causes reduced fruit production, improper ripening, leaf burn or scald, fruit drop.

OTHER PHYSICAL FACTORS AFFECTING PLANTS:

Air pollution	Causes spotting and damaged foliage and fruit, discoloration, reduced growth and yield, dieback.
Pesticide injury	Causes leaf and fruit spotting, burning, or scorching; russeting of fruit; fruit and leaf drop; delayed maturity; leaf deformity.
Hail and rain	Causes physical injury to leaves and fruit, resulting in spotting and yield reduction; leaf and fruit drop, lodging of plants.
Wind	Desiccates leaves, resulting in spotting, burning, or scorching; improper pollination; destruction of blossoms.
Fire damage	Injures bark and nutrient transport system of plant; results in stunting, poor production and growth.

OTHER PHYSICAL FACTORS AFFECTING PLANTS (continued):

Soil compaction	Limits root growth and nutrient availability, resulting in problems associated with water stress or excess and nutrient deficiencies.
Potbound root systems	Causes problems similar to soil compaction; plant exhibits weak growth, yellowed foliage, and may undergo leaf drop and twig dieback.
Improper agricultural practices	Causes physical injury to roots or trunks of plants and reduces water and nutrient uptake; provides an entry point for disease infection and insect infestation.

TABLE 2-9

Guidelines for Sampling and Sending Plants for Identification of Disease Causing Pathogens.

SAMPLING:
1. Select plants which are most representative of the observed disease symptoms. Collect several plants.
2. Include roots by digging up plant and shaking off soil.
3. Place plants in plastic or paper bags. Keep plant material cool. Put them in an ice chest while you are in the field. Store samples in a refrigerator until they can be shipped.

LABELING:
Attach a label to the outside of each sample bag. Include the following information on labels:
1. Your name, address, and telephone number.
2. Crop or plant type, including variety.
3. Location of field or property (name nearby crossroads).
4. Portion of planted area that sample represents. Include a map if necessary.
5. Brief description of the crop history (previous crops) or any information on what was planted in the area before diseased plants were grown there.
6. Observations by you and the owner or operator of the property of previous problems and of when present problem was first detected.
7. Date samples were taken.

SHIPPING:
1. Contact the person or laboratory who will receive samples to determine the best method of shipping and to inform them that samples will be arriving.
2. Pack samples in a sturdy, well-insulated container to prevent crushing or heat damage.
3. Mark package clearly and request shipper to keep it in a cool location.
4. Ship packages early in the week so they will arrive before a weekend.

through and reject some pathogens as causes of the observed symptoms. In some cases, disease-causing pathogens of specific types are more prevalent in specific locations or under certain environmental conditions; knowing these limitations will also help you to narrow the choices.

Fungi often produce distinctive structures such as molds or spore-producing bodies which can aid in identifying the organism. Also, lesions or rotted areas of infected plants may provide clues to the type of organism causing the damage. Inspect the plant's vascular system for damage. Be sure to dig up some infected plants and examine the root system.

Check to see how disease symptoms are distributed throughout a population of plants or animals. Do many individuals show signs of infection, or only a few? Are the diseased plants distributed evenly or are they confined to specific locations? Try to associate the distribution of infection with conditions that would favor specific diseases. This may provide valuable clues to the identity of the pathogen.

Bacteria

Pathogenic bacteria are small, usually single-celled organisms, although the *Streptomyces* species are many cells attached end-to-end. Bacteria cannot be seen without a microscope. Plant-infecting bacteria are rod-shaped and most have threadlike structures, called flagellae, that propel them through liquids. Most bacteria require warmth and moisture to multiply. Some bacteria will cause serious illness or death in animals and people. Only a few bacterial diseases are serious problems on plants in California because of the semi-arid climate and lack of significant summer rainfall.

Types of Bacteria. There are six genera of bacteria that have species capable of infecting plants and causing disease symptoms. These are *Pseudomonas, Xanthomonas, Agrobacterium, Erwinia, Corynebacterium,* and *Streptomyces.*

How Bacteria Are Spread. Anything that moves and comes in contact with bacteria may spread them to other areas. This includes farm equipment, rain or irrigation water, plant material, seeds, birds, insects, nematodes, and people.

Bacterial Infections. Bacteria invade plant tissues through natural openings and through wounds. They must enter tissues to infect the

TABLE 2-10

Examples of Bacterial Diseases in Plants.

SYMPTOM	EXAMPLES	PATHOGEN
galls	crown gall	*Agrobacterium* spp.
leaf spots	leaf spot of cucurbits	*Pseudomonas* spp.
	halo blight of beans	*Pseudomonas* spp.
	bacterial spot of tomato	*Xanthomonas* spp.
	walnut blight	*Xanthomonas* spp.
systemic infections	fire blight of apple, pear, and quince	*Erwinia* spp.
	bacterial wilt of cucurbits	*Erwinia* spp.
soft rot	blackleg of potato	*Erwinia* spp.
	vegetable soft rots	*Pseudomonas* spp.
scab	potato scab	*Streptomyces* ssp.
	gladiolus scab	*Pseudomonas* spp.

plant. Bacterial diseases can be seen as galls, leaf spots, soft rots, scabs, and systemic disorders. Examples of these diseases are given in Table 2-10.

Fungi

Fungi are a diverse group of microscopic, primitive plants. They must obtain their nutrients from some organic source such as living or dead plant material. Most fungi live off dead organic matter and are known as saprophytes. They are generally beneficial because they help in breaking down these materials and building up soil fertility, although some invade wood and other building materials. Certain fungi live off both dead and living plants; some of these can cause plant diseases while others may be structural pests. A few fungi require living host plants to grow and reproduce; they include the major plant pest species.

Important Characteristics. Most pest fungi have a vegetative body, called the *mycelium*, which is made up of tiny filamentous strands called *hyphae*. The mycelium grows through the tissues of an infected host. This structure is usually large enough to be seen without magnification. Reproduction in fungi is primarily by means of spores. Many species of fungi produce more than one type of spore—some act as resting structures to carry the fungus through adverse conditions, while others are responsible for the ongoing secondary spread of the organism. Fungal identification is often based on characteristics of the mycelium and structures, known as fruiting bodies, that produce spores. Diagnosis of a fungal infection is usually made by looking for these structures on or in plant tissues or infected wood.

Disease Symptoms. Plants infected by fungi may exhibit many different types of symptoms. Such symptoms are often helpful in identifying the fungal organism. Symptoms may include soft, watery rots of fruits, severe stunting of the plant, profuse gumming, smuts, rusts, leaf spots, wilting, and malformation of plant parts—such as thickening and curling of leaves. Powdery mildew, downy mildew, root and stem rots, sooty mold, and slime molds are types of fungi.

How Fungi Are Spread. Spores of fungi can be spread by wind, rain, irrigation water, insects, and cultural practices. Any practice that moves infected plant material from one location to another can also spread fungal spores. Roots of noninfected plants may grow into areas where fungal organisms are present and become infected in this manner.

Fungal Infections. Fungi enter plant tissues through wounds and natural openings and by penetrating the outer surface of the host. Usually certain environmental conditions are required for a fungal organism to begin to grow and infect; these often include high humidity or presence of water and warm temperatures.

Viruses

Viruses are very small organisms that multiply in living cells to produce disease symptoms in plants and animals. Most can only survive for short periods of time outside host plant or animal cells. Viruses alter chemical activity (metabolism) within host cells, and these metabolic changes cause disease.

Important Characteristics. Viruses are so small they can only be observed with an electron microscope. They have different shapes depending on the type. Structurally, viruses are very simple compared with other living organisms. Viruses invade cells of plants or animals, then, using their own genetic information, alter these cells to produce more viruses rather than the cell's usual proteins or nucleic acid.

Symptoms of Virus Diseases. Almost all viruses reduce plant growth, resulting in stunting. Another common symptom is a change in coloration; this is most noticeable in the leaves, and is often expressed as a mosaic pattern of light and dark blotches. Color variation may also be seen in flowers. Sometimes the veins of leaves become lighter, producing a netlike appearance. Russeting is a color change which appears on fruit. Viruses can also produce malformations or abnormal growth in various parts of affected plants, including leaves, fruits, stems, or roots. Some viruses cause parts of infected plants to die, a condition called necrosis. Severe necrosis results in complete death of the plant.

Identifying Viruses. Three methods are commonly used to identify viruses. The simplest is to compare the infected plant and its symptoms with photographs and written descriptions that are found in publications such as *Descriptions of Plant Viruses*, published by the Commonwealth Mycological Institute in England. A method called *indexing* involves inoculating a series of indicator plants with extracts from the diseased plant and observing the results. Disease symptoms, if any, are compared with the results obtained from having inoculated similar plants with known viruses. If symptoms from the unknown pathogen match any of the plants inoculated with a known virus, you can assume the unknown pathogen is that virus. A third method, known as *serological testing*, involves testing plant extracts for the presence of virus proteins using antibodies specific for known viruses.

How Viruses Are Spread. Mites and several plant-feeding insects, such as aphids, leafhoppers, and whiteflies, are the most common ways viruses are moved from plant to plant. Disease-producing viruses may also be transmitted by some nematodes and fungi, cultivation practices, by pollination, or through pruning and grafting. Some viruses can be transmitted through seeds of an infected plant. Each virus usually can be transmitted in only one or a few of these ways.

Viroids

Viroids differ from viruses in several ways, but are most distinguishable microscopically because they are much smaller and are not enclosed in a protein coat. Only a few viroid-caused plant diseases have been identified, although viroids are suspected of causing many other plant and animal disorders. Potato spindle tuber, citrus exocortis, chrysanthemum stunt, chrysanthemum chlorotic mottle, and cucumber pale fruit are diseases known to be caused by viroids. Viroids are spread primarily by infected stock. Infected plant sap can be carried on hands or tools during propagation and cultural practices. Some viroids can be transmitted through pollen and seeds; insect transmission is unknown.

Mycoplasmas

Mycoplasmas are intermediate in size between viruses and bacteria. They can survive for extended periods outside of plant cells and are the smallest known independently living organisms. Mycoplasmas are responsible for several insect-transmitted plant diseases, including pear decline, aster yellows, western-X disease of peach, and citrus stubborn disease. Most mycoplasma diseases are transmitted by insects, the most important being leafhoppers. Mites may also transmit mycoplasmas. Mycoplasmas are also readily transmitted among woody crop plants by grafting.

3 Pest Management

Pheromone traps can be used to monitor the activities of some species of insects.

Pest management is the science of safely preventing, suppressing, or eliminating unwanted organisms. Carrying out a successful pest management program involves choosing the right prevention or control methods and knowing how to use these techniques to reduce pest populations to an acceptable level. Often several methods must be used together for best control.

Besides knowing the identity of an organism, you must be able to determine its importance as a pest. Most pests may be classified as key pests, occasional pests, or secondary pests. *Key pests* are those that may cause major damage on a regular basis unless they are controlled. Many weeds, for example, are key pests because they compete with crop or ornamental plants for resources and require regular control efforts to prevent or reduce damage. *Occasional pests* are those that become intolerable only once in a while due to their life cycles, environmental influences, or as a result of.peoples' activities. For instance, ants sometimes become occasional pests when sanitation practices change, providing them with food where previously none existed; they also move into buildings after a rainfall or other event destroys an outdoor food source. *Secondary pests* become problems as a result of actions taken to control a key pest. Some weed species only become pests after key weeds, which are normally more successful in competing for resources, are controlled. Fleas often attack people after a pet dog or cat dies or is kept out of a home for a period of time. Certain species of fleas, ticks, and blood-feeding bugs attack people only when their natural hosts have been eliminated.

APPROACHES TO PEST MANAGEMENT

A pest management program may be directed at preventing, suppressing, or eradicating pest populations. Often pest management programs combine preventive and suppressive techniques.

Prevention

There are often economical and environmentally sound ways to prevent loss or damage when a pest is likely to be present. Such techniques include planting weed- and disease-free seed and growing varieties of plants that are resistant to diseases or insects. Other alternatives are using cultural controls to prevent weedy plants from seeding and choosing planting and harvesting times that minimize pest problems. Often sanitation methods will reduce the buildup of pests. Other preventive methods include excluding pests from the target area or host and employing practices for the conservation of natural enemies. It is often important to ensure that plants, poultry, or livestock receive adequate water and nutrients to reduce stress and susceptibility to diseases or pests.

Pesticides are sometimes used preventively. For instance, weeds are often treated with preplant or preemergence herbicides before they emerge, based on the knowledge that weed seeds are present. Plant disease pathogens cannot usually be prevented from causing economic damage once they have infected susceptible plants, so pesticides are normally applied before infection occurs or when environmental conditions favor infection. Likewise, pesticides are applied to structural lumber before construction to protect it from insects, fungi, or marine borers.

Suppression

Suppressive pest control methods are initiated to reduce pest population levels in order to limit damage, competition, or inconvenience. The methods chosen do not usually eliminate pests, but reduce their populations to a tolerable level or to a point below an economic injury level; additional suppressive measures may need to be taken later. Suppression sometimes lowers pest populations so that natural enemies are able to maintain control. Suppression is the goal of most pesticide applications used to manage weeds, insects and mites, and microorganisms. Other techniques such as cultivation or mowing of weeds and release of biological control agents against arthropods are also used to suppress pest populations.

Eradication

Eradication is the total elimination of a pest from a designated area. This is a common objective of pest control efforts in buildings or other small, confined spaces where, once the pest is eliminated, it can be excluded (eradicating cockroaches, rats, and mice from commercial food establishments, for example). Over larger areas, however, eradication is a radical approach to pest control which can be very expensive and often has limited success. Large eradication programs are usually directed at exotic or introduced pests posing an area-wide public health or economic threat; such programs are generally coordinated by governmental agencies. Recent Mediterranean fruit fly and hydrilla eradication efforts in California are examples of this type of pest management approach.

Which pest control strategy you choose depends on the nature of the pest, the environment of the pest, and economic considerations. Combining prevention and suppression techniques usually enhances a pest management program. Objectives might differ, however, for the same pest in different situations. For example, the Mediterranean fruit fly (Medfly) is an established pest in the state of Hawaii, so the emphasis there is both prevention and suppression to reduce crop damage. Regulatory agencies in California, however, are using eradication measures to prevent the Mediterranean fruit fly from becoming permanently established here, in hopes of preventing severe economic losses to the agricultural industry of the state.

SETTING UP A PEST MANAGEMENT PROGRAM

To establish a management program that will control an identified pest you must: (1) know what control methods are available; (2) evaluate the benefits and risks of each method; (3) select methods that are most effective and least harmful to people and the environment; (4) use several methods whenever possible; (5) use each method correctly; (6) observe local, state and federal regulations; and (7) evaluate the success of the pest management program.

Pest Management Methods

Once a pest problem or pest complex is anticipated or identified, you can begin to plan your pest management program. To carry it out, you must know what management methods are available and be able to evaluate the benefits and limitations of each. Select methods that are most effective but least harmful to people and the environment. There are usually several ways to manage pests or pest complexes; whenever possible, combine the use of several compatible methods into an integrated pest management program.

Biological Control

Most arthropod pests have natural enemies that control or suppress them effectively in some situations. Occasionally arthropods or pathogens, such as microorganisms, contribute to the control of certain weeds. Microorganisms also provide natural control of some species of plant pathogens, nematodes, birds, and rodents. Natural enemies and pathogens are being used successfully as biological control agents to manage certain insect, mite, fish, and weed pests.

Classical Biological Control. Classical biological control is directed against pests that are not native to a geographical area. These introduced pests are often problems in their new location because there are no natural enemies to help control them. Classical biological control involves locating the native home of an introduced pest and finding suitable natural enemies there. After extensive testing and evaluation, selected natural enemies are imported, reared, and released (Figure 3-1). If successful, the introduced natural enemies will become established within large areas and effectively lower target pest populations for long periods of time with no further need for intervention. The process is complicated because it is often difficult to locate the native home of some pests, and natural enemies cannot be released until it is proven that they will not become pests themselves. Laws have been enacted that strictly control the importation of all organisms, including biological control agents, into the United States. Other countries have similar restrictions.

Augmentation. Augmentation involves the mass release of large numbers of a natural enemy in a field, orchard, greenhouse, or other location for control of specific pests. It is similar to classical biological control since it requires rearing and releasing natural enemies; however, augmentation usually does not have long-term results so natural enemies must be released periodically. Several natural enemies are reared or cultured commercially; predatory mites are used for the control of plant-infesting spider mites and parasitic wasps and lacewings are used to control various insect pests. Nematodes are being studied as biological control agents for certain weeds and some insects, and a spray-on rust

FIGURE 3-1.

Classical biological control involves locating the native home of a pest (usually outside of the U.S.), finding and rearing one or more of its natural enemies, and releasing these natural enemies into the pest-infested area. In this photo the parasitic wasp Hyposoter exiguae *is laying an egg inside a beet armyworm larva.*

pathogen is under development as an early nutsedge control. General predators such as praying mantids and lady beetles are sold with claims made for biological control. In many cases, however, their effectiveness has not been established.

Naturally Occurring Control. Many factors influence pests, including natural enemies, environmental and geographical features, and climatic conditions. Maintaining populations of natural enemies by avoiding damaging cultural practices or the indiscriminate use of pesticides can be one of the most economical means of control. Select pesticides that are known to be less toxic to natural enemies or, if recommended, apply pesticides at lower-than-label rates so as not to affect natural enemy populations. Sometimes it is possible to modify certain parts of the environment, such as by planting crops or groundcovers to maintain or enhance natural enemies.

Pesticides

Pesticides often have a key role in pest management and occasionally may be the only control method available. Major benefits associated with

FIGURE 3-2.

Pesticides are an effective way of controlling pests. Usually pest damage stops within a few hours to a few days after a pesticide application, such as this fungicide being applied to carnation plants in a greenhouse.

the use of pesticides are their effectiveness, the speed and ease of controlling pests, and, in many instances, their reasonable cost compared to other control options. Usually pest damage stops or pests are destroyed within a few hours (for arthropods) to a few days (for weeds) after application of a pesticide. Plants may receive immediate, short-term protection against microorganisms after a fungicide is used (Figure 3-2). Any material that is applied to plants, the soil, water, harvested crops, structures, clothing and furnishings, or to animals to kill, attract, repel, or regulate or interrupt growth and mating of pests, or for the regulation of plant growth, is considered a pesticide. Herbicides, insecticides, fungicides, rodenticides, nematicides, and miticides are common types of pesticides. Pesticide types and classification are explained in Chapter 4.

Mechanical and Cultural Control

An important way to control pests or protect crops, livestock and poultry, people, and manufactured products is by the use of mechanical and cultural control methods. Mechanical devices and cultural techniques are used separately or together to exclude or trap pests, destroy them, alter their life cycle, encourage natural enemies, or change the environment so that it is not suitable for pest survival.

Exclusion. Exclusion consists of using barriers to prevent pests from getting into an area. Window screens, for example, exclude flies, mosquitoes, and other flying insects (Figure 3-3). Cracks, crevices, and small openings in buildings can be patched or sealed to exclude insects, rodents, bats, birds, or other pests. Fences and ditches make effective barriers against many vertebrate pests. Wire or cloth mesh excludes birds

FIGURE 3-3.

Window screens and paper tree protectors are mechanical devices that are used to exclude certain pests.

from fruit trees. Sticky material can be painted onto tree trunks, posts, wires, and other objects to prevent crawling insects from crossing.

Trapping. Traps physically catch pests within an area or building. Several types of traps are commonly used; some kill animals that come in contact with them, while others snare animals so they can then be relocated or destroyed. Traps are either mechanical devices or sticky surfaces.

Cultivation. Cultivation is one of the most important methods of controlling weeds and is also used for some insects and other soil-inhabiting pests. Mechanical devices such as plows, discs, mowers, cultivators, and bed conditioners are used to physically destroy weeds or control their growth and to disrupt conditions suitable for the survival of some microorganisms and insects.

Other Cultural Practices. Many different cultural practices have an influence on the survival of pests. In turf, mowing, irrigation, aeration, and fertilization are all important ways of preventing pest buildup and damage. In agricultural crops, the selection of crop plant varieties, timing of planting and harvesting, irrigation management and timing, crop rotation, and use of trap crops contribute to the reduction of weeds, microorganisms, insects, mites, or other pests. Weeds can also be managed with mulching (using plastic, straw, or other materials), flooding, solarization, and flaming (a process of exposing weeds to high-temperature flames).

Environment Modification. Pests which occur in enclosed areas may sometimes be suppressed by altering environmental conditions, such as the temperature, light, and humidity, of those areas. Refrigeration, for example, is used to protect stored food products, furs, and other items from insect pests; lowered temperatures either kill the insects, cause them to stop feeding, or prevent egg hatch or development. Installing bright lights in attics sometimes discourages bats from roosting there. Lowering the humidity of stored grains and other food products reduces damage from molds and some insects.

Sanitation. Sanitation practices can be used to control pests in many locations. Sanitation, or *source reduction*, involves eliminating food, water, shelter, or other necessities important to the pest's survival. In agriculture, sanitation includes removing weeds that harbor pest insects or rodents, removing weed plants before they produce seed, destroying diseased plant material, removing and destroying unharvested produce and crop residues, and keeping field borders or surrounding areas free of pests and pest breeding sites. Animal manure management is an effective sanitation practice used for preventing or reducing fly problems in poultry and livestock operations. Proper storage of feeds and preventing feed waste helps reduce rodent problems in these locations. In nonagricultural areas, certain pests can be controlled by methods such as draining standing water (Figure 3-4) and waste management. Waste management is very important; closed garbage containers and frequent garbage pickup eliminate food sources for flies, cockroaches, and rodents (Figure 3-5), while removal of soil, trash, and other debris from around and under buildings helps to reduce termite and fungal rot damage and prevent rodent nesting.

Host Resistance

Sometimes plants and animals can be bred or selected to resist specific pest problems. For example, particular livestock breeds are selected for

FIGURE 3-4.

Standing water such as this should be drained as an important cultural practice to reduce mosquito populations.

FIGURE 3-5.

Sanitation around homes, restaurants, and businesses is an important way of reducing populations of cockroaches, ants, and rodents. Garbage and trash containers must have tight lids to keep pests out.

physical characteristics that prevent attack by some pests or provide physiological resistance to disease or parasite organisms. Resistance is also influenced by the host's health and nutritional needs. Certain plant varieties are naturally resistant to insects, pathogens, or nematodes. All plants actually repel different types of pests and some contain toxic substances; pests, however, often adapt to the toxins or repellents, enabling them to utilize these plants as food.

Resistance to pests can sometimes be induced in growing plants by inoculating them with a microorganism. Also, genetic engineering research has demonstrated that it is possible to build pest resistance into the genetic makeup of some plants—this resistance is then transferred to subsequent generations.

Integrated Pest Management

Integrated pest management is an ecological approach to managing pests that often provides economical, long-term protection from pest damage or competition. Factors such as prior pest history, crop growth and development, weather, visual observations, pest monitoring information, and cultural practices are considered before control decisions are made. Integrated pest management programs emphasize prevention of weed competition and other types of pest damage by anticipating these problems whenever possible. Goals include conserving natural enemies and avoiding secondary pest problems through the use of nondisruptive and mutually compatible biological, cultural, and chemical methods. Chemical control methods are usually reserved for situations when decision-making guidelines and field-collected data indicate their use is economically and environmentally justified (Figure 3-6).

Concern about pesticides in the environment and their potential harm to users and consumers has increased in recent years. Integrated pest

FIGURE 3-6.

When using an integrated pest management approach, the use of pesticides is generally reserved for situations when decision making guidelines and field-collected data indicates that it is economically and environmentally justified.

management addresses many of the problems associated with chemical pest control because IPM relies on careful and continual pest monitoring; sometimes pesticide applications can be avoided, or, when needed, can be timed more accurately. Generally, less pesticide is required, resulting in a lower risk to people and to elements of the environment. Ongoing programs in many locations have proven IPM techniques to be economical and effective; through integrated pest management, long-term pest control has been achieved that reduces reliance on more expensive, short-term pest treatments.

Regulatory Pest Control

Some pest problems cannot be controlled successfully at a local level by individuals. These problems involve pests that seriously endanger or harm the health of the public or are likely to cause widespread damage to agricultural crops or animals, forests, or ornamental plants. The introduction and spread of such pests is prevented through quarantine or eradication programs and by abatement of pest sites as specified in state and federal laws.

Quarantine. Quarantine is a pest control process designed to prevent entry of pests into areas free of that pest. The State of California maintains inspection stations at all major entry points into the state to prevent importation of pests or materials that might harbor pests. There are 16

FIGURE 3-7.

This border inspection station located in Truckee is part of the effort to control the movement of pests into California. Inspection stations are in operation on all major highways into the state from Oregon, Nevada, Arizona, and Mexico.

border stations for inspection of vehicles entering the state from Oregon, Nevada, Arizona, and Mexico (Figure 3-7). Airports and ocean ports are monitored by the California Department of Food and Agriculture in co-operation with the U.S. Department of Agriculture and local offices of county agricultural commissioners.

Quarantine is also imposed to prevent movement of designated pests within the state. Produce and other identified items being shipped from a quarantine area must be fumigated to destroy pests before shipment.

Nursery stock, plant cuttings, and budding and grafting material are regulated to prevent the spread of pests. As part of this program the State of California maintains a registry of certified pest-free plant material that can be shipped throughout California.

Eradication. Eradication methods are used against pests that are un-acceptable at any level; these pests are usually also under quarantine restrictions. Once the need for eradication is confirmed, the geographical extent of pest infestation is determined and control measures are taken to eliminate this pest from the defined area. Procedures may include an area-wide spray program, release of sterile insects, use of mechanical and cultural practices, and intensive monitoring for pests within and around the borders of the infested area. Regulations prohibiting the planting or growing of host plants (host-free areas) or prohibiting the planting and growing of these plants during certain periods (host-free periods) may also be part of an eradication program.

Abatement. Weedy areas and neglected or abandoned crops create fire hazards and may harbor vertebrate, invertebrate, or microorganism pests or provide a reservoir of weed seeds that threaten adjoining areas or crops. Specific host plants attract or promote the buildup of certain pest invertebrates or disease-causing organisms. Some weedy plants are noxious to people or livestock. Government agencies are authorized to destroy weeds and plants that cause fire hazards, harbor harmful pathogens or animals, or are noxious to people or livestock in and around agricultural areas. Similar authority applies to diseased or infected livestock or poultry and to weeds and nuisance plants in residential, commercial, and industrial areas. Mosquito abatement is an important pest control function undertaken to protect public health. Under the authority of mosquito abatement laws, state agencies drain or treat standing water that provides breeding sites for mosquitoes.

Other. Additional areas in which regulatory activities play a role include: (1) protecting native plants and preventing them from being transported out of their native habitat; (2) regulating and protecting the honey bee industry; (3) regulating the production and sale of seeds within California; (4) regulating predatory animals that damage livestock, agricultural crops, or standing timber; and (5) regulating the disposal of vessel and aircraft garbage. All these activities have the potential of spreading pest organisms to areas where they do not presently exist, or causing injury or disease in high value crops, people, or livestock and poultry.

Due to the migratory or dispersal nature of some pests and the susceptibility of California to infestations of pests from other areas, the State of California has entered into a *pest control compact*. This is an agreement with other participating states to share in managing mutual pest problems. Cooperating states contribute money to a fund that is used to help finance pest control programs included within the compact.

4 Pesticides

Pesticides are an important part of many pest management programs.

Pesticides, as defined by the Federal Insecticide, Fungicide, and Rodenticide Act (FIFRA), are "... any substance or mixture of substances intended for preventing, destroying, repelling, or mitigating any insects, rodents, nematodes, fungi, or weeds, or any other forms of life declared to be pests; and any substance or mixture of substances intended for use as a plant regulator, defoliant, or desiccant." Materials used as pesticides include synthetically produced organic chemicals (an organic chemical is one that contains carbon and hydrogen in its basic structure), naturally occurring organic chemicals, naturally occurring inorganic chemicals, microbial agents (naturally occurring and those made through genetic manipulation), plus miscellaneous chemicals not commonly thought of as pest control agents. For instance, chlorine added to swimming pools for algae control is a pesticide, as are household disinfectants.

The selection, use, and handling of pesticides is a complex topic. Hundreds of materials are registered for use as pesticides and these act in many different ways to control or destroy pests. Some pesticides are much more toxic than others and therefore present higher risks to users as well as the environment. Most pesticides can be manufactured into several different types of formulations which also affect how they will react with target pests, nontarget organisms, and the environment. This chapter describes pesticide classification and formulation and it will help you understand pesticide toxicity and how toxicity is altered by different factors. Problems with pesticide incompatibility are also explained.

PESTICIDE TOXICITY

Toxicity, just like color or boiling point, is one of the characteristics used to describe a chemical. Toxicity is the capacity of a chemical to cause injury, sometimes referred to as *potency*. Pesticides, by their nature, must be toxic in order to destroy pests. Like most toxic chemicals, pesticides can be *hazardous* because they have a *potential* for causing harm to all living organisms. Some pesticides are more toxic or potent than others. The more toxic pesticides cause injury at smaller doses and therefore are more hazardous. One way to measure the toxicity of a pesticide is by giving test animals known doses and observing the results. In this way a lethal dose or lethal concentration is established; this information is used to predict the hazards to people and nontarget organisms. Toxicity testing is also used to determine the maximum doses people or organisms in the environment can be exposed to without injury.

Lethal Dose and Lethal Concentration

Pesticide toxicity is rated by determining the amount, or lethal dose, that will kill 50% of a test population of animals, referred to as the LD_{50}

PERCENT OF TEST ANIMALS LEFT ALIVE

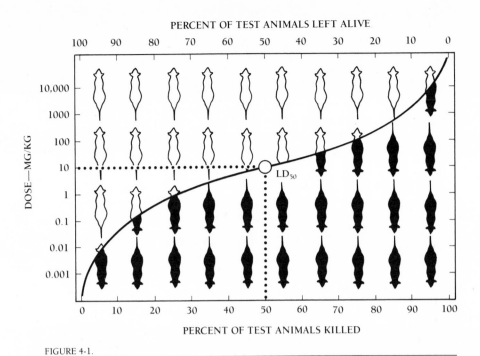

FIGURE 4-1.

The amount of pesticide that will kill half of a group of test animals is known as the LD_{50}. The smaller the LD_{50} the more toxic a pesticide is considered to be. LD_{50} values are established for both oral and dermal exposure. In this illustration the LD_{50} is 10 mg of pesticide per kg of body weight of test animal.

(Figure 4-1). Test animals are divided into groups so that different methods of exposure can be tested. LD_{50} is expressed as the milligrams (mg) of pesticide per kilogram (kg) of body weight of test animal (mg/kg). Toxicity is also rated by measuring how much pesticide vapor or dust in the air or what amount of pesticide diluted in river, stream, or lake water will cause death in 50% of a test animal population. This is the *lethal concentration* or LC_{50}. LC_{50} is expressed as micrograms (1/1,000,000 g) per liter of air or water (μg/l). Different organisms do not respond the same to toxic chemicals, however, and even organisms of the same species will sometimes respond differently because of their age, life stage, health, and the influences of environmental surroundings.

Pesticides that are more toxic present higher *hazards*, or risks of injury, to people and the environment (Table 4-1). Special precautions must be taken when handling or using these. In addition to toxicity, hazards associated with pesticides also vary according to the route of entry into the body. Normal routes of entry include the mouth, skin, eyes, and lungs. Pesticides can also be injected through the skin into a vein, artery, or tissues. The LD_{50} of a pesticide is usually established for skin applied (dermal) and ingested (oral) exposures.

Lethal dose or lethal concentration classifications do not provide information about chronic, long-term toxic effects. A pesticide that has a high LD_{50} (relatively nontoxic) may not necessarily be harmless; sublethal doses of some pesticides (doses lower than the LD_{50}) may cause skin or eye irritation, headache, nausea, or other ailments. Materials with these types of hazards require special precautions during handling and application.

TABLE 4-1.

Oral LD$_{50}$ Values for Some Pesticides. (LD$_{50}$ values may vary due to formulation types. These values are shown for comparative purposes only. Some chemicals listed may no longer be in use as pesticides.)

CHEMICAL	TRADE NAME	LD$_{50}$	TYPE OF PESTICIDE
aldicarb	Temik	0.79	insecticide
parathion		3	insecticide
azinphos-methyl	Guthion	11	insecticide
paraquat		150	herbicide
diazinon		300	insecticide
2,4-D		375	herbicide
carbaryl	Sevin	500	insecticide
copper hydroxide	C-O-C-S	1000	fungicide
pendimethalin	Prowl	1250	herbicide
malathion		1375	insecticide
ziram		1400	fungicide
propargite	Omite	2200	acaricide
iprodione	Rovral	3500	fungicide
trifluralin	Treflan	3700	herbicide
glyphosate	Roundup	4300	herbicide
simazine	Princep	5000	herbicide
captafol	Difolatan	6200	fungicide
captan		9000	fungicide
benomyl	Benlate	>10,000	fungicide
chlorothalonil	Bravo	>10,000	fungicide
oryzalin	Surflan	>10,000	herbicide
B. thuringiensis	Dipel	15,000	insecticide
methoprene	Precor	34,600	insect growth regulator

No Observable Effect Level

The no observable effect level (NOEL) is the maximum dose or exposure level of a pesticide that produces no noticeable toxic effect on test animals. NOEL is used as a guide to establish maximum exposure levels for people and for establishing residue tolerance levels on pesticide treated produce. Commonly, exposure levels and residue tolerances are set from 100 to 1000 times less than the NOEL to provide a wide margin of safety.

Threshold Limit Value

The threshold limit value (TLV) for a chemical, such as a pesticide used as a fumigant, is the airborne concentration of the chemical in parts-per-million (ppm) that produces no adverse effects over a period of time. The most common TLV is established for workers who might be exposed to low-level concentrations of a toxic chemical for 8 hours per day for 5 consecutive days. Sometimes a TLV is established for short-term exposure and would apply to workers who must briefly enter treated areas. The concentration of airborne chemical is higher than the long-term TLV but injury will not result because the exposure period is limited. The TLV is established by exposing animals to different airborne concentrations and observing and analyzing the results.

Pesticide Toxicity Classification

Pesticides are grouped into three categories by CDFA according to their toxicity or potential for causing injury to people: Category I, Category II, and Category III (Table 4-2). Category I pesticides are most toxic or hazardous and their use is normally restricted. Category III pesticides are least toxic to people and are generally less hazardous. Different label and regulatory requirements apply to each category, such as the requirement for use of closed mixing systems and other safety equipment, permits for the use and possession of some pesticides, and set-back distances between the application site and nontarget areas. The Federal Department of Transportation (DOT) also has a classification system for hazardous materials, which includes many pesticides, but should not be confused with the pesticide toxicity classification described here.

Category I Pesticides

Toxicity Category I pesticides (Figure 4-2) have an oral LD_{50} up to 50 mg/kg or a dermal LD_{50} up to 200 mg/kg. The signal word "Danger" appears on labels of pesticides in this category, along with the word "Poison" and a skull and crossbones. Category I pesticides are often the most hazardous because they are the most toxic. A few drops to a teaspoonful of a pesticide in this category could possibly cause death if taken orally. Less toxic pesticides may be included in toxicity Category I if there is a specific hazard, such as severe skin or eye injury, or a particular danger to the environment. For those, the signal word "Danger" appears on the label, but not the word "Poison" or the skull and crossbones.

Category II Pesticides

Toxicity Category II (Figure 4-3) includes pesticides that have an oral LD_{50} between 50 and 500 mg/kg or a dermal LD_{50} between 200 and

TABLE 4-2.

Pesticide Toxicity Categories.

HAZARD INDICATORS:	TOXICITY CATEGORIES			
	I *DANGER*	II *WARNING*	III *CAUTION*	IV *CAUTION*
Oral LD_{50}	Up to and including 50 mg/kg	From 50 through 500 mg/kg	From 500 through 5,000 mg/kg	Greater than 5,000 mg/kg
Inhalation LC_{50}	Up to and including 0.2 mg/liter (0–2,000 ppm)	From 0.2 through 2 mg/liter (2,000–20,000 ppm)	From 2 through 20 mg/liter (Greater than 20,000 ppm)	Greater than 20 mg/liter
Dermal LD_{50}	Up to and including 200 mg/kg	From 200 through 2,000 mg/kg	From 2,000 through 20,000 mg/kg	Greater than 20,000 mg/kg
Eye effects	Corrosive; corneal opacity not reversible within 7 days	Corneal opacity reversible within 7 days; irritation persisting for 7 days	No corneal opacity; irritation reversible within 7 days	No irritation
Skin effects	Corrosive	Severe irritation at 72 hours	Moderate irritation at 72 hours	Mild or slight irritation at 72 hours

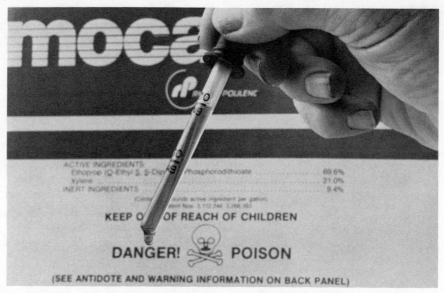

ACTIVE INGREDIENTS:
Ethoprop (O-Ethyl S, S-Dip___ Phosphorodithioate 69.6%
Xylene 21.0%
INERT INGREDIENTS 9.4%

(Cont___ pounds active ingredient per gallon)
___ent Nos. 3,112,244 3,268,393

KEEP O___ OF REACH OF CHILDREN

DANGER! ☠ POISON

(SEE ANTIDOTE AND WARNING INFORMATION ON BACK PANEL)

FIGURE 4-2.

Category I pesticides are recognized by the signal word "Danger" on the label. A few drops to a teaspoonful of Category I pesticide, taken internally, would probably cause death. These pesticides have an oral LD_{50} of 50 mg/kg or less and a dermal LD_{50} of 200 mg/kg or less.

2000 mg/kg. The signal word "Warning" is used on labels of Category II materials, indicating they are moderately hazardous. Between 1 teaspoonful to 1 ounce (6 teaspoons) of chemical in this group would probably kill an adult.

Category III Pesticides

Toxicity Category III (Figure 4-4) pesticides have an oral LD_{50} over 500 mg/kg and a dermal LD_{50} greater than 2000 mg/kg. These pesticides have the signal word "Caution" printed on their labels, which indicates they may be slightly hazardous. Taken orally, over 1 ounce of pesticide in this category would probably be required to cause death in an adult. EPA regulations provide for a fourth group of pesticides, Category IV, which include materials that have an oral LD_{50} greater than 5000 mg/kg and a dermal LD_{50} greater than 20,000 mg/kg; these must be labeled with the signal word "Caution," so are often grouped with pesticides in Category III.

Plant and Animal Testing

Pesticides are tested on living plants and animals to determine their toxicity and predict hazards to people and nontarget plants and animals. These tests also help establish exposure levels and provide information on mode of action. Different types of tests are performed depending on what kind of information is needed. In some cases, animals are fed small doses of a pesticide on a daily basis. Lower than the LD_{50}, they are known as *sublethal* doses. Such studies are used to establish *no observable effect levels* (NOEL) and to give information on the long-term, or *chronic*, effects of pesticides. Studies also assess the potential for causing sterility, birth defects, cancer, or other problems in people.

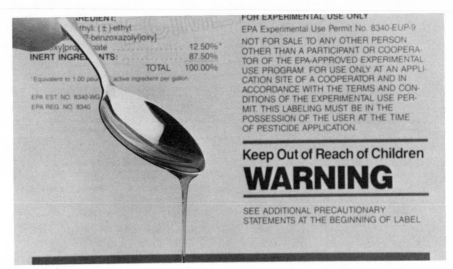

FIGURE 4-3.

Category II pesticides have the signal word "Warning" on their labels. One teaspoonful to 1 ounce of pesticide in this category would probably kill a person if taken internally. Category II pesticides have an oral LD_{50} in the range of 50 to 500 mg/kg and a dermal LD_{50} in the range of 200 to 2000 mg/kg.

FIGURE 4-4.

Category III pesticides are recognized by the signal word "Caution" on their labels. These pesticides are the least toxic. It would probably take more than 1 ounce of material in this category, taken internally, to kill an adult. Category III pesticides have an oral LD_{50} greater than 500 mg/kg and a dermal LD_{50} greater than 2000 mg/kg.

Short-term toxicity tests are used to establish LD_{50} and LC_{50} values; these are known as the *acute* effects. Groups of animals are given single, high doses of a pesticide, allowing researchers to measure immediate responses and to study pesticide mode of action. Research has shown that exposing animals for short periods of time to large doses of a chemical can also help predict human hazards from exposure to small doses over longer periods.

Pesticides are tested on mice, rats, rabbits, and dogs. Toxicity testing is also performed on nontarget plants, insects (such as bees), fish, amphibians (frogs, toads, salamanders), deer, birds, and other wildlife when these animals are at risk from pesticide exposure. Extensive tests are performed on target pests to establish dosage rates and to determine how well the pesticide works under different conditions. The effectiveness of a pesticide on its target pest is referred to as its *efficacy*.

Factors Influencing Pesticide Toxicity

Certain conditions increase or decrease the ability of a pesticide to control a pest. Some of these are temperature, humidity, and exposure to sunlight, wind, or rain. Even genetic variations cause some plants or animals to respond differently. Also, the age and general health of a pest often influence how well pesticides work. Weeds that are water-stressed or are not growing vigorously may be more tolerant to some herbicides, for example.

Factors in the environment sometimes change pesticides into different chemicals that may be less toxic or more toxic than the original pesticide. The time it takes for a pesticide to lose half of its original form is known as its *half-life*. Half-life may be affected by the chemical nature of the pesticide and its formulation, soil microbes, ultraviolet light, quality of the water used in mixing, or impurities combined with the pesticide. (Impurities can contaminate pesticides during manufacture, formulation, storage, and mixing.) Combining pesticides with other pesticides can also change toxicity or alter the half-life. Pesticides that remain in their active state in the environment for long periods of time are said to be *persistent*. Pesticide persistence is discussed in Chapter 6.

Most pesticides can be manufactured into several different types of formulations which also affect how they will react with target pests, nontarget organisms, and the environment.

HOW PESTICIDES ARE CLASSIFIED

Pesticides are classified according to their function (Table 4-3). For example, *insecticides* control insects and *herbicides* control weeds. *Plant growth regulators* are usually used to enhance growth or fruiting of cultivated plants, although they are considered to be pesticides. *Attractants* and *repellents* are also considered pesticides because they are used for pest control.

Many specific pesticides can be used to control more than one group of pests and therefore can be found in more than one pesticide class. The compound 2,4-D, for instance, when used at low rates is a plant growth regulator but is an effective herbicide when applied at higher concentrations. Oxythioquinox (Morestan) can be used as an acaricide, fungicide, or insecticide.

TABLE 4-3.

*Pesticide Classification Based on Target Pests and Pesticide Functions.
(Some chemicals listed here may not be currently registered as pesticides.)*

PESTICIDE TYPE	PESTS CONTROLLED OR FUNCTION	EXAMPLES OF PESTICIDES
acaricide	mites	propargite (Omite, Comite) fenbutatin-oxide (Vendex)
algaecide	algae	copper sulfate dichlone endothall (Hydrothol 191)
attractant	attracts pests	pheromones baits miscellaneous chemicals
avicide	birds	aminopyridine (Avitrol) starlicide Ornitrol
bactericide	bacteria	oxytetracycline (Mycoshield) copper compounds
defoliant	removes plant foliage	endothall (Accelerate) thidiazuron (Dropp) tributyl phosphorotrithioite (Folex)
desiccant	removes water from arthropod pests	boric acid powder silica gel diatomaceous earth
fungicide	fungi	benomyl (Benlate) copper sulfate captan
growth regulator	regulates plant or animal growth	gibberellic acid (Pro-Gibb) chlorocarbanilate (Sprout Nip) methoprene (Precor)
herbicide	weeds	atrazine bromoxynil (Buctril) trifluralin (Treflan) paraquat petroleum oil
insecticide	insects	diazinon permethrin (Ambush) azinphos-methyl (Guthion) parathion petroleum oils
molluscicide	snail or slugs	Snarol mesurol triphenmorph (Frescon) clonitralid (Bayluscide)

PESTICIDE TYPE	PESTS CONTROLLED OR FUNCTION	EXAMPLES OF PESTICIDES
nematicide	nematodes	carbofuran (Furadan) phosphoramidate (Nemacur) dichloropropene (Telone)
piscicide	fish	rotenone Lamprecide antimycin (Fintrol)
predacide	mammal predators	strychnine zinc phosphide compound 1080
repellent	repels animals or invertebrates	deet mesurol (methocarb) avitrol thiram
rodenticide	rodents	chlorophacinone strychnine hydroxycoumarin (Warfarin) diphacinone (Diphacin) brodifacoum (Talon) bromadiolone (Maki)
silvicide	trees and woody shrubs	tebuthiuron (Spike) petroleum oils

Pesticide Chemical Groups

Pesticides are also grouped according to chemical origin. This type of classification often reveals common characteristics such as mode of action, chemical structure, types of formulations possible, persistence in the environment, and how they may be broken down through biological processes. Table 4-8 lists the chemical groups for most commonly used pesticides and their modes of action.

Insecticides

Chemicals used as insecticides come from different chemical groups such as those listed in Table 4-8. Three of the most common insecticide groups are the organochlorines, organophosphates, and carbamates.

Organochlorines. Organochlorines (also known as chlorinated hydrocarbons) were frequently used for insect and mite control, although most early forms of these compounds have now been banned due to environmental persistence or other problems. DDT, chlordane, toxaphene, and dieldrin are some of the earlier developed organochlorine insecticides which are no longer used in agriculture. Others still in use include dicofol and methoxychlor. Newly developed organochlorines are being used for insect and rodent control. Most organochlorines do not break down rapidly in the environment, and many are stored in fat tissues of animals. Some of these materials are quite poisonous to mammals, including people, while others are reasonably harmless. The ability of organochlorine pesticides to persist for long periods of time make them useful for structural pest control and other applications in which long residuals are needed.

Organophosphates. Organophosphates are an important group of pesticides, widely used for control of insects and mites. These pesticides are derivatives of phosphorous compounds and some are among the most acutely toxic chemicals known. Many organophosphates are easily absorbed through a person's skin, lungs, or digestive tract. They interfere with animal and human nervous systems, similar to carbamates. Organophosphate pesticides usually break down rapidly in the environment once they have been applied, which is a benefit for most pest control applications. Well-known organophosphate insecticides include parathion, malathion, phosdrin, diazinon, chlorpyrifos (Lorsban), and azinphos-methyl (Guthion).

Carbamates. Carbamates are a widely used group of synthetic organic pesticides because they are highly effective, moderately priced, and, under normal conditions, short-lived in the environment. They are derivatives of carbamic acid, and include the sulfur-containing subgroups of dithiocarbamates and thiocarbamates. Besides being used as insecticides, carbamates have uses as fungicides, herbicides, molluscicides, and nematicides. In animals, some carbamates impair nerve function and some are highly toxic to mammals, including human beings. Carbamates are not stored in animal tissue, so their toxic effects are often short-lived and reversible. Examples of carbamate insecticides include carbaryl (Sevin), aldicarb (Temik), and methomyl (Lannate, Nudrin).

Herbicides

Table 4-8 lists many of the chemical groups that have compounds used to control weeds. Early herbicides were generally broad-spectrum contact materials which are now being replaced by more unique chemicals that destroy weeds through many varied and specific modes of action.

Fungicides

Fungicides include compounds such as inorganic metals and sulfur and a wide spectrum of synthetic organic chemicals. These different chemical groups are listed in Table 4-8.

Special Types of Pesticides

A few pesticides are specialized types of materials that occur naturally, are produced by living organisms, or are unique in other ways, but do not belong to specific chemical groups. These include antibiotics, anticoagulants, botanicals, insect growth regulators, inert dusts, microbials, petroleum oils, pheromones, plant hormones, and soaps.

Antibiotics

An antibiotic is a material produced by one organism that kills or inhibits another. The antibiotic penicillin, used for control of bacterial infections in people and animals, is derived from a fungus. Streptomycin, used to treat bacterial diseases in people, animals, and plants, is both a naturally and synthetically produced antibiotic. Terramycin and other similar synthetic antibiotics resemble naturally occurring antibiotics and are used to control several plant diseases caused by fungi, viruses, and bacteria.

Anticoagulants

Anticoagulants interfere with the blood-clotting mechanism of mammals, causing them to die of blood loss after sustaining an injury. Anticoagulants are used to control rodents such as rats and mice. Rodents must feed on some anticoagulants over a period of several days before accumulating enough toxic material to cause death. Other anticoagulants are effective after a single feeding.

Botanicals

Some plants contain substances that are naturally poisonous to insects and other animals. These include certain species of chrysanthemum flowers from which pyrethrum is extracted, the roots of the cube plant that supply rotenone, and species of lily plants that provide sabadilla and hellebore. Ryania is derived from a tropical South American plant. Nicotine, extracted from tobacco, was once used extensively as an insecticide, but its use has declined with the introduction of newer, synthetic pesticides. Strychnine is obtained from the dried seed of a small tree found in India, Ceylon, Australia, and French Indochina. Table 4-4 lists insecticidal chemicals derived from plants.

Inert Dusts

Inert dusts, also called *desiccants* or *sorptive* dusts, are fine powders, often low in toxicity, that are used for control of insects and other invertebrates. These dusts kill pests through a physical, rather than chemical, action; some are abrasive and scratch the pest's waxy body covering, causing them to lose water, while others remove (or *adsorb*) the protective waxy coating (Figure 4-5). Inert dusts are sometimes combined with aluminum fluosilicate to give the mixture an electrostatic charge and cause it to cling to surfaces, including the pest's outer covering. Because some inert dusts are low in toxicity, they are applied in locations where other pesticides cannot be safely used. The killing action is physical rather than chemical, so these dusts do not lose their effectiveness over time due to environmental degradation. Inert dusts are not effective once they become wet, however. Diatomaceous earth, silica gel, and boric acid powder are some of the materials used as desiccants. These pesticides leave a highly visible residue on all treated surfaces. Boric acid powder is toxic if ingested, so should not be used in areas accessible to young children. Although most inert dusts are low in toxicity, avoid inhaling them because they can cause serious lung irritation.

DESICCANT BEING APPLIED . . .

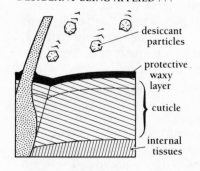

desiccant particles

protective waxy layer

cuticle

internal tissues

AFTER APPLICATION . . .

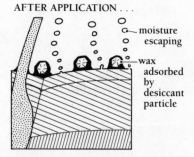

moisture escaping

wax adsorbed by desiccant particle

FIGURE 4-5.

Desiccants (inert dusts and sorptive powders) destroy insects and mites by removing or disrupting the protective outer body covering, as illustrated here. This causes the organism to lose body fluids.

TABLE 4-4.

Insecticidal Chemicals Derived From Plants.

INSECTICIDE	SOURCE	USES/COMMENTS
Pyrethrum	Extract from dried flowers of certain chrysanthemum species.	Has contact, stomach, and fumigant poisoning action on insects. Is also toxic to cold blooded animals. Kills aphids, mosquitoes, flies, fleas, mealy bugs, cabbageworms, thrips, beetles, leafhoppers, lice, loopers, and many others. Insecticidal action is degraded rapidly by sunlight. Insecticidal action is often enhanced by addition of piperonyl butoxide, a synergist.
Pyrethrins	Chemical extracts of naturally occurring pyrethrum.	Similar uses as for pyrethrum. Many different pyrethrins are derived from pyrethrum. Some may have more specific action to certain insect pests and be safer to nontarget insects.
Rotenone	Derived by grinding roots of certain legume plants (68 different species). U.S. supplies come primarily from roots of the Cube plant.	Contact and stomach poison. Used for control of beetles, weevils, slugs, loopers, mosquitoes, thrips, fleas, lice, and flies. Also used for control of unwanted fish. Also acts as a repellent and acaricide. Rotenone is slow acting and has a short residual. It is nontoxic to honey bees.
Nicotine	Nicotine is an extract from several species of tobacco, usually used as the sulfate.	Registered for use on many types of crops. Used in greenhouses and household applications for control of aphids, thrips, leaf hoppers, and other sucking insects. Kills by contact and fumigation poison activity. Toxic to people and domestic animals if used improperly. Used also to repel dogs and rabbits.
Sabadilla	Obtained from the dried, ripe seed of a South American lily plant.	Has contact and stomach poison action against cockroaches, several species of bugs, potato leafhopper, imported cabbage worm, house fly, thrips, and the cattle louse. Is also toxic to honey bees. Used on many types of tree and vine fruits, forage crops, and vegetables. Not highly toxic to mammals.
Hellebore	Made from the dried rhizomes of several species of lily plants, many of which occur naturally in the U.S.	Used against several types of insects. Hellebore is rapidly broken down by sunlight. Does not have high insecticidal activity.

INSECTICIDE	SOURCE	USES/COMMENTS
Ryania	Ryania insecticide is the powdered roots, leaves, and stems of a native South American plant. A synergist is often used to enhance its activity.	This compound is effective against corn earworm, codling moth, German cockroach, house fly, mosquitoes, European corn borer, oriental fruit moth, and the imported cabbage worm. Ryania has low toxicity to mammals.

Insect Growth Regulators

Insect growth regulators (IGRs) are chemicals used to control insects by modifying normal development (Table 4-5). Growth hormones produced by insects control how long an insect will remain in each of its larval or nymphal stages and when it becomes a reproductive adult. IGRs are synthetically produced chemicals that either mimic or interrupt the action of natural growth hormones. They block the insect's ability to change from a juvenile to an adult or force it to change into an adult before it is physically able to reproduce.

TABLE 4-5.

Insect Growth Regulators.

COMPOUND	BRAND NAME(S)	USES
hydroprene	Gencor	Used for control of cockroaches.
methoprene	Altosid, Diacon, Pharorid, Precor	Controls ants, cigarette beetles, fleas, flies, leafhoppers, lice, mosquitoes, moths, stored products pests, and others.

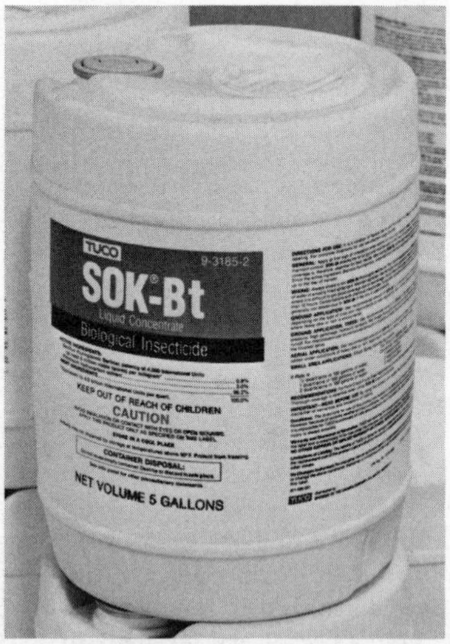

Microbials

Microbial pesticides are microorganisms that have been combined with other ingredients to form pest control products. Strains of the bacterium *Bacillus thuringiensis* (Dipel, SOK-Bt, Thuricide, Vectobac) are used to control species of moth larvae, mosquito larvae, and black fly larvae (Figure 4-6); *Agrobacterium radiobacter* is used to control the bacterium that produces crown gall in trees, shrubs, and vines. A fungus manufactured under the name of Mycar controls some pest mites. Another fungus formulated into the herbicide Collego is used in rice and soybeans for control of the northern jointvetch weed. A virus produced under the name Elcar is used for control of several moth pests, and another virus is used for control of codling moth. The plant disease *Phytophthora palmivora*, sold as the herbicide Devine, is used to control milkweed vine in citrus. There is an increasing interest in the use of microbial pesticides because of their extremely low hazard to people and nontarget organisms and their specificity to target pests. Besides naturally occurring microbial organisms, genetically altered organisms are part of a new group of materials showing promise as pest control products.

FIGURE 4-6.

Microbial pesticides contain microorganisms that kill other living organisms. The insecticide shown here contains spores of the bacterium Bacillus thuringiensis.

Petroleum Oils

Some highly refined petroleum oils are used as insecticides and acaricides. These are often used to control aphids, scales, mealybugs, eggs of these insects, and mites and mite eggs. Petroleum oils destroy plant-feeding pests through a suffocating action. Less refined petroleum oils are used as nonselective herbicides; they destroy plants by injuring cell membranes. Insecticidal and herbicidal oils are usually formulated with emulsifiers and other inert ingredients to improve mixing in water.

Refined oils used as insecticides and acaricides consist of five grades of *classified* oils (light, light-medium, medium, heavy-medium, and heavy) and two types of *unclassified* oils (supreme and narrow-range). Oils in these groups include types commonly referred to as dormant oils, summer oils, supreme oils, superior oils, and narrow-range oils. Each type has certain features important for controlling pests and reducing injury to treated plants (materials that injure plants are said to be *phytotoxic*).

Three characteristics (Table 4-6) which influence the safety and effectiveness of insecticidal oils are: (1) the *unsulfonated residue* (UR) rating; (2) the *distillation temperature and range*; and (3) the *hydrocarbon composition* of the oil.

Unsulfonated residue is a measure of purity, therefore the closer the UR rating is to 100, the purer the oil. Oils used for insecticides and acaricides must have a minimum UR rating based on the grade or type of oil. Oils with higher UR ratings are safer for use on plants.

Most oils are mixtures of molecules of different sizes and types. As these oils are heated, smaller molecules begin to distill. When temperatures increase even more, larger molecules begin to distill. Classified oils are rated at the temperature where 50% of the oil distills under a vacuum of 10 mm of mercury; as this number gets smaller the oil becomes lighter

TABLE 4-6.

Characteristics of Petroleum Oils Used as Insecticides and Acaricides.

CLASSIFIED OILS: Grade	Minimum UR*	50% Point °F**	Percent Distilled at 636° F and Atmospheric Pressure
Heavy	94	671	10 – 25
Heavy–Medium	92	656	28 – 37
Medium	92	645	40 – 49
Light–Medium	92	628	52 – 61
Light	90	617	64 – 79

UNCLASSIFIED OILS: Type	Minimum UR*	50% Point °F**	10 – 90% Range***	Minimum %Cp****
Supreme	92	490	100 – 110	60
NR 440	92	440 + 8	80	60
NR 415	92	415 + 3	60	60

*UR = *Unsulfonated Residue* (see text).
**The temperature at which half of the oil distills under a vacuum of 10 mm of mercury.
***See text for description of distillation temperature range.
****The percent of paraffin-base molecules in the oil (see text).

FIGURE 4-7.

Pheromones are unique chemicals that attract certain species of insects; they are used in pest management programs to monitor pest insect activity and to determine when insecticide applications should be made. This photograph shows codling moths being monitored in a walnut orchard.

and therefore less phytotoxic. Unclassified oils are rated by the temperature range in which 10 to 90% of the oil distills. As the temperature range becomes smaller, the oil is more highly refined and consists of more uniform molecules. Oils of this type used as pesticides are known as narrow range oils.

Oils are usually mixtures of different types of hydrocarbon molecules, including paraffins, napthenes, aromatics, and unsaturates. Effectiveness and safety to plants are influenced by the percentages of these hydrocarbons in the oil. The most suitable oils for use as insecticides and acaricides are the paraffin-based oils, since they are more toxic to pests but least phytotoxic. The percent of paraffin is represented by %Cp. Unclassified oils must have a minimum %Cp of 60.

Pheromones

Pheromones are unique chemicals produced by animals to stimulate behavior in other animals of the same species. Many insects depend on pheromones to locate mates. Artificial insect pheromones are used in pest control as tools for monitoring insect activity, for timing insecticide applications, or to attract insects to poisoned sprays. Synthetic pheromones are commonly used with sticky insect traps and play an important role in the monitoring of insect activity in integrated pest management programs and for monitoring insecticide resistance (Figure 4-7).

Plant Growth Regulators and Plant Hormones

Plant growth regulators and plant hormones are chemicals that are either derived from plants, are chemically synthesized to mimic naturally occurring plant chemicals, or are other chemicals that induce growth changes in plants. In nature, hormones control plant functions such as flowering, fruit development, nutrient storage, and dormancy. Plant growth regulators and hormones are used to regulate plant growth, enhance fruit production, remove foliage for ease in harvesting a crop, and to destroy undesirable plants. Table 4-7 lists some of the available plant hormones and chemicals that have a hormonelike action.

TABLE 4-7.

Commercially Available Plant Growth Regulators and Plant Hormones.

COMPOUND	BRAND NAMES	USES*
carbaryl	Sevin	Used as thinning agent on apples.
chlorpropham	Sprout Nip	Inhibits potato sprouting.
cytokinin	Cytex	Cytokinins are extracts of kelp and marine algae. They increase cell elongation, improve fruit set, and enhance color.
daminozide	Alar	Controls height, retards growth, promotes early budding of ornamentals. Hastens ripening and concentrates maturity of cherries, peaches, nectarines. Controls premature ripening and prevents drop in pears. Many other uses.

*Some materials listed here may not be registered for use as plant growth regulators or plant hormones. Check current labels before use.

COMPOUND	BRAND NAMES	USES*
dimethipin	Harvade	Used as a desiccant, defoliant, and abscission agent for improving cotton harvest.
ethephon	Ethrel, Florel, Prep	Hastens yellowing of tobacco. Promotes and concentrates maturity of berries, tree fruits, grapes, tomatoes, and other crops. Loosens walnuts to improve harvest. Used to eliminate unwanted fruit and leaves. Promotes flowering in certain ornamentals. Accelerates opening of mature unopened cotton bolls and enhances defoliation to improve early harvest.
gibberellic acid	Pro-Gib, Gibrel	Increases fruit set, size and yield. Controls lemon maturity. Breaks dormancy in potatoes. Produces larger and more profuse flowers in ornamentals. Many other uses in vegetable, grain, and fruit crops.
gibberellic acid and N-(phenylmethyl)-1H-purine-6-amine	Promalin	Applied to apples to improve shape, size, firmness, color, and yield.
mefluidide	Embark	Regulates growth of turfgrass, broadleaf vegetation, and ornamentals in landscaped areas.
mepiquat-chloride	Pix	Improves yield, shortens plant height, accelerates maturity, and opens canopy of cotton plants.
naphthaleneacetic acid (NAA)	Fruitone N, NAA-800, Stik, Tre-Hold	Controls drop in apples and pears. Used as apple thinning agent. Used as wound dressing to suppress sprouting from pruning cuts.
N-[phenylmethyl]-1H-purine-6-amine	Pro-Shear	Used to improve growth and fullness of white pine grown for Christmas trees.
N-(phenylmethyl)-9-(tetrahydro-2H-pyran-2-yl)-9H-purine-6-amine	Accel	Increases number of lateral branches of roses, carnations, and other ornamental flowers.
pyridazinedione	Royal MH-30, Royal Slo-Gro, Sprout Stop, Sucker-Stuff	Controls sprouting of potatoes and onions.
2,4-D	2,4-D	Increases fruit size and reduces drop in citrus. Intensifies color and improves skin appearance in potatoes. (Also used as herbicide).

*Some materials listed here may not be registered for use as plant growth regulators or plant hormones. Check current labels before use.

Soaps

Pesticide soaps are used for control of insects, mites, mosses, liverworts, algae, and lichens. Killing action is obtained because soap interferes with cellular metabolism of the target pest. Insecticidal soaps are most effective on soft-bodied insects such as aphids, scales, psyllids, and some larval stages of other insects. Pesticide soaps have the advantage of being nearly nontoxic to vertebrates, including people. Some nontarget plants may be damaged by soap sprays, however, so check the label for any restrictions before use. Use only soaps that have been labeled for pest control.

MODE OF ACTION

The way that a pesticide destroys or controls a target organism is known as its *mode of action*. Some of the different modes of action are listed in Table 4-8. Understanding modes of action makes it easier to select the right pesticide and predict which pesticides will work best in a particular situation. If pests show resistance to one pesticide, selecting a material with a different mode of action will often achieve better control.

In general, pesticides within a chemical class have the same mode of action on specific types of pests. They may also have similar characteristics such as chemical structure, persistence in the environment, and types of formulations possible.

The type of injury caused by a pesticide may be local, systemic, or both. Local injury usually involves damage to tissues that initially come in contact with the pesticide, such as leaves. Pesticides with systemic action move, after application, to other tissues where damage may occur. For instance, some leaf-applied herbicides move through the plant to growing points (roots and shoots). Ingested anticoagulant rodenticides are carried from the intestines into the blood of target rodents to interrupt the normal clotting mechanism. Organophosphate and carbamate insecticides interfere with nerve transmission at specific sites within the central nervous system of insects.

Another form of systemic activity involves the application of pesticides to host plants or animals to *protect* them from damage by pest organisms. For example, systemic insecticides may be applied to soil where cultivated plants are growing; these are picked up by plant roots and carried to the leaves, making the plants toxic to foliage-feeding insects. Pets and livestock may be fed low doses of systemic insecticides for control of internal and external parasites. Trees are sometimes injected with insecticides or antibiotics to protect them from pest organisms.

The mode of action of certain herbicides is to destroy weeds by damaging leaf cells and causing plants to dry up. Others alter the uptake of nutrients or interfere with the plant's ability to grow normally or convert light into food. The mode of action often dictates when and how an herbicide is used. Those that inhibit seed germination or seedling growth are used as *preemergent* herbicides; they are incorporated into the soil to control weed seedlings before they break through the soil surface. Other types are used as *postemergent* herbicides and are applied to the foliage

TABLE 4-8.

Pesticide Chemical Groups.

CHEMICAL TYPE	EXAMPLES	MODE OF ACTION
INSECTICIDES:		
Petroleum oils	Supreme oil, superior oil	Physical toxicants
Organochlorines	DDT, Methoxychlor	Axonic poisons
	Hexachlorocyclohexanes	Axonic poisons
	Cyclodienes (chlordane, thiodan)	Central nervous system synaptic poisons
	Polychloroterpenes (toxaphene)	Axonic poisons
Organophosphates		
Aliphatic derivatives	malathion	Central nervous system synaptic poisons
	dimethoate (Cygon)	
	disulfoton (Di-Syston)	
Phenyl derivatives	ethyl parathion	Central nervous system synaptic poisons
	methyl parathion	
	sulprofos (Bolstar)	
Heterocyclic derivatives	diazinon	Central nervous system synaptic poisons
	azinphos-methyl (Guthion)	
	chlorpyrifos (Lorsban, Dursban)	
	phosmet (Imidan)	
Organosulfurs	propargite (Omite)	
	aramite	
Carbamates	carbaryl (Sevin)	Central nervous system synaptic poisons
	methomyl (Lannate, Nudrin)	
	aldicarb (Temik)	
	propoxur (Baygon)	
	bendiocarb (Ficam)	
Formamidines	chlordimeform (Galecron)	Adrenergic insecticides
	amitraz (Baam)	
Thiocyanates	thanite	Interfere with cellular respiration and metabolism
Dinitrophenols	dinoseb	Metabolic inhibitors
	dinocap (Karathane)	
Organotins	cyhexatin (Plictran)	Metabolic inhibitors
	fenbutatin-oxide (Vendex)	
Botanicals	Nicotine	Postsynaptic poisons
	Rotenone	Metabolic inhibitors
	Sabadilla	Muscle poisons
	Ryania	Muscle poisons

CHEMICAL TYPE	EXAMPLES	MODE OF ACTION
	Pyrethrum	Axonic poisons (also inhibit mixed function oxidase when mixed with a synergist)
Pyrethroids	permethrin (Ambush and Pounce) fenvalerate (Pydrin) allethrin (Pynamin) resmethrin (Synthrin)	Axonic poisons (also inhibit mixed function oxidase when mixed with a synergist)
Inorganics	silica gel	Physical toxicants
	boric acid	Physical toxicants
	sulfur	
	arsenic	Inhibit respiration
Fumigants	methyl bromide	Narcotic and alkylating agents
	ethylene dibromide	Narcotic
	hydrogen cyanide	Narcotic and alkylating agents
	chloropicrin	Narcotic
	vapam	
	telone	
	naphthalene	
Microbials	*Bacillus thuringiensis* viruses fungi	Various
Insect growth regulators	methoprene diflubenzuron (Dimilin) other chitin synthesis inhibitors	Influence growth and development
HERBICIDES:		
Inorganics	sodium chlorate	Desiccant
Petroleum oils		Physical toxicants
Organic arsenicals	MSMA, DSMA, cacodylic acid	Interfere with cellular respiration and metabolism and other functions
Phenoxyaliphatic acid	2,4-D, 2,4,5-T diclofop methyl (Hoelon)	Multiple actions
Amides	propanil (Kerb) napropamide (Devrinol) alachlor (Lasso) metolaclor (Dual)	Inhibit root and shoot growth
Substituted anilines	trifluralin (Treflan) oryzalin (Surflan) pendimethalin (Prowl)	Inhibit root and shoot growth

CHEMICAL TYPE	EXAMPLES	MODE OF ACTION
Substituted ureas	tebuthiuron (Spike) diuron (Karmex) fenuron (Dybar)	Block photosynthesis
Carbamates	propham (Chem-Hoe) barban asulam (Asulox)	Block photosynthesis and interfere with cell division
Thiocarbamates	molinate (Ordram) cycloate (Ro-Neet) butylate (Sutan)	Interfere with cellular respiration and metabolism, block photosynthesis, and inhibit root and shoot growth
Triazines	atrazine (Aatrex) simazine (Princep) metribuzin (Lexone) cyanazine (Bladex)	Block photosynthesis
Aliphatic acids	TCA dalapon	Unknown
Substituted benzoic acids	dicamba (Banvel) DCPA chloramben (Amiben)	Unknown
Phenol derivatives	dinoseb	Destroy cell membranes, also a desiccant
Nitriles	dichlobenil (Casoron) bromoxynil (Buctril)	Interfere with cellular respiration and metabolism and inhibit carbon dioxide fixation
Bipyridyliums	diquat, paraquat	Destroy cell membranes, desiccant, and block photosynthesis
Microbials	*Phytophthora palmivora* (Devine)	
Uracils	bromacil (Hyvar-X) terbacil (Sinbar)	Block photosynthesis
Sulfonylureas	chlorsulfuron (Glean) sulfomethuron-methyl (Oust)	Interfere with cell division
Miscellaneous herbicides	endothall glyphosate (Roundup) oxyfluorfen (Goal)	Inhibits metabolism and protein synthesis
PLANT GROWTH REGULATORS: Auxins Gibberellins Cytokinins Ethylene generators Growth inhibitors and retardants	IAA, 2,4-D, VAR ethephon (Ethrel) benzoic acid daminozide (Alar)	

CHEMICAL TYPE	EXAMPLES	MODE OF ACTION
FUNGICIDES:		
Inorganic fungicides	copper	Enzyme inhibitor
	sulfur	Metabolic inhibitor
Dithiocarbamates	thiram	Enzyme inhibitors
	maneb	
	ferbam	
	ziram	
	Vapam	
	zineb	
Thiazoles	ethazol (Terrazole)	Enzyme inhibitor
Triazines	anilazine	Inhibits metabolism and protein synthesis
Substituted aromatics	hexachlorobenzene	Enzyme inhibitors
	chlorothalonil (Bravo)	
	chloroneb	
Dicarboximides	captan	Enzyme inhibitors
	folpet	
	captafol (Difolatan)	
Oxathiins	carboxin	Metabolic inhibitors
	oxycarboxin	
Benzimidazoles	benomyl (Benlate)	Inhibit metabolism and protein synthesis
	thiabendazole	
	thiophanate (Topsin)	
Acylalanines	metalaxyl (Dual)	
Triazoles	tridimefon (Bayleton)	
Piperazines	triforine	
Imides	iprodione (Rovral)	
	vinclozolin (Ronilan)	
Quinones	chloranil	Enzyme inhibitors
	dichlone	
Aliphatic nitrogen compounds	dodine	Inhibit metabolism and protein synthesis
Fumigants	chloropicrin	
	methyl bromide	
Antibiotics	streptomycin	
	cycloheximide	

or soil of growing weeds. Some postemergents have contact activity, meaning they kill the plant by destroying leaf and stem tissues. Other postemergents are *translocated* (moved within the tissues of the plant) from leaves and other green parts to growing points.

Insecticides may act as nerve poisons, muscle poisons, desiccants, growth regulators, sterilants, or may even have a purely physical effect by clogging air passages. Often an insecticide will have more than one mode of action.

Some fungicides are used as *eradicants* because they are capable of destroying fungi that have already invaded and begun to damage plant tissues. Their mode of action is to inhibit the metabolic processes of the growing fungal organisms. Others act as *protectants* to prevent fungal infections, and therefore will have a mode of action that retards fungal growth or prevents the organisms from entering a treated plant.

Factors Influencing Reactions to Pesticides

Several factors can influence how a pest reacts to a pesticide. Primary among these are the life stage of the target pest and the ability of the pesticide to reach the active site within the organism (pesticide uptake).

Life Stage

The life stage of a target organism may influence its response to a pesticide. For instance, young plants are generally more susceptible to herbicides than older ones, and perennial plants that are beginning to flower often will be more severely damaged by some herbicides than ones that have not yet begun to flower or have completed their flowering stage. Perennial weeds are more difficult to control once they have developed rhizomes or nutlets. Insects go through several life stages including eggs, nymphs or larvae, pupae (in some orders), and finally adults. Each life stage has different susceptibilities to insecticides due to biological and physical characteristics, feeding habits, and the physical location of the organism. Similarly, success of rodent control with toxic baits depends on the rodent's life stage and food preferences at different times of year.

Pesticide Uptake

Most pesticides have specific sites of action. Before they can act they must usually be taken into the tissues of intended target organisms and reach these sites. Structural differences, protective coatings, and habits of the pest may influence pesticide entry. Types of pesticide formulations as well as environmental conditions may influence uptake.

Various terms are used to describe methods and routes of pesticide uptake. Pesticides with *contact* activity pass through the target organism's outer covering (for example, plant cuticle, arthropod cuticle, or vertebrate skin). Some insecticides and rodenticides are known as *stomach poisons* and must be fed upon to cause poisoning (feeding enables the toxin to be absorbed through the lining of the pest's mouth or intestinal tract). Other pesticides have *fumigant* activity, meaning they are vaporized and taken into the tissues of the target plant or animal through respiration or breathing channels, or by passing through skin or cuticle. Certain pesticides include all of these types of uptake.

FORMULATIONS OF PESTICIDES

Pesticide chemicals in their "raw" or unformulated state are usually not suitable for use in pest control because they are highly concentrated, may not mix well with water, and may be chemically unstable. For these reasons, manufacturers add substances to improve storage, handling, application effectiveness, and to make the chemical safer to use. The final product is known as a *pesticide formulation*. This formulation consists of: (1) the pesticide active ingredient; (2) the carrier, such as an organic solvent or mineral clay; (3) surface-active ingredients, often including stickers and spreaders; and (4) other ingredients such as stabilizers, dyes, and chemicals that improve or enhance the action of the pesticide. Liquid pesticides sometimes have antifreeze added for protection against freezing. Usually a formulation is mixed with water or oil for final application, but baits, granules, and dusts are ready for use without additional dilution. Specialized pesticides, especially products designed for the homeowner, are sold in ready-to-use formulations.

The amount of actual pesticide in a dry formulation is expressed as percent of active ingredient (a.i.). For instance, a 50-W wettable powder contains 50% by weight of actual pesticide. Ten pounds of "D-Z-N Diazinon 50W" contains 5 pounds of diazinon and 5 pounds of inert ingredients. The amount of active ingredient in liquid formulations is represented by the pounds of active ingredient in 1 gallon of formulated pesticide. For example, in "Lorsban 4E," the "4" indicates that there are 4 pounds per gallon of the active ingredient *chlorpyrifos*.

Formulation type is usually indicated by letters that follow or are a part of the brand name of the pesticide (Table 4-9). In the examples above, the "W" represents a wettable powder and the "E" indicates that the pesticide is an emulsifiable concentrate. These codes may also describe how the pesticide is intended to be used, what it is used for, or will describe some special characteristics of the formulation. In some cases, they indicate that the formulation is intended for use in specific locations.

Selecting a Formulation

It is often possible to select from two or more formulations of the same pesticide for control of a target pest. When this is possible, make your selection based on the type of control desired, safety, cost, and other factors such as those listed in Table 4-10. For example, emulsifiable formulations of insecticides usually provide a faster kill but have a shorter residual action, as compared with wettable powders. Whenever a choice is available, consider the safety of pesticide applicators, helpers, persons working or living in the area of application, pets, and livestock and poultry. Also, select a formulation that is compatible with available application equipment. Evaluate the habits and growth patterns of each pest, and be sure that the formulation is suitable for the life stage that needs to be controlled. Cost can also influence the formulation selection. Environmental concerns are of equal importance; choose a formulation that

TABLE 4-9.

Suffixes of Chemical Brand Names.

SUFFIX	MEANING
DESCRIBE THE FORMULATION:	
AF	Aqueous Flowable
AS	Aqueous Suspension
D	Dust
DF	Dry Flowable
E	Emulsifiable Concentrate
EC	Emulsifiable Concentrate
ES	Emulsifiable Solution
F	Flowable
FL	Flowable
G	Granular
OL	Oil-Soluble Liquid
P	Pelleted
PS	Pelleted
S	Soluble Powder
SG	Sand Granules
SL	Slurry
SP	Soluble Powder
ULV	Ultra-Low-Volume Concentrate
W	Wettable Powder
WDG	Water-Dispersible Granules
WP	Wettable Powder
DESCRIBE HOW A PESTICIDE IS USED:	
GS	For Treatment of Grass Seed
LSR	For Leaf Spot and Rust
PM	For Powdery Mildew
RP	For Range and Pasture
RTU	Ready To Use
SD	For Use as a Side Dressing
TC	Termiticide Concentrate
TGF	Turf Grass Fungicide
WK	To Be Used with Weed Killers
DESCRIBE CHARACTERISTICS OF THE FORMULATION:	
BE	The Butyl Ester of 2,4-D
D	An Ester of 2,4-D
K	A Potassium Salt of the Active Ingredient
LO	Low Odor
LV	Low Volatility
MF	Modified Formulation
T	A Triazole
2X	Double Strength
LABEL FOR USE IN SPECIAL LOCATIONS:	
PNW	For Use in the Pacific Northwest (ie., Benlate PNW)
TVA	For Use in the Waterways of the Tennessee Valley Authority (ie., Aqua-Kleen TVA)

TABLE 4-10.

Comparisons of Pesticide Formulations.

FORMULATION	MIXING/ LOADING HAZARDS	PHYTO-TOXICITY	EFFECT ON APPLICATION EQUIPMENT	AGITATION REQUIRED	VISIBLE RESIDUES	COMPATIBLE WITH OTHER FORMULATIONS
Wettable powders	Dust inhalation	Safe	Abrasive	Yes	Yes	Highly
Dry flowables/water dispersible granules	Safe	Safe	Abrasive	Yes	Yes	Good
Soluble powders	Dust inhalation	Usually safe	Non-abrasive	No	Some	Fair
Emulsifiable concentrates	Spills and splashes	Maybe	May affect rubber pump parts	Yes	No	Fair
Flowables	Spills and splashes	Maybe	May affect rubber pump parts; also abrasive	Yes	Yes	Fair
Solutions	Spills and splashes	Safe	Non-abrasive	No	No	Fair
Dusts	Severe inhalation hazards	Safe	—	Yes	Yes	—
Granules and pellets	Safe	Safe	—	No	No	—
Microencapsulated formulations	Spills and splashes	Safe	—	Yes	—	Fair

will cause the least harm to the environment. Drift, runoff, wind, and rainfall must be considered, along with soil type and characteristics of the surrounding area.

Common Pesticide Formulations

Some pesticides, those used primarily for residential, industrial, and institutional pest control, consist of pure, technical grade, undiluted liquids that are applied in small quantities as aerosols by combining them with an inert gas. Because these pesticides are usually not flammable and will not conduct electricity, they are safer to use around sparks and electrical outlets than formulations containing petroleum solvents or water. Most pesticides, however, must be mixed and diluted with water or oil before being applied as liquid sprays. Occasionally, pest control problems lend themselves better to the use of dry pesticides. Three dry formulations—dusts, granules, and pellets—are available. In addition, several miscellaneous formulations and packaging methods are available for specific uses. Some formulated pesticides are packaged in ways that increase their safety or make them easier to use.

Wettable Powders (W or WP)

Wettable powders are formulations consisting of the pesticide combined with a finely ground dry carrier, usually mineral clay, along with other ingredients that enhance the ability of the powder to suspend in water. Most wettable powder formulations contain between 15 and 75% active ingredient. Use formulations with the highest percentage of active ingredient if visible residues are a concern, because carriers and other inert ingredients are the most common source of unsightly residues on fruit, other marketed produce, foliage, stored items, and furnishings. Also, the cost per unit of active ingredient is usually less in formulations having a high percentage of active ingredient. A higher percentage of active ingredient in the formulation, however, makes the wettable powder more hazardous and requires more care in handling and mixing.

Wettable powders are among the safest formulations to use when plant injury (phytotoxicity) is a problem because the carriers are inert minerals. These formulations are compatible with many other pesticides (especially other wettable powders) and fertilizers. A disadvantage is the abrasiveness of the inert carrier which contributes to pump and nozzle wear. Also, mixing is more difficult if you must use hard or alkaline water. Wettable powders always require agitation during application to keep the mixture suspended. When wettable powders are applied to porous materials, water in the mixture will penetrate, but the pesticide remains on the surface. On nonporous surfaces, water evaporates and leaves the pesticide as a residue.

A serious problem with wettable powders is the potential hazard of inhaling dust during handling and mixing. Dust particles usually contain a high concentration of pesticide, are very fine, and can remain suspended in the atmosphere for several hours. To overcome this hazard, some manufacturers package wettable powders in water-soluble bags that can be dropped unopened into a filled spray tank.

Dry Flowables or Water-Dispersible Granules (DF or WDG)

The active ingredient in dry flowables (also called water-dispersible granules) is incorporated with emulsifiers and other enhancers, similar to wettable powders. However, rather than being a powder, the pesticide formulation is formed into granules that must be mixed with water before use. Less carrier is used, giving this formulation a higher percentage of active ingredient per unit of weight. Dry flowables do not have the dust problems associated with wettable powders. Measuring and mixing are simple because granules are packaged in easy-to-pour plastic containers enabling them to be measured out by volume, similar to liquids, rather than by weight. Like wettable powders, dry flowables require constant agitation during application, and are abrasive to application equipment.

Soluble Powders (S or SP)

A soluble powder formulation is similar to a wettable powder except that the pesticide, its carrier, and all other formulation ingredients completely dissolve in water to form a true solution. Once dissolved, soluble powders require no additional mixing or agitation and are not abrasive to spray nozzles or pumps. Only a few pesticides are available in this formulation because most pesticide active ingredients will not dissolve in water. While mixing, soluble powders can be hazardous if the dust is inhaled.

Emulsifiable Concentrates (E or EC)

Many pesticides that are not soluble in water may dissolve in petroleum solvents; emulsifiable concentrates are petroleum-soluble pesticides formulated with emulsifying agents (soaplike materials) and other enhancers. Solvents will not mix with water, but emulsifiers enable the dissolved pesticide to form a suspension. When emulsifiable concentrates are added to water, a milky substance, known as an emulsion, is formed. After mixing, the emulsified pesticide is evenly distributed in water, although agitation during application is necessary to keep the emulsion uniform. Emulsifiable concentrates are among the most versatile of all formulations and have many applications. They penetrate porous materials such as soil, fabrics, paper, and wood better than wettable powders. Since they are liquid, they can be poured easily for mixing. Emulsifiable concentrates do present several operator hazards: (1) they are subject to spills that spread easily and are difficult to clean up; (2) they can be absorbed by porous protective clothing and leather boots; (3) they are more readily absorbed through the skin than powders; and (4) they may cause serious damage if splashed into the eyes.

Emulsifiable concentrates will often be more harmful to sensitive plants (phytotoxic) than wettable powders because they contain petroleum solvents. These solvents also contribute to the deterioration of rubber and plastic hoses, gaskets, and some pump parts and may be corrosive to painted surfaces.

Flowables (F)

A flowable formulation combines the qualities of both emulsifiable concentrates and wettable powders. This formulation is used when the active ingredient is an insoluble solid and cannot be dissolved in water or other solvents; finely ground pesticide particles are mixed with a liquid, along with emulsifiers, to form a concentrated emulsion. Flowables share the features of liquid emulsifiable concentrates, even having similar disadvantages. They require agitation to keep them in suspension and leave visible residues, similar to wettable powders. Flowables are easy to handle and apply, but because they are liquids, are subject to spilling and splashing like emulsifiable concentrates. They contribute to abrasive wear of nozzles and pumps. Flowable suspensions settle out in their containers, so should always be thoroughly shaken before mixing. Because of their tendency for settling, flowables are packaged in containers of 5 gallons or less to make remixing possible.

Water-Soluble Concentrates or Solutions (S)

Liquid pesticide formulations that dissolve in water are called water-soluble concentrates or solutions. Once dissolved, they require no further mixing or agitation. The same operator hazards are present with water-soluble concentrates as with other liquids, but they are nonabrasive to application equipment. There are only a limited number of pesticides capable of dissolving in water and being formulated as a solution or water-soluble concentrate.

Low-Concentrate Solutions (S)

Low-concentrate formulations are ready to use and require no further dilution. They consist of a small amount of active ingredient dissolved in an organic solvent. They usually do not stain fabrics or have unpleasant odors, so they are especially useful for structural and institutional pests as well as for home use. Major disadvantages to low-concentrate formulations are their limited availability and high cost per unit of active ingredient. Many organic solvents are harmful to foliage, limiting their usefulness for plant protection.

Ultra-Low-Volume Concentrates (ULV)

Ultra-low-volume concentrates are highly concentrated pesticide solutions, usually between 80 and 100% active ingredient. They are applied with little or no dilution and require application equipment specially suited to applying small quantities of pesticide over a large area. This means less frequent refilling of spraying equipment, a major advantage when large areas are treated. When ULV formulations are diluted, they are usually diluted with vegetable oil rather than water. Droplets of ULV formulations do not evaporate as rapidly as those from emulsions. Calibration and application must be done with extreme care when using ULV concentrates because of the high concentration of active ingredient.

Slurry (SL)

A slurry is a thin watery mixture of finely ground dusts. Because they are dusts before mixing, slurries have similar respiratory hazards to the operator as other powdered formulations. Slurries are usually first mixed in a small container by combining water with the powder and stirring to form a paste; this mixture is then slowly added to water in a partially filled spray tank. The mixture must be constantly agitated to prevent settling out. When all the paste has been added to the tank, it is filled to the top with water and allowed to mix thoroughly before being applied. Slurries are applied to seeds and plants for protection against insects or fungi. After application, slurries dry and leave thick residues on treated surfaces. These residues are usually highly visible. Slurries are abrasive and will contribute to pump and nozzle wear.

A commonly used slurry formulation is the Bordeaux mixture applied to plants as a fungicide. Hydrated lime and copper sulfate are combined in various proportions with water to produce a slurry. The separate dry materials are usually combined just prior to application, although some premixed dry formulations are commercially available.

Fumigants

Fumigants are used to treat pests of stored products, to treat soil pests such as weeds, insects, nematodes, and microorganisms, and to control vertebrates such as ground squirrels and gophers. They are also used for the control of structural and other pests in ships, boxcars, aircraft, trucks, residences, warehouses, greenhouses, and commercial buildings. Fumigants may be either solid, liquid, or gas in form. Solids and liquids volatilize (evaporate) into a gas after or during application. Fumigants which are gases at room temperature are packaged under pressure in steel cylinders and are metered into the treatment area through valves and hoses.

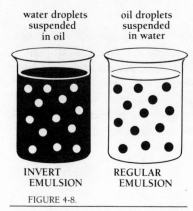

water droplets
suspended
in oil

oil droplets
suspended
in water

INVERT
EMULSION

REGULAR
EMULSION

FIGURE 4-8.

An invert emulsion consists of small water droplets suspended in oil. Compare this to the regular emulsion where oil droplets are suspended in water.

Fumigants present a serious inhalation hazard to applicators and other people in or near the treated area. Applicators often must wear supplied air breathing equipment and protective clothing. Atmosphere-monitoring equipment is often used to detect fumigant concentration and to determine when an area can be safely entered without protective equipment.

Soil fumigants may be applied through an irrigation system or injected into the soil. Usually soil moisture needs to be high to prevent fumigants from volatilizing into the atmosphere. Often soil is tarped with plastic sheeting to confine the fumigant and maintain its concentration. Texture, soil type, amount of organic material, and soil condition affect how well a soil fumigant will work; soil temperature and weather conditions during and after application may also influence the effectiveness of a fumigant.

Structures should be sealed with tarps or be otherwise air-tight to prevent fumigants from escaping. Fans help to distribute the gas evenly. Buildings must be thoroughly ventilated after fumigation before they can be reoccupied. Stored products are usually fumigated in air-tight containers or specially designed rooms or buildings. Fumigation of rodent burrows is most effective when the soil moisture level is high.

Invert Emulsions

Invert emulsions are liquid formulations having small water droplets suspended in oil, as opposed to regular emulsions where water contains small droplets of oil. Pesticides dissolve in either the oil or water phase (Figure 4-8). Invert emulsion concentrates have the consistency of mayonnaise and are usually safer for handling and mixing than other liquid formulations. Continuous agitation is required unless special application equipment is used to combine water with an invert emulsion as it is being sprayed.

Invert emulsions aid in reducing drift and are primarily used for this purpose. Drift is often a problem with regular emulsions and other formulations because water droplets begin to evaporate before reaching the target surface, causing them to become very small. Because oil evaporates slower than water, invert emulsion droplets shrink less, therefore more pesticide reaches the target. The oil helps to reduce runoff and improves rain resistance. It also serves as a sticker-spreader by improving surface coverage and absorption. Since droplets are larger and heavier, it is difficult to get thorough coverage of the undersides of foliage.

Dusts (D)

Dust formulations consist of finely ground pesticide combined with an inert dry carrier (Figure 4-9). Most dust formulations contain between 1 to 10% active ingredient, although higher concentrations are available, including 100% active ingredient in materials such as sulfur. Dusts are effective where moisture from a liquid spray can cause damage to the crop, foliage, or sprayed surfaces. Dusts can be used on many surfaces without harm, although they do leave a highly visible residue. Depending on the pesticide being used, dust formulations often provide long-term protection of treated surfaces.

Dusts are less frequently used than liquid sprays in agricultural applications due to drift hazards. They are effective in hard-to-reach indoor areas, and they are also used for seed treatment, for pests around homes and gardens, and for parasites on pets, livestock, and poultry. Fungicides,

FIGURE 4-9.

One type of pesticide formulation consists of a finely ground powder containing active and inert ingredients. These are applied dry as dusts. A common use of dusts is to control external parasites on livestock and pets. This dog is being dusted with an insecticide to control fleas and ticks.

herbicides, insecticides, and rodenticides are available in dust formulations.

In addition to hazards of drift to nontreated areas, dusts present serious inhalation hazards to applicators. To prevent poisoning, wear respirators whenever using dust formulations. Outdoor application is restricted to periods when the air is still. Application equipment is difficult to calibrate, and dusts require agitation during application to prevent settling and caking.

Tracking Powders. Special dusts, known as tracking powders, are used for rodent and insect monitoring and control. For rodent control, the tracking powder consists of finely ground dust combined with a stomach poison rodenticide. Rodents walk through the dust, pick it up on their feet and fur, and ingest it when they clean themselves. Tracking powders are especially useful when bait acceptance is poor due to an abundant, readily available food supply. Sometimes nonpoisonous powders, such as talc or flour, are used to monitor and track the activity of rodents in buildings. Control of some insects, such as cockroaches, can often be accomplished by using poisoned tracking powders; poisons contained in these dusts are either ingested during grooming or are absorbed through the insect's outer body covering. Desiccants or sorptive dusts, such as boric acid powder, diatomaceous earth, and silica gel, remove or disrupt the waxy protective coatings of insects, causing death by desiccation (loss of water) rather than by a toxic reaction.

Granules (G)

Granules consist of a pesticide and carrier combined with a binding agent. They range in size between 4 and 80 mesh, with the most common

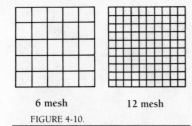

6 mesh   12 mesh

FIGURE 4-10.

The size of dust and granule particles can be measured by passing the material through screens with different sizes of mesh. Mesh is the number of wires per inch of screen. As seen here, the larger the mesh number, the finer the screen. Granules range in size between 4 and 80 mesh, while dusts are 80 mesh and finer.

formulations being in the range of 15 to 30 mesh. *Mesh* is the term used to describe the number of wires in an inch of screen (Figure 4-10). Larger numbers indicate more wires, therefore being a finer screen. (Dusts pass through 80 mesh and finer screens.) The larger size and weight of granules eliminates drift. Hazards to the applicator and environment are minimized with granular formulations due to reduced dust and lack of spray mist. Granular formulations are more persistent in the environment than other formulations because the pesticide active ingredient is released slowly.

Some granular pesticides dissolve in water and are used for control of aquatic pests such as algae, weeds, or fish. These have advantages over liquids in aquatic situations because liquids are hard to disperse; when sprayed, liquids may dry on plants or floating debris, or may not pass through the surface film of the water. Granules bounce off vegetation and easily penetrate the surface film. Some aquatic granular formulations offer sustained, controlled release of the pesticide because they dissolve slowly. Slow-dissolving granules may be hazardous to waterfowl which mistake them for food.

Granules are applied to soil for control of weeds, nematodes, and soil insects. They are also used for some systemic insecticides that are taken up through plant roots for control of leaf- and stem-feeding insects. Granular formulations may require mechanical incorporation into the soil and often need moisture to activate them. Granules are not suitable for conventional foliar application since they do not stick to leaves. However, when applied to plants like corn, the granules lodge in the leaf whorls and are effective in controlling some insect pests. For this type of application, granules may be weighted with sand.

Pellets (P or PS)

Pellets are identical to granules except they are manufactured into specific uniform weights and shapes, enabling them to be more accurately applied with equipment such as precision planters. This uniformity improves calibration accuracy, which is normally difficult to accomplish with granules.

Microencapsulated Materials

Liquid or dry pesticide particles may be surrounded by a plastic coating to produce a microencapsulated formulation (Figure 4-11). Microencapsulated pesticides must be mixed with water and are sprayed in the same manner as other sprayable formulations. After spraying, the active ingredient is released gradually as the plastic coating breaks down. There are several advantages to microencapsulated formulations: (1) highly toxic materials are safer for applicators to mix and apply; (2) delayed or slow release of the active ingredient prolongs its effectiveness, allowing for fewer and less precisely timed applications; (3) the pesticide volatilizes more slowly, so less will drift away from the application site; and (4) these formulations are not as phytotoxic to sensitive plants. In residential, industrial, and institutional applications, microencapsulated formulations offer the advantages of reduced odor, release of small quantities of pesticide over a long period of time, and greater safety. Microcapsules have less dermal hazard than ordinary formulations. There is a special hazard to bees because microcapsules are about the same size as pollen grains; bees may carry them back to their hives before the plastic coating

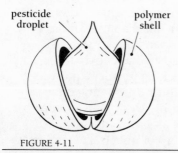

pesticide droplet   polymer shell

FIGURE 4-11.

Microencapsulated formulations consist of pesticides enclosed in tiny plastic capsules. This often makes them safer to use and increases their effectiveness.

FIGURE 4-12.

Water-soluble bags protect operators during mixing of some types of highly toxic or hazardous pesticides. A preweighed amount of formulated pesticide powder is contained in a plastic bag inside this paper envelope. The bag is removed from the envelope and dropped into a water-filled spray tank. The plastic bag will dissolve in the water and release the powder.

breaks down, resulting in poisoning of adults and brood once the pesticide is finally released.

Water-Soluble Packets

Water-soluble packets are used to reduce the mixing and handling hazards of some highly toxic pesticides (Figure 4-12). Preweighed amounts of wettable powder or soluble powder formulations are packaged in water-soluble plastic bags. When the bags are dropped into a filled spray tank, they dissolve and release their contents to mix with the water. There are no risks of inhaling or contacting the undiluted pesticide during mixing as long as the packets are not opened. Once mixed with water, pesticides packaged in water-soluble packets are no safer than other mixtures.

Baits

Baits consist of pesticides combined with food, an attractant, and/or a feeding stimulant. Baits attract target pests to a pesticide, eliminating the need for widespread pesticide application. Sometimes target pests carry baits back to their nestbound young. Baits are used indoors for control of rodents, ants, roaches, and flies. Outdoors they are used to control slugs, snails, insects, and vertebrates such as birds, rodents, and larger mammals. Dangers associated with baits include their attractiveness to nontarget animals and to children, making materials hazardous if improperly used. Baits used around children and pets should be placed in special pet- or child-proof bait stations. Some baits are colored to distinguish them as being poisonous and to make them less attractive to birds.

Attractants

Various substances are used to attract pests, including pheromones, sugar and protein hydrolysate syrups, yeasts, and rotting meat. These attractants are either used in sticky traps or are combined with pesticides and sprayed onto foliage or other items in the treatment area.

Aerosol Containers

Aerosol containers are sometimes used for packaging and dispensing pesticides, usually insecticides (Figure 4-13). The pesticide is combined under pressure with a chemical propellant in a disposable or refillable self-dispensing can. Two types of dispensers include those that emit pesticides as a fine airborne mist or fog (aerosol foggers), and those that produce a coarse spray of liquid or powder (pressure spray applicators). Pressure spray applicators allow a pesticide film to be sprayed directly onto surfaces. Aerosol foggers are generally one-time, total release units, and contain a high percentage of pesticide. Pressure spray applicators, used for intermittent sprays, may contain a lower percentage of pesticide combined with a petroleum oil carrier. Residential, industrial, and institutional pest control operators frequently use aerosol applicators, since they require no mixing or special application equipment. Pressure spray applicators have an advantage when only small amounts of pesticide are applied at any one time because the remaining pesticide does not lose its potency during storage.

FIGURE 4-13.

This aerosol container offers greater convenience without waste. It contains a low-concentration insecticide solution.

Hazards from aerosol containers include inhalation injury from breathing the spray or dust. Pressure spray applicators, because they may have a high percentage of flammable petroleum oil carrier, should not be used around open flames or other sources of ignition. It is difficult to confine the spray, fog, or dust emitted from aerosol containers.

Impregnates

Pesticides that are incorporated into household and commercial products are known as impregnates (Figure 4-14). Pet collars, livestock ear tags, adhesive tapes, and plastic pest strips contain pesticides that volatilize (evaporate) over a period of time and provide control of nearby pests. Soaps and shampoos containing pesticides are available for control of external parasites on pets, livestock, and people. Some paints and wood finishes have pesticides incorporated into them to kill insects or retard fungus growth. Special formulations of pesticides used for control of cockroaches in residences and commercial buildings consist of an insecticide incorporated into clear plastic or lacquer paint that is rolled or painted onto wall surfaces; this treatment gives 6 to 12 months of residual control. Carpeting, furniture, bedding, fabrics, and clothing are often impregnated with pesticides to prevent damage from insects and fungi. Adhesive-backed impregnated strips may be placed inside electrical boxes, electronic equipment, and appliances to provide long-term protection against insect damage.

Repellents

Insect repellents are available in aerosol and lotion formulations. They can be applied to skin, clothing, or foliage to repel biting and nuisance insects. Other repellents can be mixed with water and sprayed on ornamental plants and agricultural crops to prevent damage from deer, dogs, and other animals.

Animal Systemics

Animal systemics are pesticides used for controlling animal parasites such as insects and mites. These can be administered to pets and livestock as a food additive or given as premeasured capsules or liquids. Sometimes

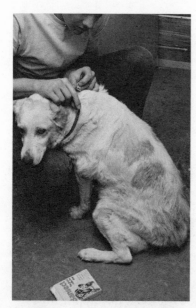

FIGURE 4-14.

Flea collars are impregnated with an insecticide that will kill fleas. A low level of this insecticide is slowly released by the collar over a period of time.

they are applied to the skin of the animal in the form of a pour-on liquid, liquid spray, or dust. These pesticides enter the animals' tissues to control pests systemically. Systemics are used against fleas and other external blood-feeding insects, as well as worms and other internal parasites. Animal systemics are used under the supervision of a veterinarian.

Fertilizer Combinations

Occasionally insecticides, fungicides, and herbicides are combined with fertilizers to provide a convenient way of controlling pests while fertilizing crops or ornamental plants. These combinations are commonly used by homeowners. The unit cost of pesticide in these preparations is very high. In commercial applications, pesticides may be custom-mixed with fertilizers by the applicator or supplier to meet specific crop requirements.

PESTICIDE MIXTURES

It is often convenient to combine two or more pesticides so they can be applied at the same time. A few pesticides are sold as premixed combinations, although most are combined at the time of application and are known as *tank mixes*. For example, fungicides are commonly mixed with insecticides as a dormant spray in deciduous crops. Two or more herbicides may be combined to increase the number of species of weeds controlled. Pesticides are mixed with micronutrients or fertilizers, saving money by reducing the time and fuel required for multiple applications. Tank mixes also reduce equipment wear, decrease labor costs, and lessen the mechanical damage done to crops and soil by heavy application equipment. If Category I pesticides are mixed with Category II or III pesticides, the mixture must be treated as a Category I pesticide. Required safety equipment and all other label restrictions must comply with the label having the greater restrictions. A special rule applies to determine the reentry interval when two or more pesticides are combined; this is discussed in Chapter 7.

clumping and separation   smooth, uniform mixture

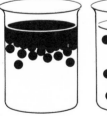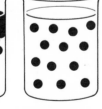

INCOMPATIBLE   COMPATIBLE

FIGURE 4-15.

Sometimes mixtures of pesticides are incompatible and may separate or curdle as shown here. An incompatible mixture will clog spray equipment and waste the pesticide material.

Incompatibility

Incompatibility is a physical condition that prevents pesticides from mixing properly to form a uniform solution or suspension. Precipitation of flakes, crystals, or oily clumps or severe separation are unacceptable (Figure 4-15). Such incompatible mixtures clog application equipment, inhibit even distribution of the active ingredient in the spray tank, and prevent effective pesticide coverage. However, if the incompatible mixture can be thoroughly mixed and kept in that condition with agitation, and without clogging nozzles, the mixture is probably safe to use.

The cause of incompatibility may relate to the chemical nature of the materials, impurities in the spray tank or water, the order in which pesticides are mixed in the spray tank, or the types of formulations being mixed. Pesticide formulations of the same type are rarely incompatible with each other because they often contain the same inert ingredients and solvents. Before preparing a tank mix, be sure the spray tank is thoroughly clean and contains no sediments or residues. Evaluate the tank mixture by performing a compatibility test.

Testing for Incompatibility. Pesticides should be mixed in small quantities to test for incompatibility problems. Table 4-11 provides instructions for a simple compatibility test that requires only a small investment of time. This test will not help you determine if the mixture has changed the chemical effectiveness of any or all of the pesticides, however.

When combining chemicals for either the compatibility test or for mixing in the spray tank, add the materials in the following order: (1) wettable powders; (2) flowables; (3) water-soluble concentrates; then

TABLE 4-11.

Compatibility Test for Pesticide Mixtures.

WARNING: Always wear a waterproof apron, gloves, eye protection, and if necessary, respiratory protection when pouring or mixing pesticides. Perform this test in a safe area away from food and sources of ignition. Pesticides used in this test should be put into the spray tank when completed. Rinse all utensils and jars and pour rinsate into the spray tank. Do not use utensils or jars for any other purpose after they have contacted pesticides.

---

1. Measure one pint of the intended spray water into a clear quart glass jar.

2. Adjust pH if necessary (see instructions under "Buffers and Acidifiers" section below).

3. Add ingredients in the following order. Stir well each time an ingredient has been added.

   a. Surfactants, compatibility agents, and activators: add 1 teaspoon for each pint/100 gallons of planned final spray mixture.

   b. Wettable powders and dry flowable formulations: add 1 tablespoon for each pound/100 gallons of planned final spray mixture.

   c. Water-soluble concentrates or solutions: add 1 teaspoon for each pint/100 gallons of planned final spray mixture.

   d. Emulsifiable concentrate and flowable formulations: add 1 teaspoon for each pint/100 gallons of planned final spray mixture.

   e. Soluble powder formulations: add 1 teaspoon for each pint/100 gallons of planned final spray mixture.

   f. Remaining adjuvants: add 1 teaspoon for each pint/100 gallons of planned final spray mixture.

4. After mixing, let the solution stand for 15 minutes. Stir well and observe the results.

COMPATIBLE
Smooth mixture, combines well after stirring. Chemicals can be used together in the spray tank.

INCOMPATIBLE
Separation, clumps, grainy appearance. Settles out quickly after stirring. Follow instructions below to try to resolve incompatibility, otherwise do not mix this combination in the spray tank.

RESOLVING INCOMPATIBILITY

1. Add 6 drops of compatibility agent and stir well. If mixture appears compatible, allow it to stand for 1 hour, stir well, and check it again. If the mixture appears incompatible, repeat one or two more times, using 6 drops of compatibility agent each time.

2. If incompatibility still persists, dispose of this mixture, clean the jar, and repeat the above steps, but add 6 drops of compatibility agent to the water before anything else is added.

3. If the mixture is still incompatible, do not mix the chemicals in the spray tank. To overcome this problem you might consider the following alternatives:

   a. Use a different water supply.

   b. Change brands or formulations of chemicals.

   c. Change the order of mixing.

4. Make only one change at a time, and perform a complete test, as described above, before making another change. Do not mix the chemicals in the spray tank if incompatibility cannot be resolved.

(4) emulsifiable concentrates. For example, when combining a water-soluble concentrate with a wettable powder, always add the wettable powder to the spray tank first. When mixing an emulsifiable concentrate with a flowable, add the flowable first.

Field Incompatibility. Sometimes tank mixes seem compatible during testing and after mixing in the spray tank, but problems arise during application. This is known as *field incompatibility* and often is related to the temperature of the water in the tank, water impurities, or to the length of time the spray mixture has been in the tank. Occasionally there are variations among different lots of pesticide chemicals that are great enough to cause an incompatibility. Usually increased agitation is sufficient to recombine the mixture.

Resolving Compatibility Problems in the Spray Tank. There are several things you should try if pesticide incompatibility develops in the spray tank. First, increase agitation and try to break up the aggregates with a water stream to get the mixture recirculating. If the material still separates, contact your pesticide dealer for an appropriate compatibility extender; add the extender to the tank and continue agitation. It may be necessary to change filter screens to a larger size and clean them frequently. When these steps do not resolve the problem, dilute the mixture with additional water and filter off larger particles. Any pesticide that cannot be sprayed on the application site must be placed in an appropriate container and disposed of in the same way as any other unused pesticide (Chapters 7 and 8 contain information on disposal of pesticide wastes).

Chemical Changes with Pesticides and Pesticide Combinations

Pesticides may mix properly in solution but the effectiveness or toxicity of the pesticides in the mixture has been changed. These *interactive effects* are due to chemical, rather than physical, reactions between combined pesticides, impurities, or the water used for mixing. Such changes are difficult to recognize because they cannot be seen.

Additive Effect

An additive effect occurs when two or more pesticides are combined, but the resulting toxicity to the target organism is no more than if an equal amount of either chemical was used alone. For example, two insecticides—Compound A and Compound B—are applied at the rate of ½ pound of each per acre. The results on the target insect are no different than if 1 pound of Compound A or 1 pound of Compound B were used. The results are greater, however, than if ½ pound of Compound A or ½ pound of Compound B were used alone (Figure 4-16).

Greater than Additive Effect

Sometimes the toxicity of pesticides being used is increased above what would be expected through just an additive effect. Three types of effects are observed (Figure 4-17):

Potentiation. Potentiation is an increase in the toxicity of a pesticide because something mixed with it lowers the pest's tolerance to that chemical. Impurities in malathion, for example, make malathion more toxic because they inactivate enzymes produced by the pest that normally detoxify malathion. In a mixture of two or more pesticides, one compound may potentiate another in a similar manner, producing an effect greater than the expected additive effect.

Synergism. Synergism is a process in which a chemical (with or without pesticidal properties) increases the toxicity of a pesticide it is

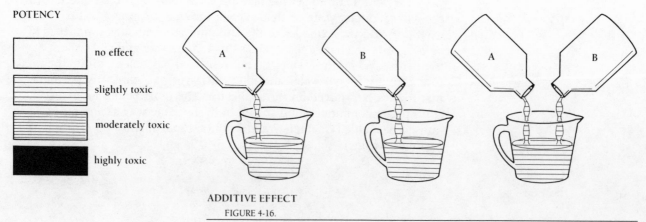

POTENCY

no effect

slightly toxic

moderately toxic

highly toxic

ADDITIVE EFFECT
FIGURE 4-16.

If two or more pesticides are combined and toxicity increases proportionally to the total amount of pesticide being used, the effect is additive.

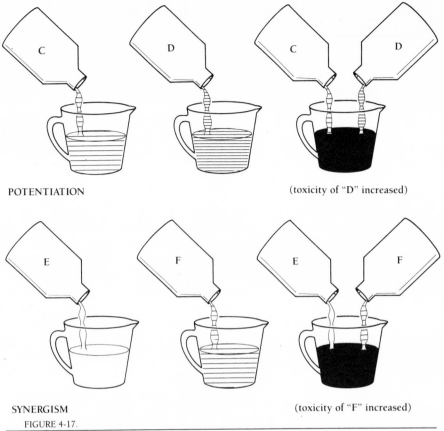

POTENTIATION

(toxicity of "D" increased)

SYNERGISM

(toxicity of "F" increased)

FIGURE 4-17.

When combined pesticides produce a response greater than what would be expected by the additive effect, the response may be due to potentiation or synergism as illustrated here.

mixed with. The synergistic chemical may slow the breakdown of the pesticide or increase its uptake by the pest. For example, the compound *piperonyl butoxide* has no insecticidal properties, but is commonly used to increase the toxicity of pyrethrum insecticides.

Coalescent Effect. A coalescent effect occurs when the toxic response (mode of action) from a pesticide mixture is unlike the expected response from either pesticide alone. The combined materials have formed a chemical with a different mode of action.

Antagonistic Effect

Antagonism occurs when the toxic effect of a pesticide is reduced by being combined with other materials. For example, when two pesticides are combined, one pesticide may enable the target organism to resist, slow down, or degrade the toxic action of the other (Figure 4-18).

Deactivation

Deactivation may take place before a spray reaches its intended destination, usually in the spray tank at the time pesticides are combined. Usually one of the pesticide compounds will alter the pH of the spray

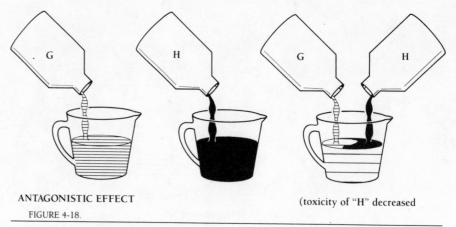

ANTAGONISTIC EFFECT   (toxicity of "H" decreased
FIGURE 4-18.

Sometimes combining pesticides will cancel or reduce the toxic effect of one or both of the components; this is known as an antagonistic effect.

mixture, causing breakdown or hydrolysis of the other pesticide. Or, one compound may alter or neutralize an electrical charge of the other to reduce its effectiveness (a problem that can be especially important when using herbicides). The quality of the water being used in the spray mixture may be the cause of deactivation of pesticides; water with a high pH (alkaline water) commonly shortens pesticide half-life.

Delayed Mixtures

If two pesticides are interactive, problems can occur even when one is applied several weeks after the other. An earlier spray may cause deactivation of the second spray. Or, the combination of the two sprays may be injurious to treated plants (phytotoxic). Check pesticide labels for this type of incompatibility.

Damage to Treated Plants or Surfaces

Pesticide combinations may actually work well controlling target organisms but still leave problems. The interaction between the chemicals may cause spotting or staining of sprayed surfaces or might cause damage to plant foliage or produce. To avoid this problem, test the mixture on a small area first and observe the results. Some surfaces are less affected by chemicals than others. Similarly, some species of plants are more sensitive than others. Plant phytotoxicity, however, may not be apparent until several weeks after application.

Sources of Information on Compatibility

Enhancing toxicity or efficacy of one pesticide by combining it with another compound is a common practice in pest control. Tank mixes also reduce the number of trips required through a treatment area to apply pesticides. However, it is important to avoid combinations that could be damaging or could cause problems with the application equipment. There are several ways to get information on pesticide mixtures:

Labels

Pesticide labels provide compatibility information on tank mixes with other pesticides. Herbicides, which are commonly tank mixed, often have lists of other materials on their labels that can be safely combined as long as all other label instructions are followed. Labels sometimes list pesticides or materials that are incompatible with the pesticide being used. These may be specific compounds or general classes of chemicals, such as "sulfur containing compounds" or "alkaline materials." Check the labels of all the pesticides you are combining to be sure they are compatible.

Compatibility Charts

Pesticide manufacturers and distributors, or publishers of agricultural chemicals handbooks, often prepare charts showing pesticide compatibility. These charts provide useful information on special mixing procedures and ways to improve compatibility. Pesticide mixtures listed on compatibility charts may be safely used as long as all label instructions are followed.

Pesticide Manufacturer

Information on compatibility can often be obtained through a telephone call or letter to the main office, district office, or field representative of a pesticide manufacturer. Manufacturer's addresses are printed on pesticide labels. Chemical dealers will usually have the names and telephone numbers of local field representatives. Most public libraries can also assist you in locating the nearest district office or corporate headquarters of a particular company.

ADJUVANTS

Adjuvants are materials that are added to a pesticide mixture in the spray tank to improve mixing and application or enhance pesticide performance. While pesticides are formulated to be suitable to many types of application conditions, they cannot be formulated for all possible situations; adjuvants are used to customize the formulation to specific needs and compensate for local conditions. Adjuvants can be used to: (1) improve the wetting ability of a spray solution; (2) control evaporation of spray droplets; (3) improve weatherability of the pesticide; (4) increase the penetration of the pesticide through plant or insect cuticle; (5) adjust the pH of the spray solution; (6) improve spray droplet deposition; (7) increase safety of the spray to target plants; (8) correct incompatibility problems; and (9) reduce spray drift. Familiarize yourself with adjuvant types to understand where and how they should be used. When selecting adjuvants, define the effect you wish the adjuvant to have, then check pesticide and adjuvant labels to make sure these materials are suitable to the application site, target pest, and application equipment (Table 4-12).

TABLE 4-12.

Comparison of Adjuvants.

FUNCTION	Surfactant	Sticker	Spreader Sticker	Extender	Activator	Compatibility Agent	Buffer	Acidifier	Deposition Aid	Defoamer	Thickener	Attractant
Reduce surface tension	■		■		■							
Improve ability to get into small cracks	■		■									
Increase uptake by target	■		■		■			■				
Improve sticking	■	■	■									
Protect against wash-off/abrasion	■	■	■	■								
Reduce sunlight degradation		■	■	■								
Reduce volatilization	■	■	■						■		■	
Increase persistence			■	■								
Improve mixing						■	■	■				
Lower pH							■	■				
Slow breakdown							■	■				
Reduce drift									■	■	■	
Eliminate foam										■		
Increase viscosity									■		■	
Increase droplet size									■		■	
Attract pests to pesticide												■

Often a single chemical will accomplish two or more separate adjuvant functions, such as a spreader-sticker, spreader-activator, or spreader-sticker-drift retardant. Some manufacturers also produce a blend of chemicals to accomplish multiple functions. The effectiveness of most adjuvants is proportional to their concentration in the spray tank mixture, however. Therefore, ready-mixed blends may limit your ability to achieve the proper concentration of some component for your application requirements unless the requirements are identical to the ratio of components in the adjuvant mixture. It is often better to use several single-active-ingredient adjuvants, at the appropriate concentrations for your specific needs, rather than to try to accomplish all the desired results with a blended, multiple function adjuvant.

Surfactants

Surfactants are surface active agents, also known as wetting agents or spreaders. They enhance spray coverage by reducing the surface tension

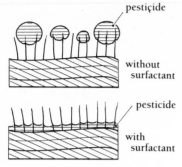

FIGURE 4-19.

Spray droplets will not spread out over waxy or hairy leaf or insect surfaces because of surface tension. Adding a surfactant, however, lowers surface tension and causes droplets to spread out and come in contact with the cuticle of the leaf or insect.

of spray droplets. Sometimes vegetable oils such as cottonseed oil and soy oil are used as surfactants. Surfactants are used for getting good coverage on waxy or hairy surfaces, such as leaves of many plants or the outer covering of insects and mites (Figure 4-19). They also help to get sprays into small cracks or openings. The amount of surface tension reduction is proportional, to an extent, to the amount of surfactant used. Surface tension is measured in dynes/cm. Water has a normal surface tension of about 72 dynes/cm, but optimum spreading is obtained when this is reduced to about 30 dynes/cm. At this point, spray is able to penetrate the small openings in leaf surfaces or insect cuticle. Increasing surfactant concentration to achieve a surface tension much below 30 dynes/cm usually causes spray materials to run off treated surfaces, resulting in reduced effectiveness and a waste of materials. Mix surfactants according to directions on their labels to achieve the appropriate surface tension reduction.

Three types of surfactants are available: *nonionic*—do not react in water; *anionic*—ionize into negatively charged ions in water; and *cationic*—ionize into positively charged ions in water. The charge or lack of charge of a surfactant is important to pesticide applications, since this affects how the spray material will react after it has been applied, dried, and exposed to environmental conditions. Emulsifiers used in the formulation of many pesticides are usually a blend of anionic and nonionic surfactants that enable petroleum or other solvent-based chemicals to break up into droplets and suspend in water.

Anionic surfactants (negative electrical charge) increase the resistance of the pesticide to washing off a sprayed plant by rain, dew, or irrigation. They prevent pesticides from being readily absorbed through plant cuticle, however, because plant surfaces are also negatively charged and like charges repel each other. Anionic surfactants are used when it is important that the pesticide remain on the outer surface of plants in the treatment area, even during adverse weather or environmental conditions. They should also be used to increase the effectiveness of insecticides and miticides that are stomach or contact poisons, because more poison remains on leaf surfaces rather than being absorbed by the plant.

Nonionic surfactants (no electrical charge) are used to increase pesticide penetration through plant cuticle. They are recommended for use with systemic herbicides such as glyphosate (Roundup) and oxyfluorfen (Goal), to improve target plant uptake of the toxicant; they are also used with insecticides and fungicides that have systemic action in plants and need to be absorbed into plant tissues so they can be translocated. Pesticides mixed with nonionic surfactants may be subject to washing off treated surfaces by rainfall, dew, or irrigation.

Cationic surfactants (positively charged) are strongly attracted to plant surfaces and, although they aid in getting pesticides through cuticle, are highly phytotoxic when not blended with other types of surfactants. Pure cationic surfactants are not used as pesticide adjuvants.

Some surfactants are a blend of anionic and nonionic surfactants and may also contain cationic surfactants. Blends are used as general purpose surfactants and usually have a wider range of application.

Selecting Surfactants. When selecting surfactants for pesticide application, consider several factors: (1) the nature of the target surface—waxy layers or fine, hairlike structures are difficult to penetrate; (2) the physical and chemical nature of the pesticide—some pesticides dissolve in water, while others are not soluble; (3) the site of action of the pesticide—some must be absorbed by the target and translocated, while other types must remain on the surface; (4) weather conditions and cultural practices, such as irrigation, which may wash pesticides off treated surfaces or influence application methods and timing; and (5) the biology or habits of the pest. Consider also the cost of the surfactant compared to the cost per unit area of treatment. Surfactants are not always pure, active ingredient; most are combined with an alcohol solvent. The percentage of alcohol varies from one brand of surfactant to another. You will need to use a larger volume of a surfactant if it contains a high percentage of alcohol.

Stickers

Stickers are substances such as latex or other adhesives that improve pesticide attachment to sprayed surfaces. They protect pesticides from washing off due to rainfall, heavy dew, or irrigation, and help prevent pesticide loss from wind or leaf abrasion. Many stickers incorporate ultraviolet inhibitors to prevent the pesticide from being degraded by sunlight. Follow label directions carefully, because too much sticker binds the pesticide so well that it may be unavailable to react with target organisms. Some pesticide formulations already contain a sticker, so additional amounts should not be used. Always read the pesticide label to be sure there are no recommendations against using a sticker.

Spreader-Stickers

Spreader-stickers are a mixture of surfactant and latex or other adhesive sticker. These are often used as a general purpose adjuvant for many pesticide applications. When using a spreader-sticker, be certain that the surfactant in the mixture is compatible with the type of pesticide being used and that the pesticide formulation does not already contain a sticker.

Extenders

Extenders are chemicals that enhance the effectiveness or effective life of a pesticide. Some extenders function by screening out ultraviolet light that decomposes many pesticides; others slow down pesticide volatilization. Stickers are used as extenders because they slow down the loss of pesticide from surfaces due to irrigation, rainfall, and abrasion. Extenders may make sprayed areas toxic for longer periods of time because they slow the pesticide breakdown or natural degradation process.

Activators

Activators increase the activity of a pesticide. Some surfactants are activators because they reduce surface tension and allow greater pesticide contact. Activators also include chemicals that speed up pesticide penetration through insect or plant cuticle. Activators must be used carefully because they may increase risk to nontarget organisms by making pesticides more toxic.

Compatibility Agents

When problems of physical incompatibility occur, compatibility agents may reduce or eliminate problems of separating or clumping. For example, one type of compatibility agent, an emulsifier, is a soaplike material that can be combined with oil to make the oil disperse in a water solution. When trying to correct an incompatibility problem with a compatibility agent, mix small quantities of the pesticides and compatibility agent in a jar. Add all components in your test in the same order that they will be mixed in the spray tank. Unless specific information on mixing order is given on the pesticide or compatibility agent labels, follow the technique described in Table 4-11 on page 108.

Buffers and Acidifiers

The term pH is a measure of the acidity or alkalinity of a solution. A neutral solution has a pH of 7. A solution with a pH of 6 is slightly acid, while one with a pH of 8 is slightly alkaline. Many pesticides are unstable in alkaline solutions, but quite stable if the solution is slightly acid. The optimum pH for most pesticides is about 6, although a solution in the range of pH 6 to 7 is usually satisfactory. Some pesticides are most effective when the solution is acidified to a pH of 3 to 3.5. High pH often causes accelerated pesticide breakdown. Table 4-13 shows some of the effects that water pH has on the activity of pesticides. Table 4-14 describes how you can measure and alter the pH of the spray solution you are using, should this be necessary.

Buffers. Buffers are substances capable of changing the pH of a water solution to a prescribed level, and maintaining this level relatively constant, even though conditions such as water alkalinity may change.

Acidifiers. Acidifiers (also called *acidulators*) are acids that can be added to spray mixtures to neutralize alkaline solutions and lower the pH. Acidifiers do not have a buffering action; therefore alkaline or acid compounds added to the spray solution after the acidifier will change the pH of the solution.

Deposition Aids

Deposition aids are adjuvants that improve the ability of pesticide sprays to reach surfaces in a treatment area. Different types of products can be used as deposition aids. Inverting agents, for instance, encapsulate the pesticide, forming oil droplets of uniform size which are then suspended in larger water droplets to form an invert suspension. Encapsulation prevents evaporation or volatilization of the pesticide before it reaches the target surface, even when some of the water carrier evaporates. Drift control agents increase droplet size by altering shear forces of the liquid as the spray is emitted from a nozzle. Because larger droplets have more momentum, they travel farther and are less influenced by air movement; the result is that more pesticide will reach target surfaces in the treatment area. Surfactants that alter the surface tension of the spray solution are considered deposition aids because they influence drop size as well as distribution on sprayed surfaces.

Defoaming Agents

Many pesticide mixtures produce copious amounts of foam as a result of the action of hydraulic or mechanical agitators. Foaming in the spray tank introduces air into the pressure system and makes it difficult to

TABLE 4-13.

Effect of Water pH on the Chemical Stability of Pesticides.

COMPOUND	BRAND NAME	HALF-LIFE AT DIFFERENT pH VALUES*
azinphos-methyl	Guthion	12 hours at pH 9.0 17.3 days at pH 5.0
benomyl	Benlate	Very stable in mildly acid to slightly alkaline solutions.
carbaryl	Sevin	24 hours at pH 9.0 2.5 days at pH 8.0 24 days at pH 7.0
chlorothalonil	Bravo	38.1 days at pH 9.0 Stable below pH 7.0
chlorpyrifos	Dursban Lorsban	1.5 days at pH 8.0 35 days at pH 7.0
diazinon	Knox-Out	37 hours at pH 6.0 Hydrolysis is very rapid in strong acid or strong alkaline solutions.
dicrotophos	Bidrin	50 days at pH 9.1 100 days at pH 1.1
dimethoate	Cygon	12 hours at pH 6.0 Maximum stability is between pH 4.0 and pH 7.0. Unstable in alkaline water.
duraphos	Phosdrin	1.4 hours at pH 11.0 3 days at pH 9.0 35 days at pH 7.0 120 days at pH 6 to pH 2
EPN	EPN	8.2 hours at pH 9.0 10 years at pH 6.0
ethoprop	Mocap	Stable in acid solutions, but hydrolyzes rapidly in alkaline solutions.
formetanate	Carzol	3 hours at pH 9.0 14 hours at pH 7.0 4 days at pH 5.0
malathion	Cythion Carbophos others	Stable in neutral or moderately acid solutions, but undergoes hydrolysis rapidly at pH values above 7.0 or below 3.0.
methomyl	Lannate Nudrin	Stable in slightly acid water. Slight hydrolysis after 6 hours in pH 9.1 solution.
monocrotophos	Azodrin	22–23 days at pH below 7.0 Hydrolysis increases rapidly at pH above 7.0.
naled	Dibrom	Undergoes 90 to 100% hydrolysis in 48 hours in alkaline water.

*These figures are generalized estimates and reflect trends, but half-life periods may vary considerably. Hydrolysis depends on other factors besides the pH of the solution, including temperature, other pesticides and adjuvants in the spray tank, and formulation of the pesticide.

COMPOUND	BRAND NAME	HALF-LIFE AT DIFFERENT pH VALUES*
phosmet	Imidan	4 hours at pH 8.0 12 hours at pH 7.0 13 days at pH 4.5
phosphamidon	Dimecron	30 hours at pH 10.0 13.5 days at pH 7.0 74 days at pH 4.0
parathion	Phoskil others	170 minutes at pH 11.0 29 hours at pH 10.0
trichlorfon	Dylox	Rapid hydrolysis under conditions greater than pH 8.5. Most stable at pH 5.0.

*These figures are generalized estimates and reflect trends, but half-life periods may vary considerably. Hydrolysis depends on other factors besides the pH of the solution, including temperature, other pesticides and adjuvants in the spray tank, and formulation of the pesticide.

maintain the even pressure required for proper mixing and uniform pesticide application, resulting in a waste of time and materials. Defoaming agents eliminate foam in the spray tank.

Thickeners

Thickeners are substances that are added to the spray mixture to make it thicker or more viscous. Although thickeners are used as deposition aids, they also assist in keeping spray mixtures in suspension and slow the separation process once these materials are applied. They help to slow the water evaporation rate, therefore extending pesticide activity and reducing drift. Thickeners are often required as drift control agents when phenoxy herbicides, such as 2,4-D, are applied.

Attractants

Attractants are food or bait, such as sugar, molasses, protein hydrolysates, or insect pheromones, that attract specific pests. These are combined with a pesticide to form a lethal mixture. Attractants allow pesticides to be applied to localized parts of the treatment area rather than the entire area. They often enhance pesticide specificity to target pests.

Spray Colorants

Spray colorants are dyes that can be added to the spray tank so an applicator can see the areas that have been treated. These are usually used in backpack sprayers when applying herbicides to turf or a landscaped area, or in rangeland areas and fence rows. Spray colorants are not suitable for food crops since the dyes may remain on produce and there may not be a residue tolerance.

TABLE 4-14.

Method of Testing and Adjusting pH of Water Used for Mixing Pesticides.

pH can be measured with an electronic pH meter, a pH test kit such as those used for testing swimming pool water, or pH test paper available from a chemical supply dealer.

TEST WATER:

1. Using a clean container, obtain a sample of water from the same source that will be used to fill the spray tank.

2. Measure exactly 1 pint of this water into a clean quart jar.

3. Check the pH of the water using a pH meter, test kit, or test paper.

pH 3.5–6.0: Satisfactory for spraying and short-term (12 to 24 hours) storage of most spray mixtures in the spray tank.

pH 6.1–7.0: Adequate for immediate spraying of most pesticides. Do not leave the spray mixture in the tank for over 1 to 2 hours to prevent loss of effectiveness.

Above pH 7.0: Add buffer or acidifier.

ADJUST pH:

1. Using a standard eyedropper, add 3 drops of buffer or acidifier to the measured pint of water.

2. Stir well with a clean glass rod or other clean, nonporous utensil.

3. Check pH as above.

4. If further adjustment is needed, add 3 drops of buffer or acidifier, stir well, then recheck pH. Repeat until pH is satisfactory. Remember how many times 3 drops were added to bring the solution to the proper pH.

CORRECT pH IN SPRAY TANK:

1. Before adding pesticides to the sprayer, fill the tank with water.

2. For every 100 gallons of water in the spray tank, add 2 ounces of buffer or acidifier for each time 3 drops were used in the jar test above. Add buffer or acidifier to water while agitators are running. If tank is not equipped with an agitator, stir or mix well.

3. Check pH of the water in the spray tank to be certain it is correct. Adjust if necessary.

4. Add pesticides to spray tank.

ORGANIC PEST CONTROL MATERIALS

In 1979, an "Organic Foods Act" was made part of the California Health and Safety Code. This law specifies types of pesticides that can be applied to agricultural crops, livestock, poultry, or dairy products sold as organic, organically grown, naturally grown, wild, ecologically grown, or biologically grown (Figure 4-20). The law requires that, in the case of perennial crops, no synthetically compounded fertilizers, pesticides, or growth regulators be applied to a field where the commodity is grown

FIGURE 4-20.

Organic produce must be grown without the use of synthetic pesticides or synthetic soil amendments. Only microorganisms, microbiological products, and materials derived or extracted from plant, animal, or mineral-bearing rock substances may be used for the control of pests.

for 12 months prior to the appearance of flower buds, and none can be applied throughout the growing and harvest season. In the case of annual and two-year crops, no synthetically compounded fertilizers, pesticides or growth regulators can be applied for 12 months prior to seed planting and throughout the entire growing and harvest season.

Organically grown animals used for food (including poultry and fish) may not receive pesticides in any form except for treatment of a specific disease or malady. Pesticides cannot be administered or introduced within 90 days of slaughter under any circumstances. Feed given to livestock or poultry must be produced using only approved pesticides. Similar restrictions are imposed on milk-producing animals.

Pesticides allowed for use on organically produced foods include microorganisms, microbiological products, and materials derived or extracted from plant, animal, or mineral-bearing rock substances. Bordeaux mixes, trace elements, soluble aquatic plant products, botanicals, limesulphur, naturally mined gypsum, dormant oils, summer oils, fish emulsion, and insecticidal soaps are permitted. Items used as pesticides must be labeled for such use. Table 4-15 lists many of the allowable organic pesticides.

In general, pesticides approved for use on organically grown produce break down rapidly and therefore are often less disruptive to natural enemies and other organisms in the environment. The modes of action of many chemicals acceptable under the Organic Foods Act are similar to synthetically manufactured pesticides. For example, copper materials and rotenone interfere with cell respiration; pyrethrins interfere with nerve transmission, similar to DDT; nicotine blocks nerve and nerve-muscle connections; petroleum oils interfere with cell membrane activity;

and ryania affects the nervous system. These pesticides have LD_{50} values similar to Category II and Category III organophosphate and carbamate pesticides (Table 4-16).

TABLE 4-15.

Pesticides Accepted for Use On or Around
*Some Types of Organically Grown Produce.**

COMPOUND	TYPE	USE
INSECTICIDES:		
Bacillus thuringiensis	microbial	Controls many species of lepidopteran larvae, mosquito larvae (depending on the variety of the B. *thuringiensis* used).
boric acid	inorganic	Boric acid is a sorptive dust having a desiccant action. Controls cockroaches, ants, other household pests. Ineffective if dust gets wet.
cryolite	inorganic	Controls mites, moth larvae, beetles, weevils, and thrips.
diatomaceous earth	inorganic	Diatomaceous earth is a sorptive dust derived from the skeletons of microscopic marine organisms. As a desiccant, controls household pests such as cockroaches and ants. Also controls some plant pests.
granulosis virus	microbial	Controls codling moth.
lime	inorganic	Controls mites, some plant-sucking insects.
lime sulfur	inorganic	Controls mites and psylla.
nicotine sulfate	plant derivative	Controls aphids, thrips, leaf hoppers, other sucking insects. Toxic to mammals.
petroleum oils	hydrocarbon	Controls aphids, psylla, scale insects, mites, aphid and mite eggs. May provide some control of other overwintering insects.
pheromones	attractants	Used mainly for monitoring to time other control measures. Sometimes used to confuse insects in localized area to disrupt mating. Occasionally used to catch large numbers of specific insects to reduce future generations.
pyrethrum	plant derivative	Broad spectrum of pests are controlled, including mosquitoes, flies, aphids, beetles, moth larvae, thrips, and mealy bugs. Provides rapid knockdown of flying pests.

*Many materials listed in this table may not be currently registered as pesticides or their labels may restrict their use to specific pests, crops, or sites. Use all pesticides only in accordance with current federal and state labels.

COMPOUND	TYPE	USE
rotenone	plant derivative	Contact and stomach poison. Controls beetles, weevils, slugs, loopers, mosquitoes, thrips, fleas, lice, and flies. Also used for control of unwanted fish. Acts as a repellent and acaricide. Rotenone is slow acting and has a short residual. It is nontoxic to honey bees.
ryania	plant derivative	Controls codling moth, thrips, and the European corn borer.
sabadilla	plant derivative	Has contact and stomach poison action against cockroaches, several species of bugs, potato leafhopper, imported cabbage worm, house fly, citrus thrips, and the cattle louse. Is toxic to honey bees. Not highly toxic to mammals.
soaps	soap	Controls mites, aphids, and other plant-sucking arthropods. Can be phytotoxic to plants under certain conditions. Soap must be specifically labeled for use as an insecticide.
sulfur	inorganic	Controls mites.
vegetable oils	plant derivative	As a contact spray, controls scale insects, aphids, and mites.
FUNGICIDES:		
basic copper sulfate	inorganic	Controls early and late blight, scab, blotch, bitter rot, fire blight, downy mildew, black rot, leaf spot, melanose, greasy spot, brown rot, anthracnose, angular leaf spot, and others.
Bordeaux mix	inorganic	Bordeaux is a slurry made of hydrated lime and copper sulfate. Controls brown rot and shot hole diseases in tree fruit. Controls some grape diseases. Also controls apple scab, blotch, apple black rot, melanose, anthracnose, early and late blight of potatoes and tomatoes, downy mildew, fire blight, leaf spot, peach leaf curl, and many other fungal diseases.
copper ammonium carbonate	inorganic	Controls angular leaf spot, alternaria leaf spot, cercospora leaf spot, early and late blight, bacterial blight, common blight, anthracnose, melanose, powdery mildew, downy mildew, and others.
copper hydroxide	inorganic	Controls cercospora leaf spot, bacterial blight, septoria, leaf blotch, anthracnose, halo blight, helminthosporum, downy mildew, leaf curl, early and late blight, angular leaf spot, melanose, scab, walnut blight, and others.

*Many materials listed in this table may not be currently registered as pesticides or their labels may restrict their use to specific pests, crops, or sites. Use all pesticides only in accordance with current federal and state labels.

COMPOUND	TYPE	USE
copper oxychloride sulfate	inorganic	Controls peach blight, peach leaf curl, damp-off, anthracnose, fire blight, shot hole fungus, pear blight, bacterial spot, walnut blight, brown rot, celery blight, downy mildew, early and late blight of vegetables, cherry leaf spot, septoria leaf spot, powdery mildew, melanose, scab, and others.
copper sulfate	inorganic	Suppresses development of fungal and bacterial organisms such as fire blight, cercospora leaf spot, early and late blight, bacterial blight, and others.
lime sulfur	inorganic	Controls powdery mildew, anthracnose, apple scab, brown rot, peach leaf curl, and others.
sulfur	inorganic	Controls brown rot, peach scab, apple scab, powdery mildew, downy mildew, rose black spot, and others.
terramycin	antibiotic derived from a fungus	Controls certain bacterial diseases in plants.

*Many materials listed in this table may not be currently registered as pesticides or their labels may restrict their use to specific pests, crops, or sites. Use all pesticides only in accordance with current federal and state labels.

TABLE 4-16.

Oral LD$_{50}$ Values for Some Pesticides Used to Control Pests on Organically Grown Produce.

CHEMICAL	TRADE NAME	LD$_{50}$	TYPE OF PESTICIDE
nicotine	Black Leaf 40	55	insecticide
rotenone		132	insecticide
Bordeaux		300	fungicide
copper hydroxide	Kocide	1000	fungicide
copper oxychloride sulfate	C-O-C-S	1000	fungicide
ryania		1200	insecticide
pyrethrum		1500	insecticide
silica aerogel	Dri-Die	3160	insecticide
sabadilla		4000	insecticide
cryolite	Kryocide	10,000	insecticide

5 Pesticide Laws and Regulations

Laws and regulations governing the use of pesticides are made by federal and state agencies.

State and federal laws and regulations control many aspects of the manufacture, sale, transportation, and use of pesticides. At the national level, the Environmental Protection Agency (EPA), established in 1970, regulates the use of pesticides in the United States under mandate from the Federal Insecticide, Fungicide, and Rodenticide Act (FIFRA). Originally passed in 1947, this law has been amended and updated several times. The EPA establishes regulations for pesticide registration and labeling, pesticide residue tolerance levels in foods, control of restricted-use pesticides, and standards for pesticide applicator certification. Other departments of the federal government, including the Department of Agriculture (USDA), the National Institute of Occupational Safety and Health (NIOSH), and the Fish and Wildlife Service (FWS), are also involved in monitoring and regulating pesticides and pesticide use. Users in all states must comply with federal laws. California laws are designed to make pesticide use safer under the special conditions existing in California and to protect people and the environment within the state (Table 5-1). California laws are sometimes more restrictive than federal laws but cannot permit uses or activities prohibited by federal law.

In California, laws regulating pesticide use and pest control are part of the California Food and Agricultural Code. These laws are made by the California Legislature in response to needs arising within the state or from federal requirements. Regulations are the working rules needed to interpret and carry out laws. Regulations pertaining to pest control and pesticide use are part of the California Administrative Code; these are written by the California Department of Food and Agriculture (CDFA) after input from public hearings. Several other state agencies—the Department of Health Services, Air Resources Board, State Department of Fish and Game, State Department of Forestry, the Occupational Safety and Health Administration of the State Department of Industrial Relations, the Waste Management Board, the Department of Water Resources, the Water Resources Control Board and Regional Water Control Boards, and the Structural Pest Control Board of the Department of Consumer Affairs—are also involved in monitoring and regulating pesticide use. Agricultural commissioners in each county develop pesticide use policies or conditions specific to the needs of their counties that must be approved by the Director of the Department of Food and Agriculture before they become effective. Counties and local governments also may create laws and ordinances governing the use and storage of hazardous materials, including pesticides, within their jurisdiction.

ENFORCEMENT

The Director of the Department of Food and Agriculture and county agricultural commissioners have the responsibility for enforcing state pesticide laws and regulations. The U.S. Environmental Protection Agen-

TABLE 5-1.

Reasons for Pesticide Laws and Regulations.
(Taken from the CDFA "Laws and Regulations Study Guide")

1. To provide for the proper, safe, and efficient use of pesticides essential for the production of food and fiber and for protection of public health and safety.

2. To protect the environment from environmentally harmful pesticides by prohibiting, regulating, or controlling uses of these pesticides.

3. To assure the agricultural and pest control workers of safe working conditions where pesticides are present.

4. To permit agricultural pest control by competent and responsible licensees and permittees under strict control of the Director of the California Department of Food and Agriculture and the local agricultural commissioners.

5. To assure the users that economic poisons are properly labeled and are appropriate for the use designated by the label.

6. To encourage the development and implementation of pest management systems, stressing application of biological and cultural pest control techniques with selective pesticides when necessary to achieve acceptable levels of control with the least possible harm to nontarget organisms and the environment.

cy enforces federal regulations. Failure to comply with federal and state pesticide laws and regulations will subject a violator to fines and/or imprisonment, as well as the possible loss or suspension of the violator's applicator license or certificate. The office of the State Attorney General or local district attorneys are charged with prosecuting violators. County agricultural commissioners may issue citations and levy fines on violators of certain pesticide-use regulations. Table 5-2 summarizes the responsibilities of the CDFA, county agricultural commissioners, and other state and federal government agencies.

Laws and regulations change periodically when new situations arise that are not addressed by existing laws and regulations. For example, pesticides and pesticide application equipment are constantly being improved and changed, and pest problems and pest management techniques may be different from year to year. As new health and environmental problems are identified, regulations must be created or modified to deal with them. Obtain a copy of the publication "Laws and Regulations Study Guide for Agricultural Pest Control Adviser, Agricultural Pest Control Operator, Pesticide Dealer, Pest Control Aircraft Pilot Examinations" from the California Department of Food and Agriculture, Pesticide Enforcement, 1220 N Street, Sacramento, California 95814. This study guide covers current information on laws and regulations that you will need for the Qualified Pesticide Applicator License (QL) or the Qualified Pesticide Applicator Certificate (QC) examinations.

TABLE 5-2.

Responsibilities of Government Agencies in
California's Pesticide Regulatory Program.

REGULATORY PROGRAM	WHO DOES IT	WHAT CAN IT DO?
Registration of pesticides	EPA, CDFA	Refuse or accept registration; suspend, cancel, or reregister pesticides.
Classification of pesticides	EPA, CDFA	EPA classifies pesticides as restricted or nonrestricted use in United States; CDFA may impose more stringent restrictions for California, based on special conditions existing in the state.
Permitting	CAC, CDFA	Issue, revoke, or refuse restricted-use pesticide permits (with use conditions) to growers, other private applicators, or certified applicators.
Licensing of commercial applicators, advisers, pesticide application businesses, dealers, and maintenance gardeners	CDFA*	Issue licenses and (in some cases) administer tests to applicants, including agents of businesses; revoke, suspend, or refuse licenses upon violation of pesticide laws.
Registering applicators and advisers, certifying private applicators	CAC	Register agricultural pest control businesses, aerial pest control operators, licensed pest control advisers, and maintenance gardeners. Through oral interview, certifies private applicators. Provide applicators and advisers with information on local pesticide use conditions. Inspect pesticide use records and pest control recommendations to verify proper pesticide use.

*In California, Structural Pest Control Operators are licensed by the Structural Pest Control Board, California Department of Consumer Affairs. Vector Control Certificates are issued by the Department of Health Services.

REGULATORY PROGRAM	WHO DOES IT	WHAT CAN IT DO?
Monitoring of pesticide residues on food and feed	USDA, CDFA, CAC**	Test food and feed for pesticide residues; quarantine or destroy illegally contaminated commodities; bring cases of violation to county district attorney or State Attorney General for prosecution.
Regulations governing pesticide use and worker safety	CDFA, CAC, DHS	General authority to regulate pest control operations, including restrictions on the time, place, and manner of application; various warning and enforcement powers.
Pesticide illness investigation	CDFA, CAC, DHS	Participate in pesticide illness investigations and in development of worker safety regulations.
Pesticide disposal and storage	CDFA, DHS, WRCB, WQCB, ARB, CAC	Regulates hazardous waste storage and disposal, pesticide container disposal sites, and water quality standards.
Protection of wildlife	EPA, FWS, DFG, CAC, CDFA	Investigate fish and wildlife losses. Identify and monitor endangered species. Restrict pesticide use to protect endangered species and other wildlife.
Citing or prosecuting violators	EPA, CDFA, CAC, SPCB, State Attorney General, Local District Attorneys	Agricultural commissioner may levy civil penalties with fines. CDFA may request Attorney General to take civil action. Attorney General may file accusation. CDFA may suspend or revoke applicator's certificate. CAC may suspend, revoke, or refuse permits and county registration.

**13 other state and 5 federal agencies monitor various parts of the environment for pesticides and other substances.

ARB: California Air Resources Board

CAC: County Agricultural Commissioner

CDFA: California Department of Food and Agriculture

DFG: California Department of Fish and Game

DHS: California Department of Health Services

EPA: U.S. Environmental Protection Agency

FWS: U.S. Fish and Wildlife Service

SPCB: Structural Pest Control Board California Department of Consumer Affairs

USDA: U.S. Department of Agriculture

WRCB: California Water Resources Control Board

WQCB: California Water Quality Control Board

PESTICIDE REGISTRATION AND LABELING

Every pesticide must be registered with the federal Environmental Protection Agency as well as with the California Department of Food and Agriculture before it can be sold or used in California. Individual pesticide *products* are registered, not generic pesticides. Labels may differ among products containing the same pesticide active ingredient. The registration procedure is necessary to provide for the proper and safe use of pesticides that protects people and the environment from ineffective or detrimental chemicals. Table 5-3 lists some of the information that a manufacturer provides in order to register a product for use as a pesticide. The registration procedure includes an evaluation of each chemical by the Pesticide Regulation Division and Pesticide Tolerance Division of the Federal Environmental Protection Agency. This evaluation establishes whether the material will be classified at the federal level as a *general-use pesticide*—one that can be sold without permit and can be used by the general public, or as a federally *restricted-use pesticide*—one that can only be sold to and used by qualified pesticide applicators or by persons hold-

TABLE 5-3.

Information Manufacturers Must Provide to CDFA to Register a Pesticide in California. (From CDFA "Laws and Regulations Study Guide").

1. *Exposure Information*: data on risks of exposure and how people can be protected.
   a. Dermal absorption.
   b. Mixer, loader, applicator exposure.
   c. Management of poisoning.
   d. Toxicology of adjuvants and other components of the formulation.
   e. Indoor exposure information.
   f. If material is a rodenticide, metabolic pathway and mode of action.
   g. Foliar residue and field reentry data.

2. *Residue test method.*

3. *Residue data.*

4. *Efficacy.*

5. *Hazard to bees.*

6. *Closed-system compatibility.*

7. *Effects on pest management.*

8. *Inert ingredient hazard.*

9. *Volatile organic compounds*: their relationship to air quality.

10. Other data as requested by the Director of the Department of Food and Agriculture, such as:
    a. Information on drift potential.
    b. Phytotoxicity.
    c. Contaminants or impurities in the product.
    d. Analytical and environmental chemistry.
    e. Effects of tank mixes on the product (compatibility).

FIGURE 5-1.

Supplemental labels are often attached to pesticide packages. Before purchasing a pesticide, make sure you have a complete set of labels.

ing a valid permit from a county agricultural commissioner. In addition to federally restricted materials, the State of California sometimes designates certain general-use pesticides as restricted-use pesticides in California due to local hazards or specific health problems. A permit from the county agricultural commissioner is required for all California restricted-use pesticides, except for certain use exemptions allowed. As part of the pesticide registration procedure in California, registration information is reviewed by representatives of the California Departments of Food and Agriculture, Fish and Game, Health Services, and Industrial Relations as well as the Air Resources Board and Water Resources Control Board. In addition, all decisions to register, renew, or reevaluate a pesticide are subject to a 45 day public review and comment period before final action can be taken. This provides an opportunity for interested persons to provide input into registration decisions. To complete registration, the manufacturer must supply a label meeting all federal and state requirements. Labels are legal documents that contain important information for the user. Labels may also refer to other documents, such as endangered species range maps, that must be considered part of the label. Obtain, read, and understand all the information on a label and referenced documents before making a pesticide application.

Pesticide Labels

Regulations establish the format for pesticide labels and prescribe what information they must contain. Some packages are too small, however, to have all this information printed on them, so manufacturers are required to attach supplemental labels (Figure 5-1); on metal and plastic containers, supplemental labels are enclosed in a plastic pouch and glued to the side of the container, while paper packages usually have supplemental labels inserted under the bottom flaps.

FIGURE 5-2.

The pesticide label is a complex legal document that must be read and understood before making a pesticide application. Pesticide applications must be made in strict accordance to the label instructions.

When to Read the Pesticide Label

Read the pesticide label (Figure 5-2):
1. *Before purchasing the pesticide.* Make sure it is registered for your intended use. Confirm that there are no endangered species restrictions or other conditions that prohibit the use of this pesticide at the application site. Be certain it can be used under current weather conditions and against the pest life stage you are trying to control. Find out what protective equipment and special application equipment is required.
2. *Before mixing and applying the pesticide.* Understand how to mix and safely apply the material. Learn what precautions are needed to prevent exposure to people and nontarget organisms. Find out what first aid and medical treatment is necessary should an accident occur.
3. *When storing pesticides.* Find out how to properly store the pesticide to prevent breakdown or contamination. Understand the special precautions to prevent fire hazards. Be sure storage areas are properly posted.
4. *Before disposing of unused pesticide and empty containers.* Learn how to prevent environmental contamination and hazards to people.

Before disposal, check with the agricultural commissioner in your area for local restrictions and requirements.

Labels on specific pesticides may change from time to time as regulations on pesticide use are put into effect. If you are using a pesticide that was purchased several months earlier, be sure to check with the local agricultural commissioner or obtain a copy of the most recent label for information on proper methods of use and disposal of unused material.

What Pesticide Labels Contain

Brand Name. A brand name is the name the manufacturer has given to the product and is the name used for all advertising and promoting.

Chemical Name. Chemical names describe the chemical structure of a pesticide and are derived by chemists based on international rules for naming chemicals.

Common Name. Chemical names of active ingredients in a pesticide formulation are often complex. For clarity, most pesticides have been assigned official common, or generic, names. For example, 0,0-diethyl 0-(2-isopropyl-6-methyl-4-pyrimidinyl) has an official common name of diazinon. Common names and brand names are not the same and not all labels will list a common name for the pesticide.

Formulation. Pesticide labels always list the formulation type, such as emulsifiable concentrate, wettable powder, or soluble powder. Manufacturers may include this information as a suffix in the brand name of the pesticide. For example, in the name *Princep 80W*, the "W" indicates a wettable powder formulation. Table 4-9 on page 97 lists definitions for suffixes used with brand names.

Ingredients. Pesticide labels list the percentage of active and inert ingredients by weight. Inert ingredients are all components of the formulation that do not have pesticidal action, although they may be toxic, flammable, or pose other safety or environmental problems—or are ingredients that are totally harmless. If a pesticide contains more than one

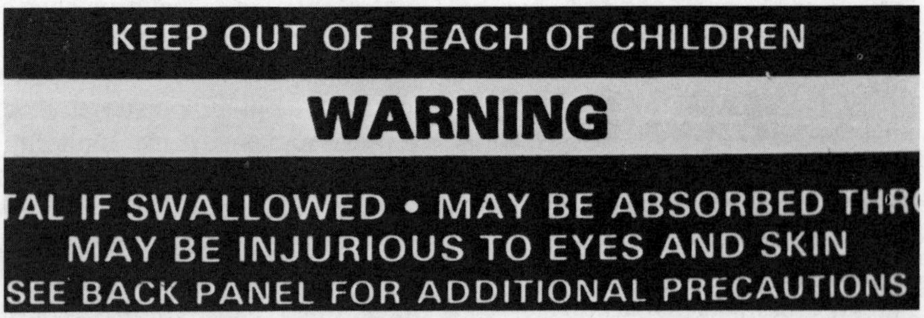

KEEP OUT OF REACH OF CHILDREN

WARNING

TAL IF SWALLOWED • MAY BE ABSORBED THR(
MAY BE INJURIOUS TO EYES AND SKIN
SEE BACK PANEL FOR ADDITIONAL PRECAUTIONS

FIGURE 5-3.

The signal word "Warning" on this label indicates that the pesticide has moderate toxicity and should be handled with care.

PRECAUTIONARY STATEMENTS
Hazards to Humans and Domestic Animals
**MAY BE FATAL IF SWALLOWED • MAY BE ABSORBED
THROUGH SKIN • MAY BE INJURIOUS TO EYES AND SKIN
Do Not Take Internally • Do Not Get In Eyes, on Skin or on
Clothing • Avoid Breathing Vapors and Spray Mist • Wash
Thoroughly After Handling**
Statements of Practical Treatment
If Swallowed, do not induce vomiting. Contains aromatic petroleum solvent. Call a physician immediately.
If On Skin: In case of contact, remove contaminated clothing and immediately flush skin with soap and water. Wash contaminated clothing before reuse.
If In Eyes: Flush eyes with plenty of water for 15 minutes. Call a physician.
NOTE TO PHYSICIAN: Chlorpyrifos is a cholinesterase inhibitor. Treat symptomatically. Atropine only by injection is an antidote.

FIGURE 5-4.

The "Precautionary Statement," "Statement of Practical Treatment," and "Statement of Use" on pesticide labels provide important information on hazards you must be aware of. Read these statements thoroughly before using the material.

active ingredient, the percentage of each will be given, but all inert ingredients may be grouped together and not specified. In the example given above, the name Princep 80W indicates that there is 80% by weight of the active ingredient 2-chloro-4,6-bis(ethylamino)-s-triazine. Labels on liquid formulation containers indicate how many pounds of active ingredient are contained in one gallon of formulated pesticide.

Contents. Labels list the net contents, by weight or liquid volume, contained within the package.

Manufacturer. Pesticide labels always contain the name and address of the manufacturer of the product. Use this address if you need to contact the manufacturer for any reason.

Registration and Establishment Numbers. The Environmental Protection Agency and the state of California assign numbers to each pesticide as it is registered. In addition, the EPA establishment number is a code which identifies the site of manufacture or repackaging of a pesticide.

Signal Word. An important part of every label is the signal word (Figure 5-3). The word "Danger," accompanied by the word "Poison" and a skull and crossbones, or the word "Danger" used alone indicates that the pesticide is highly toxic or poses a dangerous health or environmental hazard (Toxicity Category I). "Warning" indicates moderate toxicity (Toxicity Category II) and "Caution" means low toxicity (Toxicity Category III). (The "Toxicity Classification" section in Chapter 4 explains these different toxicity categories.) Part of the registration process assigns each pesticide to a toxicity category and prescribes which signal word must be used on the label.

Precautionary Statements. Precautionary statements are used to describe the hazards associated with a chemical (Figure 5-4). Always read and follow the instructions given in a precautionary statement. Three areas of hazard may be included. Most important are the hazards to people and domestic animals. This section tells why the pesticide is hazardous, what adverse effects may occur, and describes the type of protective equipment that one must wear while handling packages and mixing and applying the pesticide.

The second part of a precautionary statement gives information on environmental hazards. It indicates if the pesticide is toxic to nontarget organisms such as honey bees, fish, birds, or other wildlife, and may contain information on how to avoid environmental contamination.

The third part of the precautionary statement explains special physical and chemical hazards, such as risks of explosion if the chemical is exposed to sparks, or hazards from fumes in the case of a fire.

Statement of Practical Treatment. The statement of practical treatment tells what to do in case anyone has been exposed to the pesticide through an accident. It describes what emergency first aid measures to take when the pesticide contacts skin, splashes into eyes, or if dust or vapors have been inhaled. This section tells you when to seek medical attention.

Statement of Use Classification. Pesticides are classified by the Environmental Protection Agency as either "General-Use" or "Restricted-Use," based on the potential of the pesticide to cause harm to people, animals, or the environment. EPA restricted-use pesticides have a special statement printed on the label in a prominent place. Pesticides that do not contain this statement are considered general-use pesticides, although special state restrictions may apply; for this information, check the CDFA list of state restricted-use pesticides, available from the county agricultural commissioner. Labels may also have a *restrictive* statement indicating that they are for agricultural or commercial use only. A restrictive statement is different from a statement of use classification.

Directions for Use. The directions for use list all the target pests that the pesticide has been registered to control, plus the crops, plant species, animals, or other sites where the pesticide may be used (Figure 5-5). The directions may also include special restrictions that must be observed, such as crops that may or may not be planted in the treatment area (plantback restrictions) and restrictions on feeding crop residues to livestock or grazing livestock on treated plants. These instructions tell how to apply the pesticide, how much to use, where to use the material, when it should be applied, and also include the preharvest interval for all crops whenever appropriate. The preharvest interval is the time, in days, required after application before an agricultural crop may be harvested. Always follow these directions; it is a violation to use pesticides in a manner inconsistent with the label unless federal or state laws specify acceptable deviations from label instructions.

Misuse Statement. The misuse statement reminds users to apply pesticides according to label directions.

Reentry Statement. Sometimes restrictions apply to the time that must elapse before a person can enter an area treated with a pesticide. This reentry interval is included on the pesticide label or in state regulations. Reentry intervals may vary according to the toxicity and special hazards associated with the pesticide and the type of crop being treated

ORNAMENTAL INSECT CONTROL

To control certain insects on Arborvitae, Azalea, Birch, Boxwood, Camellia, Carnation, Chrysanthemum, Douglas Fir, Elm, Gladioli, Hawthorn, Holly, Juniper, Lilac, Locust, Maple, Oak, Pine, Plum, Poplar, Rhododendron, Rose, Spruce, and Willow, apply the recommended rates indicated below.

Insects	Rate per 3 gals. water	Rate per 100 gals. water
Aphids, Bagworms, Carnation Bud Mites, Carnation Shoot Mites, Clover Mites, Cyclamen Mites, Dipterous Leafminers, European Pine Shoot Moths, European Red Mites, Flea Beetles, Holly Bud Moths, Leafhoppers, Obscure Root Weevils, Omnivorous Leaftiers, Privet Mites, Scale	½ fl. oz.*	1 pt.*

FIGURE 5-5.

Pesticides must be applied only in the manner described in the "Directions for Use."

and may even vary from county to county. If no reentry interval is given, the treated area can usually be entered once the spray dries or dust settles, but check first with the county agricultural commissioner to be sure no state or local restrictions apply. For example, in California all Category I pesticides have a minimum reentry interval of 24 hours. A special calculation is used to determine a reentry interval when two or more pesticides are mixed (see Chapter 7).

Storage and Disposal Directions. Directions for proper storage and disposal of the pesticide and empty pesticide containers are another important part of the label. Some pesticides have special requirements. Improper storage causes the pesticide to lose its effectiveness or may even cause an explosion or fire. Pesticides must always be stored out of the reach of children and animals, in locked and posted areas. Proper disposal of unused pesticides and pesticide containers is essential to reduce human and environmental hazards. Federal, state, and local regulations control how pesticides may be disposed of; if specific information is not included on the pesticide label, it may be obtained from your county agricultural commissioner.

Warranty. Manufacturers usually include a warranty and disclaimer on their pesticide labels. This information informs you of your rights as a purchaser and limits the liability of the manufacturer.

DEVIATIONS FROM LABEL DIRECTIONS

There may be specific instances when regulations allow the use of pesticides in a manner inconsistent with label directions. These deviations generally involve using the pesticide in a *safer* manner. The University of California pest management guidelines sometimes make recommendations for pesticide use that may be different from label instructions. Following are the only label deviations allowed by California law. These exceptions may change at any time or may not apply in certain instances; always check with the Department of Food and Agriculture or with your local agricultural commissioner before using a pesticide in any manner inconsistent with label directions.

Decrease in dosage rate per unit treated. At times it may be recommended to use less pesticide than called for on the label because a lower rate will be less disruptive to natural enemies. For example, University of California pest management guidelines for control of spider mites in almonds recommend growers apply as little as $\frac{1}{10}$ of the label rate of specific acaricides *when there are adequate levels of beneficials present and other conditions are met.* This lower rate reduces pest spider mite levels without harming predators, allowing predators to maintain control. However, sometimes the use of lower rates speeds up the development of pesticide resistance in the target organism. To avoid possible problems associated with using lower rates of pesticides, check first with the local farm advisor or county agricultural commissioner.

Under no circumstances is it legal for an applicator to increase the amount of pesticide per unit of treatment area above the rates printed on the label.

Decrease in the concentration of the mixture applied. Label instructions usually recommend the amount of water that must be used when preparing a spray mixture. It is always possible to use more water than recommended, although excess water may cause problems of excessive dilution and runoff, resulting in not enough of the pesticide getting to the target pest. In most cases, use only as much water as necessary to obtain thorough coverage, as long as this is not less than label recommendations.

The only time it is possible to use less water than recommended by the label is when the increase in concentration corresponds with published recommendations of the University of California.

Increase in concentration as long as it corresponds with published recommendations of the University of California. There may be instances in which it is practical to use a more concentrated mixture of pesticide than specified on the label. The labeled amount of pesticide must be applied per unit of area, however less carrier (water or oil) is used. An example involves the use of a controlled droplet applicator for applying herbicides; a more concentrated mixture is needed because of the nature of the application equipment. The pesticide application must be done in accordance with published UC recommendations, and all other label instructions should be followed. Verbal recommendations of any type are not acceptable. Hazards associated with increased concentrations include possible phytotoxicity, spotting, or staining of sprayed surfaces.

Application at a frequency less than specified. Label instructions often prescribe how often to apply a pesticide in order to maintain adequate control of the pest being treated. It is permissible under the law to make applications less frequently than the label recommends. If pest problems are being monitored and pests are found to be adequately controlled by less frequent (or even single) pesticide applications, there would be no need to make additional treatments. Additional applications sometimes increase other pest problems by harming beneficials. Besides, it is expensive to use pesticides when they are not needed.

Conversely, reducing the frequency of pesticide application below label recommendations may result in problems if adequate pest control is not achieved. Proper monitoring and follow up must be done to prevent control failure.

There are no circumstances when it is permissible to apply a pesticide more frequently than the recommended interval listed on the label. If the label does not prescribe a treatment interval, or if terms such as "apply as needed to control" are used, you may apply the pesticide as often as necessary. Be sure, however, that repeated applications are needed.

Use to control a target pest not on label when commodity or site is on label and use against unnamed pest is not expressly prohibited. You may wish to use a particular pesticide on a commodity or site listed on the label, but find that the target pest is not on the label. As long as the label does not specifically forbid use of this pesticide against the target pest on the proposed commodity or site, it is possible to use it in this manner. The commodity or target site must be listed (for other pests) and all other label instructions must be followed.

Use of any method of application not prohibited, provided other label directions are followed. Most label recommendations do not specify exactly how the pesticide must be applied. Should this be the case, it is possible to use any method, as long as the method used allows the application to be done consistent with all other label directions.

Application of a pesticide by ground or by air is an example. If there is no prohibition against aerial application on the label, either method may be used provided all label directions can be complied with, even though aerial application may not be specified. (It may not be possible to apply a pesticide by air if the label prohibits the lower dilution rates that are required for aerial applications.)

Mixing with another pesticide or fertilizer, unless prohibited. You may want to combine one pesticide with one or several others or with fertilizers; this type of application saves time and reduces application costs. Unless specifically prohibited by directions on any of the labels, it is permissible to apply pesticides in combination.

Even though they may not be prohibited, there can be incompatibility problems with certain combinations of pesticides or pesticides and fertilizers. Check for incompatibility before mixing large volumes. Refer to Chapter 4 for methods of determining incompatibility and resolving incompatibility problems.

No pesticide should be mixed with another pesticide or fertilizer or any other material if the label prohibits such a mixture. Label restrictions may specify general classes of chemicals, such as sulfur-containing materials, alkaline chemicals, or oils.

EMERGENCY EXEMPTIONS
AND SPECIAL LOCAL NEEDS

Occasionally pest problems arise that cannot be controlled with currently registered pesticides, or the commodity, target, or site is not on the registered pesticide label. In some of these situations, an emergency registration exemption or a special local needs registration (SLN) can be issued by the California Department of Food and Agriculture.

Emergency registration exemptions are allowed under Section 18 of the Federal Insecticide, Fungicide, and Rodenticide Act (FIFRA) and must be granted to the California Department of Food and Agriculture by the Environmental Protection Agency. The basis for such an exemption would be an emergency condition, such as a pest situation that threatens the health of the public. If an emergency exemption has been allowed by the Environmental Protection Agency, the California Department of Food and Agriculture then issues an authorization for use of the nonregistered pesticide. The authorization for use serves as a supplemental label; it must be in the possession of the user at the time the pesticide is applied and the application must be performed in accordance with its restrictions. Pesticides used under the provisions of an emergency registration exemption are considered restricted-use materials. A permit is required from the local agricultural commissioner prior to the possession or use of these pesticides.

When a pest problem exists that requires a pesticide treatment and no registered pesticide is available, a special local needs registration may sometimes be issued. The registrant of the pesticide may request a special local needs registration, or an individual, organization, or government agency may apply for this type of registration. A special local needs registration cannot be issued if: (1) adequate registered products are available that will meet the need; (2) the registration is being requested for a use that has been canceled, suspended, or denied by the EPA; or (3) the pesticide will be used on a food or feed crop where a residue tolerance has not been established or there is no exemption from tolerance. Special local needs registrations will specify conditions that must be followed when using the pesticide.

RESTRICTED-USE PESTICIDES

State and federal laws restrict the sale and use of Toxicity Category I materials or other pesticides that meet certain hazard criteria (Figure 5-6) or are being used under the provisions of an emergency registration exemption. In addition to enforcing federal and state restrictions, counties may place further restrictions on some pesticides. For agricultural crops, a *Restricted Material Use Permit* must be obtained by the person growing the crop (or their authorized representative); permits are obtained from a county agricultural commissioner. Licensed agricultural pest control businesses do not need a permit to possess or apply federally restricted pesticides on agricultural crops unless the material is also a California restricted pesticide. However, they must have a written recommendation from a licensed pest control adviser and a copy of the restricted-use

FIGURE 5-6.

This is a restricted-use pesticide. Unless proper precautions are used, restricted-use pesticides may be hazardous to people or the environment. You may need to obtain a "Restricted-Use Permit" before applying these pesticides in agricultural areas.

permit issued to the grower or person responsible for the crop. For all other applications, only certified applicators are allowed to purchase and use federally restricted pesticides. Permits are not required for federally restricted pesticides used in nonagricultural areas unless there are additional state restrictions. Materials classified as restricted in California require a permit for *any* use other than use by structural pest control operators. Obtain information on federal, state, and county use restrictions from the county agricultural commissioners' offices in counties where you make applications. Note that laws and restrictions vary from county to county.

Criteria for California Restricted-Use Pesticides

The California Department of Food and Agriculture has defined several criteria for restricting the use of certain pesticides (Table 5-4). A current list of state restricted materials may be obtained from any county agricultural commissioner or from CDFA.

Restricted materials may often be used only at certain times of the day or season or only under special conditions specified by the county permit. Restricted-use pesticides may be used only where it is reasonably certain that no injury will result and where no nonrestricted material or procedure is equally effective and practical.

Restricted-Use Permits

A certified commercial pesticide applicator may purchase federal restricted-use pesticides without a permit as long as there are no additional state restrictions on that material. A grower, private applicator, or person responsible for the property (or an authorized representative such as an employee or licensed pest control adviser) must obtain a written restricted-use permit (Figure 5-7) before either a state or federal restricted-use pesticide may be purchased or applied to agricultural crops. Permits must be obtained from the county agricultural commissioner in the county where the use is intended; each permit is issued for a specific agricultural site and crop. Permits for California restricted materials used in nonagri-

TABLE 5-4.

Criteria for Restricting the Use of Certain Pesticides in California.
(Taken from CDFA "Laws and Regulations Study Guide").

1. Danger to public health.

2. Hazard to applicators and farmworkers.

3. Hazard to domestic animals, honey bees, and
   crops from direct application or drift.

4. Hazard to the environment from drift onto
   streams, lakes or wildlife sanctuaries.

5. Hazards related to persistent residues in the
   soil resulting in contamination of the air,
   waterways, estuaries or lakes, causing damage
   to fish, wild birds, and other wildlife.

6. Hazards to subsequent crops due to persistent
   soil residues.

7. A federally restricted pesticide.

cultural areas are issued *only* to certified applicators; these permits are
not site-specific. A valid pesticide applicator license or certificate must
be presented to the agricultural commissioner at the time the permit
is being applied for. The commissioner, before issuing a permit, will
consider local conditions, including whether the pesticide will be applied
in the vicinity of dwellings, schools, hospitals, recreational areas, or livestock or poultry, and if there are any adjacent crops or plantings that
will be adversely affected by this pesticide. The commissioner must also
consider weather conditions and how they might affect the pesticide application, what measures are being taken to protect honey bees, and how
the pesticides will be stored and containers disposed of. Table 5-5 describes how to obtain a restricted-use permit. Permits are usually issued
each year for an entire season. Before an application of a restricted material is actually made, however, the agricultural commissioner who issued
the permit must be notified at least 24 hours in advance by filing a
"Notice of Intent" (Figure 5-8).

Exceptions. Permits are not required for the agricultural use of pesticides that are not federal or state restricted materials, unless the Director
of the California Department of Food and Agriculture or local agricultural
commissioner determines that their use will present an undue hazard
under local conditions. Structural pest control operators, licensed pesticide dealers, and commercial warehouses do not need permits for pos-
sessing restricted pesticides.

Exempt Pesticides. Some pesticides that may ordinarily be included
as restricted-use materials are exempted from permit requirements by
the Department of Food and Agriculture when it is determined that further restrictions, beyond those imposed by the label, are not necessary.
The Department of Food and Agriculture maintains a list of exempt materials; a permit is not required to use materials on this list only as long
as county requirements do not impose restrictions.

APPLICATION—RESTRICTED MATERIALS PERMIT

☐ FOR POSSESSION ONLY ☐ FOR POSSESSION AND USE

PERMITTEE _____

PERMIT NO. _____

| PERMITTEE ADDRESS | CITY | ZIP | PHONE | TYPE OF PERMIT ☐ SEASONAL ☐ JOB | EXPIRATION DATE _____ |

☐ PRIVATE APPLICATOR ☐ STRUCTURAL PCO ☐ AGRICULTURAL PCO ☐ COMMERCIAL APPLICATOR

NOTICE OF INTENT REQUIRED ☐ MUST BE SUBMITTED AT LEAST _____ HOURS PRIOR TO APPLICATION. METHOD:

A. PESTICIDES/PESTS

1. _____ 7. _____ 13. _____
2. _____ 8. _____ 14. _____
3. _____ 9. _____ 15. _____
4. _____ 10. _____ 16. _____
5. _____ 11. _____ 17. _____
6. _____ 12. _____ 18. _____

B. LOCATION	SEC	TWN	RNG	MAP ID	COMMODITY	ACRES/ UNITS	PESTICIDES	PESTS	F*	M**	RATE	DILUTION/ VOLUME	APPL	DATE/ TIMING
1.														
2.														
3.														
4.														
5.														
6.														
7.														

| PCO NAME | ADDRESS | PHONE | PCO NAME | ADDRESS | PHONE |

C. JUSTIFICATION FOR NON-AG USE:

D. CONDITIONS:

I understand that this permit does not relieve me from liability for any damage to persons or property caused by the use of these pesticides. I waive any claim of liability or damages against the County Department of Agriculture based on the issuance of this permit. I further understand that this permit may be revoked when pesticides are used in conflict with the manufacturer's labeling or in violation of applicable laws, regulations and specific conditions of this permit. I authorize inspection at all reasonable times and whenever an emergency exists, by the Department of Food and Agriculture or the County Department of Agriculture of all areas treated or to be treated, storage facilities for pesticides or emptied containers and equipment used or to be used in the treatment.

*FORMULATION: L—LIQUID B—BAIT D—DUST F—FUMIGANT G—GRANULES WP—WETTABLE POWDER O—OTHER

**METHOD: A—AIR GR—GROUND F—FUMIGATION O—OTHER

PERMIT APPLICANT _____ SIGNATURE _____ TITLE _____ DATE _____

☐ RESTRICTED MATERIAL PERMIT IS HEREBY GRANTED FOR THE ABOVE MATERIALS.
DISTRIBUTION: WHITE & YELLOW—COUNTY; PINK & GOLD—PERMITTEE

☐ APPLICATION DENIED.
87 82647

BY _____ DATE _____
PESTICIDE ENFORCEMENT BRANCH FORM 33-125 (REV. 7-87)

FIGURE 5-7.

A restricted-use permit must be issued by the county agricultural commissioner before most federal restricted and all California restricted pesticides can be applied to agricultural areas. Permit holders are required to comply with all the conditions of the permit, including notifying the commissioner before beginning application.

Reports and Records. Within 7 days after each use of a restricted material, a *Pesticide Use Report* (Figure 5-9) must be prepared by the operator of the property or the licensed pest control operator and be submitted to the agricultural commissioner of the county where the pesticide was used. This report is not required if restricted-use pesticides are applied by a public agency, vector control agency, maintenance gardener, agricultural pest control business performing residential or right-of-way applications, or structural pest control businesses and the use is included in their monthly report to the county agricultural commissioner. Businesses involved in selling restricted-use pesticides must maintain records of the amount and type of every sale or transfer of any restricted material. These records must be kept for 2 years, and should include a copy of either the pest control adviser's written recommendation or a notice that no recommendation was made.

TABLE 5-5.

How to Obtain a Restricted-Use Permit.

TO OBTAIN A RESTRICTED-USE PERMIT:

1. Telephone the agricultural commissioner's office in the county where the restricted material will be used. Make an appointment for issuance of a pesticide restricted-use permit.

2. Prepare a list of the restricted materials that will be used during the calendar year.

3. Prepare a list of pests to be controlled by the restricted materials.

4. (For agricultural use only) Prepare a map of each location where restricted-use pesticides will be used. Include the following information:
   a. Legal location of property (section, township, range).
   b. Physical location of property (nearest crossroads) and nature of surrounding properties.
   c. Number of acres to be treated.
   d. Indicate location and types of adjoining crops.
   e. Indicate location of residences, roads, schools, hospitals, recreation areas, farm labor housing, feed lots, bee colonies, game preserves, and waterways within ½ mile.

5. If applicable, the name, address, and telephone number of pest control operator(s) who will be applying these pesticides.

6. Take all this information to the agricultural commissioner's office when applying for a restricted-use pesticide permit.

WHAT WILL HAPPEN:

1. You will be asked to fill out an application and to verify that you hold a valid pesticide applicator's license or certificate. If you are a private applicator (such as a grower), the commissioner must certify you as being qualified to safely handle and use restricted-use pesticides. This certification may require an oral interview.

2. The commissioner or a biologist will review the following information with you:
   a. Pesticide use requirements.
   b. Methods of transportation, storage, and disposal of pesticides.
   c. Bee protection, if applicable.
   d. Environmental precautions.
   e. Equipment care and calibration.
   f. Posting requirements, if needed.
   g. Mixing and loading.
   h. Employee training and supervision.
   i. Protective clothing and equipment.
   j. Emergency medical requirements.
   k. Record keeping and reports.
   l. County policies and requirements.

AFTER THE PERMIT IS ISSUED:

1. File Notice of Intent with commissioner's office at least 24 hours before making pesticide application. (This may be done in writing, in person, or by telephone. Some offices have 24-hour telephones with an answering device to receive messages.) Provide the following information:
   a. Your name, date, and time.
   b. Name on permit if different from your name.
   c. Your telephone number.
   d. Permit number.
   e. Location of application, including field number if applicable.
   f. The pesticide to be used, its formulation, amount to be applied and dilution.
   g. Method of application (ground sprayer, air, etc.).
   h. Legal location of application, if applicable.
   i. Pests to be controlled.
   j. Commodity and acreage (if agricultural application) or what pesticide will be used on.
   k. Name of applicator of pesticide.
   l. Beginning date and time of application.
   m. Location of any hazards that were not listed on the permit (livestock, poultry, crops, residences, etc., adjacent to treatment area).

2. When applicable, notify beekeepers at least 48 hours before application of the pesticide.

3. If necessary, notify all persons living or working in or near treatment area. In agricultural situation notify supervisors of field-workers and make sure they are informed of symptoms of pesticide poisoning. Supervisors must verbally notify all field-workers.

4. Post treated area if necessary.

AFTER PESTICIDE IS APPLIED:

1. Dispose of pesticide containers according to instruction from county agricultural commissioner's office.

2. File Pesticide Use Report with agricultural commissioner within 7 days of application.

3. Remove posted signs at the appropriate time.

STATE OF CALIFORNIA
DEPARTMENT OF FOOD AND AGRICULTURE
PESTICIDE ENFORCEMENT
33-126X (REV. 3/85)

**NOTICE OF INTENT TO APPLY
RESTRICTED MATERIALS**

1956478

| COUNTY NO. | SECTION | TOWNSHIP | RANGE | BASE & MERIDIAN | APPLICATION METHOD | COMMODITY OR SITE TREATED | APPLICATOR FIRM, (NAME AND ADDRESS) |
| 1 | 2 | 3 N S | 4 E W | 5 S H M | AIR ☐ GROUND ☐ 6 OTHER ☐ | 7 | |

MAP ID/DESCRIBE LOCATION

	DATES APPLIED		ACRES OR UNITS TREATED		
	PROPOSED	ACTUAL	PROPOSED	ACTUAL	
	10		11		8

| PERMITTEE/CUSTOMER | USE PERMIT NO. | APPLIED/SUPERVISED BY (PERSON'S NAME) |
| 9 | 12 | 13 | 14 |

| MFG. AND NAME OF 15 PRODUCT APPLIED | 16 | REGISTRATION NO. FROM LABEL: INCLUDE ALPHA CODE | 17 RATE | 18 DILUTION/ VOLUME | 20 TARGET PEST(S) |

ENVIRONMENTAL CHANGES/COMMENTS:

21

| SUBMITTED BY | DATE | TIME | PCA NAME |
| 22 | 23 | 24 | 25 |

| RECEIVED BY | BOX NO. | DATE | TIME | ☐ APPROVED |
| 27 | 28 | 29 | 30 | 31 ☐ DENIED |

NW — N — NE
W — TREATMENT AREA — E
SW — S — SE

26 ADJACENT CROPS, SCHOOLS, DWELLINGS, ETC.

(1) CAC   WHITE   Submit to AGRICULTURAL COMMISSIONER at least 24 hours before application.   86 97019

FIGURE 5-8.

For agricultural uses, you must file a "Notice of Intent to Apply Restricted Materials" with the county agricultural commissioner at least 24 hours before a restricted-use pesticide is applied.

STATE OF CALIFORNIA
DEPARTMENT OF FOOD AND AGRICULTURE
PESTICIDE ENFORCEMENT
33-126X (REV. 3/85)

PESTICIDE USE REPORT

1956480

| COUNTY NO. | SECTION | TOWNSHIP | RANGE | BASE & MERIDIAN | APPLICATION METHOD | COMMODITY OR SITE TREATED | APPLICATOR FIRM, (NAME AND ADDRESS) |
| 1 | 2 | 3 N S | 4 E W | 5 S H M | AIR ☐ GROUND ☐ 6 OTHER ☐ | 7 | |

MAP ID/DESCRIBE LOCATION

	DATES APPLIED		ACRES OR UNITS TREATED		
	PROPOSED	ACTUAL	PROPOSED	ACTUAL	
	10		11		8

| PERMITTEE/CUSTOMER | USE PERMIT NO. | APPLIED/SUPERVISED BY (PERSON'S NAME) |
| 9 | 12 | 13 | 14 |

MFG. AND NAME OF 15 PRODUCT APPLIED	16	REGISTRATION NO. FROM LABEL: INCLUDE ALPHA CODE	17 RATE	18 DILUTION/ VOLUME	19 TOTAL PRODUCT USED (CIRCLE UNIT OF MEASURE)	20 TARGET PEST(S)
					LB OZ PT QT GA	
					LB OZ PT QT GA	
					LB OZ PT QT GA	
					LB OZ PT QT GA	

ENVIRONMENTAL CHANGES/COMMENTS:

21

| SUBMITTED BY | DATE | TIME | PCA NAME |
| 22 | 23 | 24 | 25 |

| RECEIVED BY | BOX NO. | DATE | TIME | ☐ APPROVED |
| 27 | 28 | 29 | 30 | 31 ☐ DENIED |

NW — N — NE
W — TREATMENT AREA — E
SW — S — SE

26 ADJACENT CROPS, SCHOOLS, DWELLINGS, ETC.

(2) CAC   YELLOW   Submit to AGRICULTURAL COMMISSIONER within 7 DAYS after application.   86 97019

FIGURE 5-9.

The application of restricted-use pesticides must be reported to the local agricultural commissioner on this form, or the monthly form shown on page 147. Agricultural applications must be reported within 7 days; nonagricultural applications must be reported monthly.

State of California
DEPARTMENT OF FOOD AND AGRICULTURE
INFORMATION SERVICES UNIT
MONTHLY SUMMARY PESTICIDE USE REPORT

39-060 (REV. 3/84)

DATE SUBMITTED_____

OPERATOR (FIRM NAME)		COUNTY (WHERE APPLIED)	COUNTY NO.	MONTH/YR. OF USE	NAME OF PERSON RESPONSIBLE FOR REPORT
ADDRESS		CITY		ZIP CODE	**SUBMIT IN DUPLICATE TO THE COUNTY AGRICULTURAL COMMISSIONER BY THE 10TH OF THE MONTH FOLLOWING THE APPLICATION.**
PHONE NUMBER ( )	LICENSE NUMBER	RESTRICTED MATERIAL PERMIT NO.	TOTAL # OF APPLICATIONS		

– COMPLETE COLUMNS A, B AND C FOR ALL USES –

IF USE IS: CODE

– STRUCTURAL PEST CONTROL.......... 1
– LANDSCAPE MAINTENANCE 3
– RIGHTS OF WAY................................ 4
– PUBLIC HEALTH PEST CONTROL.... 5
– VERTEBRATE PEST CONTROL......... 8
– REGULATORY PEST CONTROL.........10

THEN:
• ENTER CODE IN COLUMN D
• LEAVE COLUMNS E & F BLANK

IF USE IS:

– COMMODITY FUMIGATION
– SPOT TREATMENT
– SEED TREATMENT
– OTHER (AS PERMITTED BY THE COUNTY AG. COMM.)

THEN:
• LEAVE COLUMN D BLANK
• FILL IN COLUMNS E & F

USE DEFINITIONS

STRUCTURAL PEST CONTROL – any pest control work performed within or on buildings and other structures. This includes work done by a licensed structural pest control operator.

LANDSCAPE MAINTENANCE – any pest control work performed on landscape plantings around residences, or other buildings, golf courses, parks, school grounds, cemeteries, etc.

RIGHTS-OF-WAY – any pest control work performed along or on roadsides, power lines, ditch banks and similar sites.

PUBLIC HEALTH PEST CONTROL – any pest control work performed by or under contract with state or local public health or vector control agencies.

REGULATORY PEST CONTROL – any pest control work performed by public employees or contractors in the control of regulated pests.

	A	B	C	D	E	F
	MANUFACTURER AND NAME OF PRODUCT APPLIED	REGISTRATION NUMBER FROM LABEL INCLUDE ALPHA CODE	TOTAL PRODUCT USED (Circle One Unit of Measure)	CODE	COMMODITY OR SITE TREATED	ACRES/UNITS TREATED
1			LB OZ PT QT GA			
2			LB OZ PT QT GA			
3			LB OZ PT QT GA			
4			LB OZ PT QT GA			
5			LB OZ PT QT GA			
6			LB OZ PT QT GA			
7			LB OZ PT QT GA			
8			LB OZ PT QT GA			
9			LB OZ PT QT GA			
10			LB OZ PT QT GA			

6 Hazards Associated with Pesticide Use

Many pesticides are hazardous materials that may harm people or the environment if improperly used.

There are several types of potential hazards associated with pesticide use. People exposed to some types of pesticides may suffer short-term or long-term health problems. Excessive residues in the environment may lead to loss of water quality or injury to nontarget vegetation, honey bees, birds, or other wildlife. Pesticides may have phytotoxic—that is, injurious to plants—effects on crops or ornamental plants with economic or esthetic costs. Improperly applied pesticides also may cause damage to treated surfaces or, through drift, surfaces near the treatment area. In addition, indiscriminate use of pesticides may result in pest resistance to certain compounds or disruption of biological control through destruction of natural enemies.

HUMAN INJURY

Breathing or swallowing pesticides or spilling or splashing pesticides into the eyes or onto the skin may result in some type of injury. Poisonous chemicals injure or kill people by interfering with biochemical and physiological functions; the nature and extent of injury depends on the toxicity of the chemical as well as the dose (amount of material) that enters the body's tissues. Some pesticides are very toxic and cause poisoning at low doses (a few drops of these are capable of causing severe illness or death). Other pesticides are so mildly toxic that several pounds must be consumed before signs of illness can be detected. Because potential hazards exist, however, anyone working with pesticides should avoid exposure to their skin, lungs, digestive tract, and eyes. Treat all pesticides with respect. It is impossible to accurately predict what effects can result from long-term repeated exposures to even the least hazardous pesticides.

How People Are Exposed

There are several ways people come in contact with pesticides, but the most common are during mixing and application and when entering or working in treated areas soon after application. If proper protective clothing and other protective equipment are worn during mixing and application, the amount of exposure to pesticides will be very low. Following label guidelines for reentry and harvest intervals will protect workers and consumers.

People also may be exposed to small doses if they live in the vicinity of the use of pesticides, eat contaminated food, or touch recently treated livestock or poultry, foliage, stored products, clothing, or furnishings.

Spills or accidents may result in exposure to large amounts of pesticides. Protective clothing and prompt emergency procedures can greatly

FIGURE 6-1.

The agricultural industry has the highest number of pesticide-related accidents. Accidents that result in pesticide injury or poisoning occur mainly during mixing and application, and are often the result of carelessness or improper techniques. This person is wearing protective clothing and equipment that will prevent injury or poisoning in case of an accident.

reduce the chances of serious injury when a person is involved in an accident.

Poisoning symptoms or injury sometimes result from a single exposure to a large quantity of pesticide; in other cases, symptoms do not occur until a person has been exposed repeatedly to small doses over a period of time. It is quite common for individuals to vary in their sensitivity to the level of pesticide exposure. Some people may show no reaction to a dose that causes severe illness in others. A person's age and body size often influences their response to a given dose, thus infants and young children are normally affected by smaller doses than adults. Also, adult females often are affected by lower doses than adult males.

Accidental Exposure

Usually the most harmful levels of pesticide exposure result from accidents, some of which are caused by carelessness. Each year the agricultural industry has the highest number of pesticide-related accidents, with most of these occurring during mixing and application (Figure 6-1). Many of these accidents result in injury or poisoning.

Spills, explosions, or similar accidents during manufacturing and packaging of pesticides have the potential to seriously injure plant employees and people living or working near these facilities. Persons involved in transporting pesticides risk possible injury if pesticide containers rupture and spill their contents or are involved in a fire. Pesticide spills during transport also can pose hazards to the public. Spills, fires, or explosions in pesticide warehouses may seriously endanger employees, emergency workers, and others.

Work-Related Exposure

Although applicators and loaders are most at risk, farm field-workers, tractor drivers, and irrigators may be injured if they are not wearing the proper protective equipment. The establishment of reentry intervals following application of highly toxic or hazardous pesticides has been an important step in reducing farm worker injury (Figure 6-2). Techniques such as reducing drift and making spray applications at times when workers are not present in adjacent fields have also helped. An important part of field-worker safety involves the training of applicators and field-workers and their supervisors on how to avoid contact with pesticide residues.

Persons performing maintenance or repair of application equipment may come in contact with pesticide residues on that equipment. Oil-soluble pesticides are a major concern because they accumulate more in grease deposits and on oily surfaces and may be difficult to remove. Thorough cleaning of application equipment, performed on a regular basis, lowers risks to maintenance workers and operators because cleaning removes grease deposits as well as pesticide residue. If equipment cannot be cleaned before repairs or maintenance, mechanics should wear protective clothing to avoid unnecessary exposure.

Workers in packing sheds and food processing plants may unknowingly contact contaminated produce or soilborne residues, especially when persistent pesticides have been used. Harvest intervals for treated produce are established to protect consumers from illegal pesticide residues but also may help reduce field, packing shed, and processing plant worker exposure by providing more time for pesticides to be degraded.

FIGURE 6-2.

Establishment of reentry intervals following the application of highly toxic or hazardous pesticides has helped to reduce farm worker injury. Treated fields like the one shown here are often posted to warn workers not to enter during the reentry period.

Persons working in greenhouses and nurseries often come in close contact with treated foliage due to the density of plants and the narrow aisles. Greenhouses have limited ventilation, increasing the potential for breathing sprays or getting dusts or mists onto the skin or into the eyes during an application. Similar conditions exist for pest control operators working in enclosed areas of homes, warehouses, factories, and offices. Workers in confined spaces should always wear protective clothing and, when necessary, respirators during pesticide applications or if they need to enter recently treated areas.

Exposure in Homes

Excessive or improper use of pesticides in the home by homeowners subjects inhabitants to possible injury. *Accidental ingestion of pesticide products by children accounts for the major portion of nonagricultural pesticide poisoning* (Figure 6-3). Proper storage can prevent accidental ingestion.

Food Contamination

Illegal pesticide residues found on food are a rare but possible source of pesticide ingestion. Incidents in which people have become poisoned by illegal residues are few and have always been the result of pesticide misuse.

Residues of some pesticides can remain on edible produce for varying periods of time. It is possible for soil residues of some pesticides to be picked up by plants and accumulated in edible parts. Hence, residue tolerance standards have been established limiting the amounts and types of pesticides allowable on human and animal food. The pesticide label

FIGURE 6-3.

Children are the major group of nonagricultural pesticide poisoning victims. Improper storage of pesticides in the home is the prime reason why children find and ingest pesticides.

states the quantity which may be used and the frequency of use to prevent tolerances from being exceeded. Preharvest intervals, established to reduce pesticide residues on produce, require that a specific time elapse between pesticide application and harvest. For the safety of the consumer, food crops are routinely inspected and tested to monitor for illegal residues.

Occasionally, food products become contaminated by improper use of pesticides after harvest. Pesticides are used to prevent damage to foodstuffs during storage, often through fumigation. The food industry may use pesticides during packing, processing, and packaging to protect produce and to prevent pest problems in and around processing plants. Pesticides often are applied in food warehouses and retail food stores to protect processed and fresh food and to control pests that infest these facilities. Restaurants may have rodent or insect pests that require control with pesticides. Pesticide labels provide specific instructions on how, when, and where to make applications to control pests yet avoid contaminating food products. To monitor for illegal pesticide residues on foodstuffs, health codes and other regulations provide for product testing and governmental inspections of processing, storage, and retail food establishments.

Pesticides applied in homes may be another possible source of food contamination. Pesticides must never be applied to foods or to food preparation surfaces, dishes, or utensils. After a pesticide application is made in kitchens or food storage areas, food preparation surfaces should be thoroughly washed. Volume 2 discusses many ways to prevent contaminating food while controlling pests in or around food preparation areas. Always follow pesticide label directions.

Animals raised for meat, dairy products, or eggs can become contaminated if excessive amounts of pesticides are used directly on the animals to control parasites, or if pesticides are used improperly in and around animal housing areas. Animals also may ingest pesticide residues on treated or contaminated feed. To prevent illegal residues in meat or animal products, federal and state laws may prohibit animal feed or crop residues treated with certain pesticides from being fed to livestock or poultry. In other cases, regulations may mandate a time period after a pesticide application before animal feed or crop residues can be given to livestock or poultry and when and if livestock may graze in treated pastures; pesticide labels provide this information. For food animals, regulations also establish a time period following pesticide application before slaughter can occur. Pesticide labels prescribe dosage, methods of treatment, and preslaughter or milk discard times aimed at reducing pesticide residues in food animal products.

Drinking water contamination offers another potential way for people to ingest pesticides. Improper use or disposal of pesticides may contaminate groundwater through wells or by leaching through the soil. Refer to the "Groundwater" section of this chapter for ways to protect groundwater from pesticide contamination.

Pesticide Exposure Through Other Sources

People may also be exposed to pesticides through contact with nonfood items. Residues remaining after fumigation of homes or work areas are a source of possible exposure to people entering these areas too soon after application. Clothing, furniture, and carpeting are sometimes treated to prolong useful life by protecting them from insect damage or reducing the buildup of fungi or bacteria, and, in some instances, may subject people to low levels of exposure. Pets and other animals treated with pesticides may be a low-level source of exposure if people come in close contact with recently treated animals. Lawns, shrubs, and other areas of home, industrial, and public landscape that have been treated for pests sometimes can be sources of pesticide exposure for people.

How Pesticides Enter the Body

Pesticides can enter the body through several different routes. The usual ways entry may occur are *dermal*—through the skin, *oral*—through the mouth, *respiratory*—through the lungs, and *ocular*—through the eyes (Figure 6-4).

Dermal Exposure

Dermal, or skin contact, is the most frequent route of pesticide exposure. Pesticides on the skin may cause a skin rash or mild skin irritation (known as dermatitis), although more severe skin injury may be caused by some types of pesticides or prolonged exposure. Internal poisoning may result when a sufficient quantity of pesticide is absorbed through the skin into the blood and is transported to other organs within the body.

The ability of a pesticide to be absorbed through the skin depends on chemical characteristics of the pesticide and its formulation. Pesticides that are more soluble in oil or petroleum solvents penetrate skin more

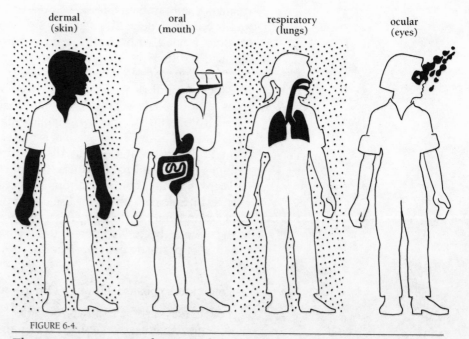

FIGURE 6-4.

The most common ways for pesticide exposure to occur are through the skin (dermal), through the mouth (oral), through the lungs (respiratory), and through the eyes (ocular).

easily than those that are more soluble in water. To prevent dermal exposure, always wear protective clothing when working with pesticides and avoid contact with plants, animals, and commodities that have been treated recently. Wash thoroughly if you accidentally get a pesticide on your skin, however *washing is not an alternative to preventing exposure.*

Oral Exposure

Oral ingestion may occur by accidentally drinking a pesticide, by splashing spray materials or pesticide dust into your mouth during mixing or application, or by eating or drinking contaminated foods or beverages. Smoking while handling pesticides may also cause ingestion. Chemicals are absorbed into the blood through the linings of the mouth, stomach, and intestines. If sufficient quantities are ingested, illness may result. Protective equipment, such as a respirator or faceshield, minimizes the risk of pesticides getting into the mouth. Before eating, drinking, or smoking, be sure to wash thoroughly and change clothing. Keep food and beverages away from areas where pesticides are being applied or mixed, and *never put pesticides into anything that can be mistaken for a food or beverage container.* Keep all pesticides in their original packages. Do not mix or measure pesticides with utensils that will be used later for food preparation or serving.

Respiratory Exposure

Pesticide handlers and others may become poisoned if pesticide dusts or vapors enter their lungs. In the lungs, pesticides are quickly absorbed and transported to all other body areas. Breathing dusts or vapors while mixing or during application is difficult to avoid unless one uses appropriate respiratory equipment. *Always* wear respirators during mixing and application when working with hazardous dusts or vapors; be sure filters are clean. Check the pesticide label for special instructions regarding the use of protective respiratory devices.

Pesticide Contact With Eyes

Serious damage can result if chemicals enter your eyes. Besides eye injury itself, the eyes provide another route for entry of pesticides into your body through the blood system. Your eyes are easily protected by wearing a faceshield or goggles. California law requires eye protection be worn: (1) during all mixing and loading activities; (2) while adjusting, cleaning, or repairing contaminated mixing, loading, or application equipment; and (3) during most types of ground application. If pesticides get into your eyes, flush them with a gentle stream of clean water for 15 minutes, holding the lids apart. Seek immediate medical attention if irritation persists.

Effects of Exposure

The type and severity of injury or poisoning depends on the toxicity and mode of action of the pesticide, the amount absorbed by the body, how fast it is absorbed, and how fast the body is able to break it down and excrete it. The severity of injury may also be reduced by prompt first aid and medical treatment. Very small doses usually produce no injury or poisoning symptoms. Depending on the toxicity of the pesticide, larger doses may cause severe illnesses. Effects of exposure may be localized—such as irritation of the eyes, skin, or throat, or generalized—when pesticides are absorbed into the blood and distributed to other parts of the body. Pesticides may affect several different internal systems at the same time. The extent of involvement and damage is related to the characteristics of the pesticide and the dose.

If you have been in a situation in which pesticide exposure could occur, and you suspect that you are experiencing pesticide poisoning symptoms, seek medical attention. Also, be aware of these problems in co-workers, since individuals may fail to recognize pesticide poisoning in themselves. If you suspect poisoning has occurred, make every effort to locate the source of exposure and take steps necessary to prevent it from happening again.

Symptoms

Symptoms are any abnormal conditions that a person sees or feels or that can be detected by examination or laboratory tests that indicate the presence of an injury, disease, or disorder. When a person becomes exposed to a large enough dose of pesticide to produce injury or poisoning, there may be either an immediate or delayed appearance of symptoms (Table 6-1). Immediate symptoms are those observed soon after expo-

TABLE 6-1.

Common Pesticide Poisoning Symptoms.

POSSIBLE SYMPTOMS RELATED TO SKIN CONTACT WITH
PESTICIDE DUST, LIQUID, OR VAPORS:
  Staining of the skin.
  Reddening of skin in area of contact.
  Mild burning or itching sensation.
  Painful burning sensation.
  Blistering of the skin.
  Cracking and damage to nails.

POSSIBLE SYMPTOMS RELATED TO EYE CONTACT
WITH PESTICIDE DUST, LIQUID, OR VAPORS:
  Discomfort, including watering and slight burning.
  Severe, painful burning. (Permanent eye damage may occur)

POSSIBLE SYMPTOMS RELATED TO INHALING OR
SWALLOWING PESTICIDE DUST, LIQUID, OR VAPORS:
  Sneezing.
  Irritation of nose and throat.
  Nasal stuffiness.
  Swelling of mouth or throat.
  Coughing.
  Breathing difficulties.
  Shortness of breath.
  Chest pains.

sure—known as *acute onset*. Sometimes symptoms from pesticide exposure may not show up for weeks, months, or even years. These delayed, or *chronic onset*, symptoms may either come on gradually or appear suddenly; they may be difficult to associate with their cause because of the lapse of time between exposure and observable effect.

Poisoning symptoms vary between classes of pesticides and pesticides within a class. The presence and severity of symptoms usually is proportional to the amount of pesticide (dosage) entering the tissues of the exposed person. Common symptoms include skin rashes, headaches, or irritation of the eyes, nose, and throat; these types of symptoms may go away within a short period of time and sometimes are difficult to distinguish from symptoms of allergies, colds, or flu. Other symptoms, which might be due to higher levels of pesticide exposure, include blurred vision, dizziness, heavy sweating, weakness, nausea, stomach pain, vomiting, diarrhea, extreme thirst, and blistered skin. Poisoning may also result in apprehension, restlessness, anxiety, unusual behavior, shaking, convulsions, or unconsciousness. Although these symptoms can indicate pesticide poisoning, they also may be signs of other physical disorders or diseases. Usually diagnosis requires careful medical examination, laboratory tests, and observation.

Types of Injuries

Injuries may be caused either by a single massive dose being absorbed during one pesticide exposure, or from smaller doses absorbed during repeated exposures over an extended period of time. The illness or damage may be *acute*—having a sudden onset and lasting for a short duration, or it may become *chronic*—persisting for a long time. Injuries caused by pesticides usually are *reversible*; that is, they can either be repaired by the body itself or through some form of medical treatment. Accidental exposure to some pesticides, however, may cause *irreversible* or permanent damage which can result in a chronic illness, disability, or death.

OTHER EFFECTS ON PEOPLE

Allergies

Certain individuals occasionally may exhibit allergic reactions when using some types of pesticides or when these pesticides are applied in or around their homes or places of work. The material causing the reaction may be the pesticide or one of the ingredients in the pesticide formulation. Symptoms often include breathing difficulties, sneezing, eye watering and itching, skin rashes, apprehension, and general discomfort. Sensitive persons should avoid exposure to pesticides that produce allergic reactions. Different types of pesticides or formulations should be tried or alternate, nonchemical control methods used.

Anxieties

Some people are very concerned about the possible dangers of pesticides and may exhibit symptoms of illness if they suspect or know they have been exposed. The illness is real, although there may be no actual pesticide injury. This type of reaction may be triggered by odors. Odors are often mistakenly associated with toxicity, with the belief that the stronger the odor, the more toxic the pesticide. (Obviously this is not true. For example, methyl bromide is extremely toxic but usually has no detectable odor. Other pesticides that have strong, objectionable odors may be far less hazardous.)

Anxiety over pesticide exposure may be partially a result of insufficient information or misinformation regarding the potential hazards of pesticide use. Accidents, poisonings, and pesticide residues in food are publicized regularly and continue to receive much media attention. Little information is provided to the public, however, regarding the benefits of judicious use and how pesticides can be used safely and risks minimized. It is important to provide factual information regarding risks and how they can be avoided when applying pesticides in public areas, workplaces, and homes. However, do not belittle people's genuine concerns and make every effort to minimize public exposure to pesticides.

GROUNDWATER CONTAMINATION

Potential contamination of groundwater with pesticides is a serious concern. Roughly 97% of the total water in the world is saltwater in the oceans. Less than 1% of all the water on the planet is in a form available to people for drinking, irrigation, household, and manufacturing purposes. Two-thirds of this fresh water (or about 2 million cubic miles) is groundwater, trapped beneath the soil; the rest is surface water in lakes, ponds, streams, and rivers. Water in the form of polar ice caps and glaciers is also surface water but is frozen and unavailable for use. Groundwater provides 40% of California's water needs. Cities obtain 46% of their water, including drinking water, from groundwater sources; 93% of the water used in rural areas is groundwater. Groundwater, therefore, is our most important freshwater source.

The extent of the groundwater contamination potential from pesticides is just beginning to be realized. Originally it was believed that pesticides were not a threat to groundwater since available testing techniques did not detect contamination. Studies suggested that microorganisms, environmental factors, and soil degraded or adsorbed most pesticides before they could reach groundwater sources. Pesticides that did enter groundwater were believed to be rapidly decomposed. More recently, however, newer detection methods have shown that small amounts of chemicals, including certain pesticides, can be found in some groundwater locations. Hazardous levels of toxic chemicals in groundwater could affect much of the state's population, therefore stringent laws have been put into effect to protect this resource. Pesticide use and disposal are regulated by these laws.

How Groundwater Accumulates

Water applied to the soil surface slowly filters down through soil particles, a process known as *percolation*, until it reaches goundwater basins. Common water sources are rainfall, surface irrigation, and melting of ice, snow, and glaciers. Some of the water contained in lakes and ponds

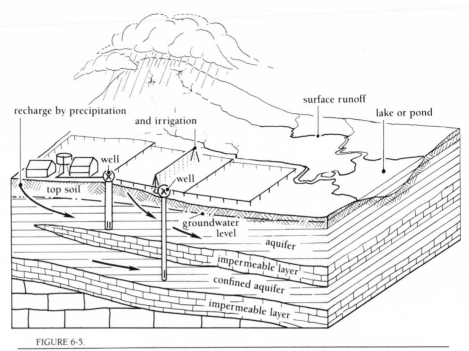

FIGURE 6-5.

Water is present in underground reservoirs called aquifers, diagrammed here. This water is contained in sand and gravel formations. Impermeable layers of clay and silt or solid rock prevent water from leaving the aquifer.

or flowing in rivers and streams also percolates into groundwater basins, although much also flows out to the oceans. The amount and rate of percolation depends on the size of soil particles. Fine particles, such as clay and loam, retain more water than coarse, sandy soil, therefore percolation is slower as soil texture becomes finer.

Groundwater becomes trapped in large underground masses known as *aquifers*. Aquifers are usually areas of water-saturated sand and gravel having one or more impermeable layers of rock or clay and silt which confine the water. Aquifers may be hundreds of miles in length and width and range in depth to as much as several thousand feet (Figure 6-5). In California, nearly half of the landmass overlies groundwater basins. It is common to find several layers of aquifers in the same area, with each being separated from the others by impermeable layers, or connected because of interruptions in the layers. Water trapped in aquifers flows toward the sea, but much more slowly than surface water; water in a stream or river has a flow rate measured in feet-per-second, while underground water may only flow a few feet per year.

Groundwater in California is especially vulnerable to contamination because of the vast size of this resource and because it is located directly below so much of the cultivated, industrial, and residential land area. Contamination, when it occurs, may be difficult or impossible to contain. Because the water flows so slowly, it may take several hundred years before contaminants are removed from the system through natural means.

Ways Pesticides Enter Groundwater

Pesticides enter groundwater in two different ways: by leaching out of the soil and by direct entry through wells or other structures that are in

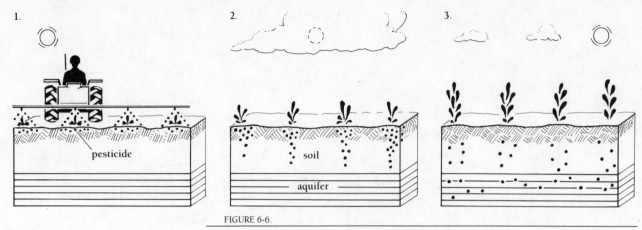

FIGURE 6-6.

Water enters aquifers by percolation through the soil. As water passes down through the soil it may dissolve some pesticides and carry them into the aquifer. This process is called leaching.

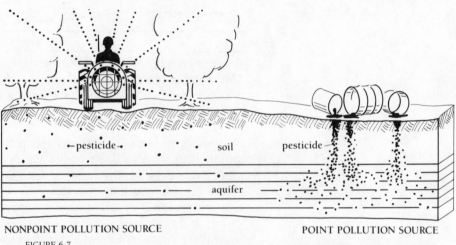

NONPOINT POLLUTION SOURCE   POINT POLLUTION SOURCE

FIGURE 6-7.

Point pollution sources are areas where large quantities of pesticide or other pollutants are discharged into the environment, such as from spills, waste discharge pipes, or dump sites. Nonpoint pollution sources are those arising from normal application where the pesticide or other material is applied over a large area.

contact with aquifers. Leaching occurs as rainwater or irrigation water percolates through the soil, dissolving water-soluble chemicals—including some pesticides (Figure 6-6).

Several types of pesticides which are incorporated into the soil by mechanical means or by irrigation or rainfall can, under certain conditions, be leached out and eventually may be carried to groundwater sources. Those applied to crop surfaces are subject to being washed off by water or abraded by wind and may enter the soil where they can leach into groundwater. Pesticides entering groundwater from normal application practices are said to come from *nonpoint pollution sources*. These sources are widely distributed over a large area, with only extremely small amounts of pesticide having the potential to enter the groundwater from any one location. Pesticides and other chemicals can also contaminate soil or surface water as a result of manufacturing accidents, storage and handling, transportation, or from improperly constructed disposal sites or holding facilities. In these situations, larger quantities of contaminants have the potential of entering the groundwater at small, defined locations, known as *point pollution sources* (Figure 6-7).

Wells are direct channels into an aquifer and may provide a connection between several aquifers (Figure 6-8). Activities that involve the use of pesticides around water wells furnish opportunities for these chemicals to enter the groundwater directly through the well. Mixing pesticides near a well, pumping water into pesticide application equipment, or injecting pesticides into an irrigation system are some of the ways that, if performed carelessly or improperly, could contaminate groundwater. Disposing of surplus pesticides and washing contaminated equipment in the vicinity of a well are other means. Abandoned wells that have been improperly sealed provide possible routes for pesticides and other contaminants into the underground system. Occasionally pesticide waste or runoff may enter groundwater through direct channels such as sinkholes or exposed shallow aquifers.

Factors Influencing Groundwater Contamination

Many factors influence the potential for groundwater contamination; an important one is the nature of the pesticide itself. Pesticides that were used in years past included many chlorinated hydrocarbon compounds that were often highly persistent but immobile. The chemical structure of these compounds causes them to bind tightly to soil particles. Such pesticides do not move down through the soil profile but remain concentrated near the soil surface and, therefore, usually do not pollute groundwater. Recently, however, pesticide use trends have been toward less persistent but more highly mobile chemicals, including some of the phenoxy, urea, and carbamate compounds. These products do not bind

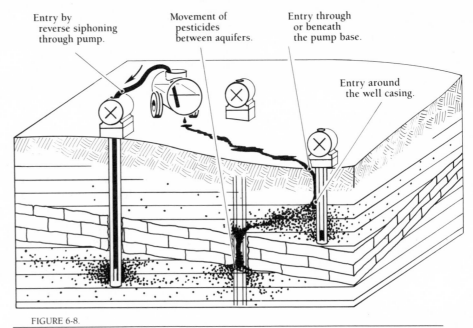

FIGURE 6-8.

As this illustration shows, water wells are direct channels into an aquifer, and may provide a connection between several aquifers. Pesticides and other contaminants can enter groundwater directly through wells.

TABLE 6-2.

Mobility of Different Pesticides, Illustrating Ability to Leach in the Soil. (Some chemicals listed here are no longer used as pesticides).

MOBILITY CLASS 1 *very unlikely to leach in soil*	MOBILITY CLASS 3 *moderately able to leach through soil*
aldrin	alachlor
benomyl (Benlate)	atrazine
chlordane	diphenamid
dacthal (DCPA)	endothall
DDT	fenuron
dieldrin	prometone
diquat	propham (IPC)
disulfoton (Disyston)	simazine
endrin	terbacil
ethion	2,4,5-T
heptachlor	
lindane	
morestan	**MOBILITY CLASS 4** *leaches readily through soil*
paraquat	
parathion	amitrole
phorate (Thimet)	bromacil
trifluralin (Treflan)	2,4-D
	fenac
	MCPA
MOBILITY CLASS 2 *slightly able to leach through soil*	picloram (Tordon)
azinphos-methyl (Guthion)	**MOBILITY CLASS 5** *Very mobile in the soil*
bensulide (Betasan)	
carbaryl (Sevin)	chloramben
chlorpropham	dalapon
diazinon	dicamba
diuron	TCA
linuron	
molinate (Ordram)	
prometryn (Caparol)	
propanil	
pyrazon (Pyramin)	

tightly with soil particles and some are soluble in water, enabling them to leach more easily. Table 6-2 compares the soil mobility of several different chemicals which have been or are currently being used as pesticides.

The potential to contaminate groundwater is also influenced by soil type and geological formations in the area where pesticides are used. Liquids percolate faster through sand or gravel soils than they do through denser silt, loam, and clay soils. The amount of organic material in the soil influences pesticide retention and may speed up its breakdown. Other factors such as soil temperature, pH, moisture content, dissolved salts, and quantity and type of soil organisms may influence the stability of pesticides and can affect their ability to reach groundwater. If there is no aquifer beneath the treatment area, or if the aquifer is very deep, risks are minimized. Shallow water tables beneath treated areas are more susceptible to contamination because pesticides pass through less soil and are therefore not subject to as much degradation. Some locations have layers of impermeable subsoil that prevent leaching of pesticides or other contaminants into the groundwater. The slope of the treated land often

TABLE 6-3.

Factors Influencing Pesticide Leaching.

AGRICULTURAL PRACTICES:

1. Amount and type of pesticide used.

2. Method of pesticide application.

3. Irrigation practices at treatment site.
   a. Frequency of irrigation.
   b. Timing of irrigation in relation to pesticide application.

GEOLOGIC CONDITIONS OF TREATMENT AREA:

1. Slope.

2. Underlying formations.

3. Proximity of surface water channels such as ponds, lakes, rivers.

INTERACTION OF PESTICIDES IN SOILS:

1. Properties of pesticides:
   a. Water solubility.
   b. Volatility.
   c. Soil adsorption.
   d. Decomposition

2. Soil influence on pesticides:
   a. Soil texture.
   b. Soil organic matter content.
   c. Soil water content.

---

determines whether contaminated water runs off or is leached. Table 6-3 summarizes factors that influence groundwater contamination by pesticides.

Pesticides that decompose quickly are, in general, less apt to cause groundwater contamination. Recent research, however, suggests that groundwater environments may retard the breakdown of some pesticides. Groundwater is not exposed to air or ultraviolet light, it is cooler than soil surfaces, and it lacks much of the biological activity found in soil.

How to Prevent Pesticide Contamination of Groundwater

The following techniques will help prevent pesticides from contaminating groundwater:

Storage. Store pesticides in enclosed areas, on an impermeable surface and protected from rain. In case of fire or rupture of storage containers, be sure runoff is contained and remove contaminated soil.

Mixing and Loading. Avoid spilling pesticides. If a spill occurs, clean up and dispose of wastes as described in Chapter 8; remove contaminated soil. Triple rinse liquid containers and pour rinsate into the spray tank for application to the target site. Take rinsed containers to a designated disposal site. Do not overfill spray tanks. Use a check valve or air gap on filling pipes to prevent backflow of contaminated water into water supplies.

Application. Whenever possible, select pesticides with low soil mobility and use materials that degrade rapidly in the soil. Avoid drift of pesticides off the target site through application techniques and by making applications during optimum weather conditions. In areas where high risks to groundwater exist: (1) reduce pesticide use by integrating chemical control with other methods of control; (2) use pesticides only when necessary; (3) apply only the amounts that will adequately control pests; and (4) reduce the frequency of application whenever possible.

Disposal. Never dump excess pesticide or pesticide mixtures onto the soil or into sewers, drains, or septic systems. Unused pesticide waste must be transported to a Class I disposal site.

Cultural Practices. Avoid excessive use of irrigation water after a pesticide application and prevent irrigation water runoff.

Records. Maintain records of the quantity and type of pesticides applied to an area. Records can be helpful in planning future pest control measures to prevent pesticide accumulation in the soil and to avoid plantback restrictions.

IMPACT ON NONTARGET ORGANISMS

Nontarget organisms are all plants and animals in or near a treated area that are not intended to be controlled by a pesticide application. Some types of pesticide use may have a drastic influence on the local and surrounding environment, including living organisms, the air, soil, and water. As much as 55% of an applied pesticide may leave the treatment area due to spray drift, volatilization, leaching, runoff, and soil erosion (Figure 6-9). Pesticides that drift or move onto adjacent areas, including residences, crops, pastures, lakes, streams, and forests, may result in inadvertent exposure of people, pets, livestock, or wildlife. Pesticides that leave the treatment area may also cause damage to crops, plants in landscaped areas, and native trees and shrubs. For example, some herbicides in concentrations as low as $1/1000$ of a pound per acre cause observable damage to crops and other plants; concentrations of some herbicides in the range of $1/100$ of a pound per acre may reduce yields. Under certain weather conditions, and if large acreages are being treated, pesticide concentrations in this range can drift out of the treatment area and move for several miles before settling to the ground.

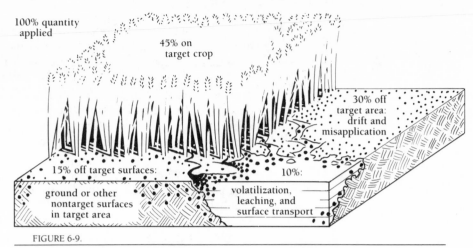

100% quantity applied

45% on target crop

30% off target area: drift and misapplication

15% off target surfaces: ground or other nontarget surfaces in target area

10%: volatilization, leaching, and surface transport

FIGURE 6-9.

As much as 55% of an applied pesticide may leave the application site due to spray drift, volatilization, leaching, runoff, and soil erosion.

Honey Bees

Certain types of insecticide and fungicide applications may kill honey bees, causing severe economic hardship to beekeepers and loss of certain crops due to poor pollination. Honey bees are most severely damaged if harmful pesticides are applied when bees are foraging for nectar and pollen (Figure 6-10), so avoid using materials toxic to bees when crops or weeds are in bloom. Sprays applied early in the morning, late in the afternoon, or during the night cause less harm because bees do not forage at these times. Selecting pesticides that are less harmful to bees is essential for protecting these very beneficial insects. Obtain a copy of the publication "Reducing Pesticide Hazards to Honey Bees with Integrated Management Strategies" (University of California Publication #2833) for information on pesticides which are highly toxic, moderately toxic, and relatively nontoxic to honey bees and for ways you can predict pesticidal hazards to bees. There are legal restrictions controlling the application of certain pesticides while bees are foraging; check with country agricultural commissioners for pesticide application cut-off dates. Beekeepers should be notified 48 hours before a spray application is intended so they have time to remove hives from the area if necessary.

Natural Enemies and Other Beneficials

Besides honey bees, pesticides can harm other beneficial organisms. Some beneficial species feed on or parasitize pest insects and mites; these beneficials are called natural enemies and include well-known types and many less known species. Other beneficials, such as beetles, flies, and soil insects and mites, help decompose dead plant and animal material. Beneficial fungi and nematodes also play an important part in decomposition and long-term natural control of pests, but also can be destroyed by soil fumigation or other pesticide applications. Disrupting the natural control of pests often promotes an increased dependency on the use of pesticides. Reduce damage to beneficials by: (1) choosing pesticides that are less toxic to beneficial species; (2) applying pesticides at times when natural enemies are least likely to be harmed, such as during dormant periods; (3) lowering dosages (when possible); and (4) using spot treatments or other methods of placing pesticides to minimize harmful effects.

FIGURE 6-10.

Honey bees may be poisoned if certain pesticides are applied while they are foraging for nectar or pollen. Avoid using materials toxic to bees when crops or weeds are in bloom. Apply sprays early in the morning, late in the afternoon, or at night to reduce chances of killing foraging bees. Also, use pesticides that have a low toxicity to bees.

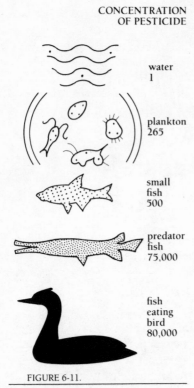

CONCENTRATION
OF PESTICIDE

water
1

plankton
265

small
fish
500

predator
fish
75,000

fish
eating
bird
80,000

FIGURE 6-11.

Some pesticides can be accumulated through the biological food chain. Microorganisms and algae containing pesticides are eaten by small invertebrates and hatchling fish. These animals are in turn eaten by larger fish and birds. Each passes greater amounts of pesticide to the larger animal.

Pest Resurgence and Secondary Pest Outbreak. Insect and mite pests often are associated with a complex of natural enemies that help control the size of the pest population; this natural control is a type of biological control. Unless a pesticide is highly selective, it will kill not only the pest but many of the pest's natural enemies. If natural enemies are not destroyed, they die off anyway because their food supply—the pest—has been eliminated. Since natural enemies depend on pest species for food, they require more time than pests to increase their population size. Therefore, after a pesticide application, pests can move back into the treated area where there will be fewer natural enemies; this enables the pest population to grow rapidly and sometimes become bigger than it was before the pesticide treatment. This phenomenon is known as *pest resurgence*.

Another problem associated with pesticide use is that of secondary pest outbreak. Secondary pests are those which are kept in control by natural enemies or by competition with a primary pest. Elimination of natural enemies or of primary pests often results in an increase in secondary pest populations to a point at which they cause economic damage.

Wildlife

Wildlife may be harmed accidentally by pesticide use either directly, when they are poisoned, or indirectly, when their food sources or habitats are altered. Vertebrates, including birds, often feed or nest in areas where pesticides are used. Sometimes these animals are the unintended victims of baits used for control of target pests or they may be injured by feeding on poisoned pest animals. Although a pesticide dose may not directly cause death, the effect might weaken a nontarget animal and indirectly cause illness or death due to its inability to get food and water or protect itself from natural enemies. Some pesticides may have an impact on the ability of wildlife to reproduce.

Fish are susceptible to many pesticides that get into waterways, even at low concentrations. Pesticides enter waterways through drift, by direct spraying, by leaching from the soil through runoff of irrigation and rainwater, through erosion of treated soil, and through accidents and illegal dumping. Some persistent pesticides can be concentrated rapidly in an aquatic environment through normal food chains—a process called *bioaccumulation* (Figure 6-11). Waterfowl may be poisoned by pesticides present in flooded fields (rice fields, alfalfa fields, etc.) or in areas where pesticides are applied through irrigation water.

Endangered Species. Federal and state laws have been enacted to help protect and prevent the extinction of certain rare or very vulnerable wildlife and plant species; severe fines and imprisonment can be imposed on people who break these laws.

As part of the enforcement of endangered species laws, the use of certain pesticides in areas where endangered species are known to exist may be highly restricted. Before using any pesticide, check the pesticide label for precautions for protection of endangered species and locations of restricted areas. For further information, consult with your local University of California Cooperative Extension office. Farm advisors in these offices may also be able to provide you with information on alternative pest control methods for use in areas where identified endangered species exist.

Check with the nearest Endangered Species Office of the U.S. Fish and Wildlife Service and the local or regional office of the California Department of Fish and Game for laws and information on the protection of endangered species. (Table 6-4 provides addresses of the Endangered Species Offices and local and regional offices of the California Department of Fish and Game).

TABLE 6-4.

Locations of the U.S. Fish and Wildlife Service Endangered Species Offices and the Local and Regional Offices of the California Department of Fish and Game.

U.S. FISH AND WILDLIFE SERVICE
Federal Building
24000 Avila Road
Laguna Niguel, CA 92656
(714) 643-4270

This office serves the counties of Imperial, Los Angeles, Orange, Riverside, San Bernardino, San Diego, Santa Barbara, and Ventura

U.S. FISH AND WILDLIFE SERVICE
Sacramento Endangered Species Office
2800 Cottage Way, Room E-1823
Sacramento, CA 95825
(916) 978-4866

This office serves all the remaining counties in California

CALIFORNIA DEPARTMENT OF FISH AND GAME
1416 Ninth Street, 12th Floor
Sacramento, CA 95814
(916) 445-3531

REGIONAL OFFICES:

REGION I
601 Locust Street
Redding, CA 96001
(916) 225-2300

REGION IV
1234 E. Shaw Avenue
Fresno, CA 93710
(209) 222-3761

REGION II
1701 Nimbus Road
Rancho Cordova, CA 95670
(916) 355-0978

REGION V
330 Golden Shore, Suite 50
Long Beach, CA 90802
(213) 590-5126

REGION III
7329 Silverado Trail
Napa, CA 94558
Mailing address:
P.O. Box 47
Yountville, CA 94599
(707) 944-5500

MARINE RESOURCES REGION
330 Golden Shore, Suite 50
Long Beach, CA 90802
(213) 590-5189

BRANCH OFFICES THAT CAN ALSO PROVIDE INFORMATION:

407 Westline Street
Bishop, CA 93514
(619) 872-1171

2201 Garden Road
Monterey, CA 93940
(408) 649-2870

619 Second Street
Eureka, CA 95501
(707) 445-6493

1350 Front Street
Room 2041
San Diego, CA 92101
(619) 237-7311

Nontarget Plants

Herbicides used to control weeds and undesirable plants in forests, along roadsides, and on rangelands may have detrimental effects on nontarget plants. Many species of plants are important in natural and undeveloped areas because they protect the watershed, reduce erosion, provide food and shelter for wildlife, and are part of the native flora. Plants in natural areas are usually part of an ecological balance; disruption of this balance by any means may favor the increase of undesirable plants or plants having minimal benefit to wildlife and watershed.

Phytotoxicity. Phytotoxicity may be a problem with a given pesticide on certain crops or ornamental and landscape plants. Phytotoxicity is not always caused by the active ingredient of the pesticide, but may result from solvents in the formulation or impurities (such as salts) in the water mixed with the pesticide. Excessive rates of application or improper pesticide dilution may also cause plant damage. Environmental conditions, such as temperature and humidity at the time of application, can often influence phytotoxicity. Plants stressed for water or nutrients may be more susceptible to injury.

PESTICIDE RESISTANCE

Resistance is a condition in which pests become tolerant to a pesticide that once controlled them. At first, higher rates or more applications of a certain pesticide are necessary to achieve the same amount of control. Finally this pesticide has little effect, no matter how much is used. Switching to a different pesticide may help, but sometimes when pests develop resistance to one chemical they also become resistant to others, even from a different chemical class. This phenomenon is called *cross-*

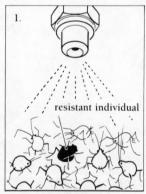  

1.	2.	3.
resistant individual		susceptible individual
Some individuals in a pest population have genetic traits that allow them to survive a pesticide application.	A proportion of the survivors' offspring inherit the resistance traits. At the next spraying these resistant individuals will survive.	If pesticides are applied frequently, the pest population will soon consist mostly of resistant individuals.

FIGURE 6-12.

This drawing illustrates how pesticide resistance can build up in a pest population. Resistance to pesticides involves a change in the genetic characteristics of pest populations which are inherited from one generation to the next. Increased or frequent use of the pesticide often hastens resistance.

resistance. Resistance involves a change in the genetic characteristics of pest populations and is inherited from one generation to the next. Initially, a pest population may possess a few individuals that are able to break down or chemically modify a pesticide, and these individuals survive when that pesticide is used. When the resistant individuals reproduce, most of their offspring are also resistant (Figure 6-12).

Insect resistance to lime-sulfur was first reported in 1914. Resistance to arsenicals (arsenic containing pesticides) began causing problems in the 1920s. Pest resistance to DDT, the first synthetic pesticide to be manufactured, developed in the 1950s. Increased use of pesticides during the past 40 years and cross-resistance of old pesticides with newer pesticides have magnified the problem of pesticide resistance. Today, every major class of pesticide has some pest which is resistant to it. Pesticide resistance has developed in insect, mite, plant pathogen, weed, and rodent pests.

Pesticide resistance is a manageable problem in many cases. How manageable it is depends on genetic and biological factors (which we cannot control) and operational factors (which we *can* control).

Genetic Factors

Genetic factors which influence the development of resistance are: (1) how the resistance is inherited; and (2) how many of the individuals in the population have the genes for resistance. If resistance genes are common in the population they will be inherited easily and resistance will spread quickly and may be difficult to manage.

Biological Factors

Biological factors which influence the development of resistance are the unique characteristics of the pest and its habits. Biological factors include: (1) the lifespan of the pest; (2) the number of offspring it produces over a period of time; (3) its ability to move large distances; and (4) its food requirements. A population of short-lived, rapidly developing, immobile pests with many offspring will develop resistance rapidly. Individuals susceptible to pesticides in this type of population will be eliminated quickly, leaving only resistant individuals to breed with each other and rapidly build up in numbers. In addition, if these pests do not move around much, very few susceptible individuals will migrate in to breed with the resistant group. The rapidly reproducing mites, aphids, fungi, cockroaches, and rodents are pests of this type and all have frequently been associated with pesticide resistance.

Operational Factors

Operational factors are the unique characteristics of the pesticide that either favor or reduce resistance. These factors can often be used to manage resistance and include: (1) the type of pesticide used; (2) the persistence of pesticide residues; (3) the pesticide application rate; (4) the life stage of the pest that is treated; and (5) whether the pesticide is combined with other pesticides.

Manufacturers are working to develop new pesticides and new formulations to replace chemicals with resistance problems. Chemicals with new modes of action are becoming harder to discover, however, and there are very high development and registration costs associated with producing new pesticides. More emphasis, therefore, should be placed on limiting opportunities for resistance development by reducing pest population exposure to pesticides and increasing mortality due to other factors.

It is the responsibility of every pesticide user to reduce the potential for pest resistance. Pesticides are important management tools when used properly. When used improperly, pesticides may become ineffective due to resistance. Pest resistance cannot be confined to a localized area; therefore irresponsible actions in one location may affect a much larger area.

Resistance Management

Pests develop resistance to pesticides fastest when control is based entirely on the use of closely related pesticides. Resistance management utilizes as many different control options as possible; pesticides will eliminate the susceptible individuals and the resistant individuals are destroyed by one or more of the nonchemical factors.

The first step in resistance management is to try to choose a chemical that selectively kills the pest and not its natural enemies, enabling natural enemies to continue to exert their control. Second, choose a nonpersistent pesticide that allows the next stage or next generation of pest to survive, therefore allowing some susceptible individuals to survive. Third, alternate selective pesticides with ones having different modes of action, or use a pesticide which has more than one mode of action—it is difficult for the pest to develop resistance in two different ways at the same time. Fourth, apply the pesticide to the damaging life stage of the pest so that it cannot use the genetic and physiological defenses of other life stages to combat the pesticide. Apply the chemicals over a limited area, such as spot treatments or strip treatments, so that some susceptible, untreated pests survive. Consider treating only alternating generations. Finally, if resistance develops to the point at which the pesticide is no longer effective at controlling the pest, stop using it.

RESIDUES

Whenever pesticides are used, they remain as residues on treated surfaces for a period of time. Residues are related to the nature of the pesticide or to the type of formulation (persistence), frequency and amount of pesticide used (accumulation), and interaction with the environment (breakdown or recombination). Residues are important and necessary in some types of pest control because they provide continuous exposure to the pest, improving chances of control. However, residues are undesirable when they expose people, domestic animals, and wildlife to unsafe levels of pesticides. Pesticide materials that miss the treatment surface can remain as residues in soil, water, or on surfaces in nontarget areas. Empty pesticide containers hold small amounts of residues that require proper disposal to prevent environmental contamination (Figure 6-13). The allowable pesticide residues remaining in the environment or on produce is restricted by law to provide a wide margin of safety; when

FIGURE 6-13.

Pesticide wastes include partial containers of pesticide that are not used, leftover mixtures in spray tanks, rinse water from pesticide containers, rinse water from inside and outside of spray equipment, and, as shown here, empty pesticide containers.

pesticides are properly applied and wastes properly disposed of, residues usually do not exceed legal tolerances.

Acceptable pesticide residue tolerances are established by government agencies; these are the maximum amounts of pesticide chemicals that may remain on or in raw agricultural commodities. Tolerances are established through laboratory and animal testing and are amounts of pesticide considered harmless to consumers. A generous margin of safety is always included in the tolerance level. Tolerances are established for each pesticide and levels vary depending on the toxicity and hazard of the pesticide and on the type of commodity. The total diet of the consumer is considered when establishing tolerance levels. State and federal agencies analyze produce to ensure that pesticide residue tolerances are not exceeded; any found having greater than maximum allowable pesticide residues are usually seized and may be destroyed if residues cannot be reduced to tolerance levels.

In the United States, hazardous pesticide residues on produce are occasionally caused by misuse. Residues exceeding tolerances can occur during crop production if the crop accumulates pesticides from the soil, if a pesticide is applied to an unregistered crop, if too much pesticide is applied to the crop, if the pesticide is applied too close to harvest, or if the crop received pesticide drift from another area. The postharvest use of pesticides also may leave residues on produce; tolerances can be exceeded by not following pesticide label directions or by using improper fumigation techniques. Illegal residues may be deposited on food when pesticides are improperly used in warehouses, markets, and restaurants.

Although hazardous pesticide residues on produce grown in the United States is a rare occurrence, it is still a major public concern. For instance, a 1984 consumer survey showed that of all the possible harmful products found in food, the public's primary concern was pesticide residues. Other concerns listed in this survey, in order of importance, were cholesterol, salt, food additives and preservatives, sugar, and artificial coloring (Table 6-5).

TABLE 6-5.

*Survey of the Public's Concern Over Possibly Harmful
Products Found in Food*.*

CONTAMINANT OR FOOD ADDITIVE	SERIOUS HAZARD	SOMEWHAT A HAZARD	NO HAZARD	NOT SURE
Pesticide residues	77%	18%	2%	3%
Cholesterol	45	48	5	2
Salt in food	37	53	9	1
Additives and preservatives	32	55	8	4
Sugar in food	31	53	15	1
Artificial coloring	26	53	17	5

*Source: Food Marketing Institute survey, "Public Attitudes Toward Food Safety." FDA Consumer, July–August 1984.

Pesticide Persistence

The amount of time from application until a pesticide has been degraded is known as its persistence. Chlorinated hydrocarbon insecticides and certain classes of herbicides do not break down rapidly in the environment and are considered to be highly persistent. Toxic effects against insects or weeds may remain from 1 to several years while the pesticide is present in the soil. Pesticide persistence is a benefit in situations in which long-term pest control is desired, such as the treatment of building foundations to prevent termite entry or season-long control of weeds. A material that has a lengthy persistence can have some drawbacks, however, because chances of poisoning are greater, longer reentry intervals may be required, and chances of the chemical leaving the treatment site through water movement, soil leaching, and wind and rain erosion are increased, especially if the material is soluble in water.

Plantback Restrictions. Persistence characteristics of certain pesticides in the soil may restrict the types of plants or crops that can legally be grown in the treated area for a specified period of time. Plantback restrictions have been established to prevent injury to subsequent crops and to avoid residues of some pesticides from occurring in nonregistered crops. When plantback restrictions apply, the pesticide label will specify which crops can be planted and how long after application a grower must wait before planting other types of crops. Plantback restrictions are an important reason for keeping accurate pesticide application records.

Accumulation

Accumulation is the buildup of a persistent pesticide resulting from repeated applications or exposures. Accumulation may occur in the soil, in groundwater, in plant and animal tissues (bioaccumulation), or in ponds and lakes. An example of accumulation is copper used as a fungicide in deciduous fruit and nut orchards. Copper may be sprayed several times each year in some orchards for suppression of fungal infections.

Copper degrades very slowly, accumulating over a period of years to a degree of possible toxicity to organisms living or growing in the soil.

Breakdown and Recombination

Special hazards may exist when pesticides break down into different toxic compounds, fail to break down properly, or possibly recombine with other chemicals in the environment into unknown materials. Each location in the environment has unique characteristics that may influence the way chemicals are broken down or combined with other chemicals, or how they react with target and nontarget organisms. Soil types and moisture, air flow, temperature, rainfall, and the presence of plants and animals are some of the variables that influence chemical efficacy, breakdown, and recombination. For example, soil texture and organic matter content have an influence on the activity and breakdown of some herbicides; application rates, therefore, must be adjusted to correspond to each type of soil where these herbicides are being applied.

Avoiding Hazardous Residues

Hazardous pesticide residues can be avoided. Follow label instructions for timing, placement, and rate of application. Avoid incompatible mixtures that cause waste of pesticides. Whenever possible, apply pesticides during dormant or fallow periods to prevent spraying edible produce. Avoid pesticide spills. Fill application equipment in ways that prevent pesticide mixtures from siphoning back into wells. Calibrate application equipment properly and make an accurate measurement of the area to be sprayed to prevent mixing too much material. Finally, when appropriate, select pesticides that break down rapidly and use formulations that reduce problems of drift.

For outdoor applications, use cultural practices, such as managing soil and water movement, to reduce pesticide residues in the environment. Control the amount and timing of irrigation water to eliminate runoff and slow the rate of percolation. In agriculture, use practices such as reduced tillage, contour farming, terracing, grass-lined waterways, and subsurface drainage to reduce soil erosion. Collect and reuse tailwater from irrigated fields to keep residues within the treatment site.

DAMAGE TO TREATED SURFACES

Pesticide treated surfaces or surfaces exposed to pesticide drift, such as painted walls and woodwork, fabrics, floor coverings, automobile finishes, wallpaper, and counter tops may sometimes become spotted, pitted, or stained by pest control chemicals, formulation solvents, or dissolved salts in the water used with the spray mixture. Application rates and the concentration of the spray mixture can influence spotting, pitting, or staining. Follow label instructions for mixing and application. When in doubt, apply a small amount to a test area to be sure the pesticide mixture will not damage treated surfaces.

7 Protecting People and the Environment

Specially designed equipment can protect applicators from excessive exposure to pesticides.

This chapter discusses ways to prevent accidental exposure to pesticides. Pesticide-related problems can be avoided by: (1) learning to use pesticides safely in order to prevent overexposure and resulting short-term or long-term illnesses or even death; (2) avoiding unsafe practices that may cause injury to plants and animals in the environment; (3) obeying all laws that apply to pesticide handling, storage, and application in your work situation; and (4) developing safe work habits.

PESTICIDE APPLICATOR SAFETY

You, the pesticide applicator, are the key to preventing pesticide accidents. If you work carefully, take the precautions described on the pesticide label, and obey the laws and regulations dealing with pesticides, you will avoid most problems. When mixing pesticides together or with other materials, confirm that they can be used safely in combination. Be sure your equipment is running properly; accidents can also be caused by faulty, broken, or worn equipment. Finally, never use alcohol or drugs before, during, or immediately after applying pesticides.

Medical Check-Up and Blood Tests

If you apply pesticides regularly, you should have periodical medical checkups. Locate a medical facility staffed with trained specialists who can help if you need medical attention. These specialists will become acquainted with you and with the nature of the pesticides you work with. They may be able to detect health problems that will limit or impair your ability to perform your work, and they can inform you of specific hazards that might affect your health in the future.

CDFA Pesticide Worker Safety Regulations require a special blood test for persons who are exposed to organophosphate or carbamate pesticides for more than 30 hours within a 30-day period. (Exposure begins the moment a pesticide container is opened for mixing or application and continues until you have bathed and changed clothing after completing the application. In case of a spill or other emergency, exposure begins at the onset of the incident and continues until you have bathed and changed clothing.) This blood test, called a red cell and plasma cholinesterase determination, establishes a baseline for measuring exposure to organophosphate and carbamate pesticides. These pesticides block the production of *cholinesterase*, the enzyme that destroys a chemical, called *acetylcholine*, that normally transmits nerve signals (Figure 7-1). Each person has a unique baseline level of cholinesterase in their blood and in their blood plasma. When a person's cholinesterase goes below their baseline level, pesticide poisoning may have occurred; the test is so sensitive it can often measure exposure well below levels that make peo-

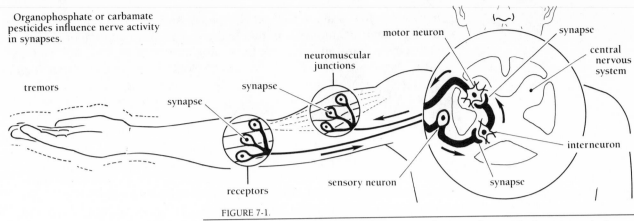

FIGURE 7-1.

Organophosphate and carbamate pesticides interfere with the transmission of nerve signals across a synapse. This may result in uncontrolled shaking or tremors.

ple sick. If tests indicate abnormal changes in your cholinesterase level, your physician may advise you to avoid applying organophosphates and carbamates for a period of time until future blood tests indicate that your cholinesterase level has returned to normal.

Different medical procedures, such as urine analyses and other types of blood tests, can be used to detect poisoning symptoms resulting from exposure to some classes of pesticides besides organophosphates and carbamates. These tests, however, are not mandated by law.

Regular (yearly or more often) follow-up visits to the medical facility should be planned to monitor your exposure and your health. The frequency of these visits depends on the amount of pesticides you are applying, the type and toxicity of the chemicals you are exposed to, and the nature of your work. Your employer or physician may advise you on the frequency of medical examinations.

Training

California's pesticide laws establish minimum standards of training required for persons applying pesticides for hire or as part of their normal work duties. This training must include information on reading and understanding pesticide labels, the proper methods of mixing and applying pesticide chemicals, handling and disposing of pesticides and pesticide wastes, recognizing pesticide poisoning symptoms, and the types of protective equipment that you should use and how to check out, fit, and properly wear this equipment. Your employer is responsible for providing the equipment and necessary training you need. If you are self-employed and require training information, contact your local agricultural commissioner.

Continuing Education. Being employed in the pesticide application profession gives you an obligation to keep current and increase your knowledge of pesticide use, pesticide safety, and pest management. Pesticide laws and regulations change or are revised quite frequently. Pesticide chemicals are constantly being improved, with several new chemicals introduced each year. Pesticide use changes with increased knowledge of pests and their habits. Techniques of pesticide application often reflect new technologies in application equipment and formulation types. Pest management techniques are modified as new information about pests is discovered and understood. Even pests are changing; new pests are introduced and some older pests are developing resistance to pesticides used previously for their control. Without continuing education, a pesticide applicator can quickly lose touch with modern pest control practices and pesticide application techniques.

One of the best ways to keep pace with changes in pesticides and pesticide application is to attend meetings of professional organizations affiliated with your work. Most occupational areas involved in pesticide use are represented by a professional organization; these groups may also publish newsletters that contain important pest management and pesticide use information. The University of California Cooperative Extension Service and local agricultural commissioners often sponsor workshops and seminars on aspects of pest control, pest management, and pesticide use. Courses are also available through local community colleges, state colleges and universities, and the University of California.

Planning

Planning for pesticide safety helps to prevent accidents and enables you to do a better job. Find out about the pesticides you use by studying material safety data sheets and pesticide labels. Learn about the dangers and what precautions you need to take. Inspect areas where you will be working to locate potential hazards that may affect your safety. Finally, plan what you need to do if an accident happens.

Material Safety Data Sheet. Material Safety Data Sheets (MSDSs) provide valuable information about pesticide hazards (Figure 7-2). These sheets are prepared by manufacturers and must be made available to

FIGURE 7-2.

Material Safety Data Sheets provide valuable information about pesticide hazards.

every person selling, storing, or handling pesticides. Ask your employer, or, if self-employed, obtain them from the chemical manufacturer or pesticide supplier. MSDSs can be obtained for every labeled pesticide; they describe the chemical characteristics of active and other hazardous ingredients, and list fire and explosion hazards, health hazards, reactivity and incompatibility characteristics, and types of protective equipment needed for handling. Storage information and emergency spill or leak clean-up procedures are described. LD_{50} and LC_{50} ratings are given for various test animals. Emergency telephone numbers of the manufacturer are also listed on MSDSs.

Pesticide Label. Thoroughly read and understand the *entire* pesticide label. The label gives specific information relating to the type of application you will be making. It is also the legal document that you are obligated to follow for any pesticide use. Look for signal words (Danger—Category I, Warning—Category II, Caution—Category III). Check to see what personal protective equipment is required and make sure you have and use this equipment and that it is in good condition. Study the label for any special environmental precautions. Always be sure that the intended application site is listed. Finally, consult the label for information on how you will dispose of unwanted pesticide and empty pesticide containers.

Hazards. After reading the Material Safety Data Sheet and pesticide label, you will have an awareness of the problems that might be encountered, including the dangers to you, to people and animals occupying the area where the pesticide will be applied, and to the environment. Carefully inspect the treatment area to look for conditions or objects that may influence the safety of the application:

(1) Evaluate weather conditions and make sure they are acceptable. If appropriate, choose a time of day when the pesticide application will be least disruptive.

(2) Become familiar with the boundaries of the treatment site. When appropriate, determine soil type and variations in soil types; these might influence the efficacy of herbicides or other soil-applied pesticides or influence percolation into groundwater. Look for environmentally sensitive areas such as streams, irrigation ditches, ponds, lakes, homes, schools, or parks.

(3) Make arrangements to protect or remove pets, livestock, honey bees, or other animals in the area that could be injured by a pesticide application.

Special hazards may exist if pesticides are being applied inside and around homes, businesses, offices, and other buildings. Be careful around food, food preparation areas and utensils, bedding, pets, and surfaces contacted by people. Extreme care must be exercised when infants or toddlers live in a treatment area. Table 7-1 is a checklist for planning a pesticide application.

Planning for Accidents. Prepare yourself by planning for the possibility of an accident. After an accident occurs is a poor time to select medical care or to look for someone to clean up a spill; always gather this information before you begin. Carry with you the telephone numbers of medical facilities closest to the application site where you are working so you can quickly obtain emergency treatment of poisoning, burns, or injuries. Also have the telephone numbers of the local fire department,

TABLE 7-1.

Checklist for Planning a Pesticide Application.

PERSONAL:
☐ Medical checkup and necessary blood tests?
☐ Properly trained for this type of application?

PESTICIDE:
☐ Read and thoroughly understood label?
☐ Checked to be sure use consistent with target pest and application area?
☐ Read Material Safety Data Sheet for information on hazards?
☐ Obtained necessary permits?
☐ Know proper rate of pesticide to be applied?

EQUIPMENT:
☐ Proper personal protective equipment (boots, gloves, respiratory equipment, protective clothing, eye protection, headwear)?
☐ Necessary measuring and mixing equipment?
☐ Suitable application equipment for this job (tank capacity, pressure range, volume of output, nozzle size, pump compatible with formulation type)?
☐ Application equipment properly calibrated?
☐ Emergency water and first aid supplies?
☐ Necessary supplies to contain spills or leaks (absorbent materials, cleaning supplies, holding containers)?

TRANSPORTING:
☐ Can transport pesticides safely to application site?
☐ Pesticides and containers secured from theft or unauthorized access?
☐ Vehicles properly marked and permits obtained, if necessary, for transporting hazardous materials and hazardous wastes?

MIXING AND LOADING:
☐ Safe mixing and loading site located?
☐ Obtained clean water for mixing?
☐ Water pH tested?
☐ Proper adjuvants obtained for correcting pH, preventing foaming, and improving deposition?
☐ Checked compatibility of pesticide tank mixes or fertilizer-pesticide combinations?
☐ Liquid containers triple rinsed with rinsate put into spray tank?

TREATMENT SITE:
☐ Boundaries of treatment site inspected?
☐ Environmentally sensitive areas within and around treatment area identified
☐ Notified people working or living in or near treatment area, including field-workers and their supervisors in agricultural applications?
☐ Treatment site properly posted with required signs?
☐ Soil types determined and noted, if these are factor in pesticide efficacy?
☐ Livestock, pets, honey bees, other animals properly protected?
☐ Aspects of groundwater determined, if applicable?
☐ Hazards within treatment site identified, including electrical wires and outlets, ignition sources, obstacles, steep slopes, and other dangerous conditions?
☐ Plants in treatment area in proper condition for pesticide application (correct growth stage, not under moisture stress, other requirements as specified on pesticide label)?

WEATHER CONDITIONS:
☐ Weather is suitable for application (low wind, proper temperature, lack of fog or rainfall)?

APPLICATION:
☐ Application pattern established suitable for treatment area, hazards, and prevailing weather conditions?
☐ Application rate selected which will give most uniform coverage?
☐ Equipment frequently checked during application to assure that everything worked properly and provided a uniform application?

CLEAN-UP:
☐ Application equipment properly cleaned and decontaminated after application?
☐ Personal protective equipment safely stored and then cleaned or laundered according to approved methods?
☐ Disposable materials burned or disposed of in approved way?

DISPOSAL:
☐ Paper pesticide containers burned or disposed of according to local regulations.
☐ Plastic and metal containers triple rinsed?
☐ Plastic and metal containers properly stored until disposed of in suitable disposal area?

STORAGE:
☐ Unused pesticides returned to supplier or stored in locked facility for later use?
☐ Storage facility is suitable for pesticides?

REPORTS:
☐ Necessary reports filed with requesting agency?

FOLLOW-UP:
☐ Treatment areas inspected after application to assure that pesticide controlled target pests without causing undue damage to nontarget organisms or surfaces of items in treatment area?

DAMAGE:
☐ Damage, if it occurred, promptly reported?

sheriff, and highway patrol (Table 7-2)—the emergency number "911" usually provides immediate access to medical help, local fire services, and law enforcement agencies. Plan what to do in the event of a pesticide spill and be prepared to protect the public from danger. Know the proper first aid to be administered to victims of pesticide exposure. Consider the four types of exposure (dermal, oral, respiratory, and ocular) and plan how to reduce injury to yourself and others in case of an accident. Be sure you have emergency water for washing eyes and skin.

Understanding Pesticide Combinations

The toxicity, mode of action, and efficacy of pesticides change when they are combined with other pesticides and other chemicals. Combinations can alter the toxicity of pesticides to people as well as to target pests; the ease with which pesticides enter your body and the speed of entry may be influenced, altering your body's ability to deactivate toxic

TABLE 7-2.

Emergency Numbers for Pesticide Accidents and Spills.

WHEN PEOPLE HAVE BEEN EXPOSED TO PESTICIDES:

Dial 911 for emergency medical assistance. Notify the operator that the problem is a pesticide exposure. Provide an accurate location and information on the type of pesticide involved.

After obtaining medical treatment for exposed persons, determine if a spill has taken place. Follow instructions below for a spill.

Contact the nearest agricultural commissioner's office to report the incident. Find the telephone number in the white pages of the local telephone book, under the county where the accident occurred.

FOR PESTICIDE SPILL ON STATE OR FEDERAL HIGHWAYS:

Notify local office of the California Highway Patrol and the local fire department (dial 911). Inform operator that a pesticide spill has occurred; provide accurate location and type of pesticide.

Contact CHEMTREC at 800-424-9300 for assistance in cleaning up a pesticide spill.

Contact State of California Office of Emergency Services. Usually a written report will need to be filed. Check white pages of telephone directory under "State Government Offices" section or contact the main office in Sacramento: Office of Emergency Services, 2800 Meadowview Road, Sacramento, CA 95832. (916) 427-4990.

Contact local agricultural commissioner's office. Find the telephone number in the white pages of the local telephone book, under the county where the accident occurred.

**FOR PESTICIDE SPILL ON LOCAL CITY
OR RURAL ROADS OR PRIVATE LAND:**

Contact local police or sheriff and local fire department (dial 911). Inform operator that a pesticide spill has occurred; provide accurate location and type of pesticide.

Contact CHEMTREC at 800-424-9300 for assistance in cleaning up a pesticide spill.

Report spill to Office of Emergency Services. Contact local agricultural commissioner's office. Find the telephone number in the white pages of the local telephone book, under the county where the accident occurred.

---

materials. Adjuvants that enhance penetration or toxic action sometimes increase hazards.

Avoiding Medications, Alcohol, and Drugs

Medications, alcohol, and drugs cause drowsiness, impair judgment, and often influence your ability to apply pesticides safely. These substances may also alter the toxicity of pesticides within your body. For example, a severe illness may result if alcohol is consumed shortly after exposure to thiram. If you are taking any medication at all, including nonprescription preparations, consult with a physician before handling, mixing, or applying pesticides. Do not use alcohol or drugs before, during, or immediately after a pesticide application.

PERSONAL SAFETY EQUIPMENT

The purpose of personal safety equipment is to prevent pesticides from contacting your body or clothing. This equipment also protects your eyes and prevents inhaling toxic chemicals. Nevertheless, personal safety

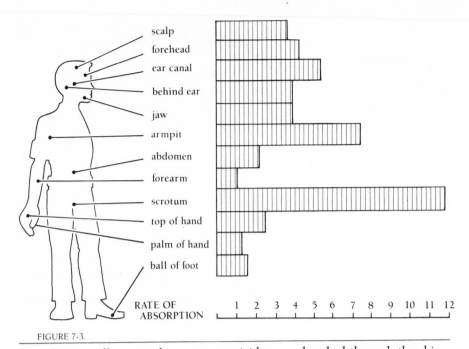

scalp
forehead
ear canal
behind ear
jaw
armpit
abdomen
forearm
scrotum
top of hand
palm of hand
ball of foot

RATE OF
ABSORPTION

1 2 3 4 5 6 7 8 9 10 11 12

FIGURE 7-3.

This drawing illustrates how some pesticides are absorbed through the skin at different rates, depending on the area of the body exposed (the rates illustrated here are based on studies using the insecticide parathion). Never handle or use pesticides without adequate protective clothing. Wear protective clothing suitable to the type of work being performed.

equipment is effective only if it fits correctly, is used properly, and is kept cleaned and maintained. Select equipment that offers maximum protection.

The greatest risk of pesticide poisoning comes from pesticides contacting your skin. Pesticides highly soluble in oil pass through skin faster than pesticides highly soluble in water. In addition, some parts of your body absorb pesticide more quickly than other areas (Figure 7-3). In a test using parathion, for example, it was found that the forearm is one of the least susceptible areas of the body for pesticide absorption; the palms of the hands and soles of the feet absorb parathion slightly faster than the forearm; the top of the hand is almost 2½ times more susceptible to absorption than the forearm; the scalp, face, and forehead are 4 times more susceptible; the ear canal absorbs at a rate almost 5½ times faster than the forearm; absorption in the armpit is nearly 7½ times greater than the forearm. The scrotum is the most susceptible area of the body to parathion absorption, being nearly 12 times more absorptive than the forearm.

The way pesticides are applied also influences the exposure you receive. For instance, persons applying pesticides in an orchard with an air blast sprayer (Figure 7-4) are usually subjected to much greater exposure than those applying similar chemicals to a row crop with a low-volume boom sprayer. Applying chemicals in enclosed areas such as greenhouses, warehouses, or residences usually subjects a person to higher levels of exposure than when applying pesticides in outdoor landscape or right-of-way areas.

The part of your body receiving the greatest amount of exposure is related to the nature of your work and the position you assume while making the application. Choose personal safety equipment that affords

FIGURE 7-4.

The type of pesticide application being performed dictates the amount and type of protective equipment you should wear. Pesticide labels are a guide for the minimum protection. Some applications, such as spraying orchards as shown here, expose the operator to higher levels of pesticide and require maximum protection.

the maximum protection for the type of work you are performing, and not just to comply with minimal label requirements; adequate equipment in one situation may not be satisfactory in another. Table 7-3 is a protective clothing and equipment guide developed by the California Department of Food and Agriculture. In most instances, using safer equipment than what is recommended increases your protection. The only exception

PROTECTIVE CLOTHING AND EQUIPMENT REQUIREMENT

 Waterproof apron made from rubber or synthetic material. Use for mixing liquids.

 Waterproof boots or foot coverings made from rubber or synthetic material.

 A daily change of clean coveralls or clean outer clothing. Wear waterproof pants and jacket if there is any chance of becoming wet with spray.

 Faceshield, goggles, or full face respirator. Goggles with side shields or a full face respirator is required if handling or applying dusts, wettable powders, or granules or if being exposed to spray mist.

 Waterproof, unlined gloves made from rubber or synthetic material.

 Waterproof, wide-brimmed hat with nonabsorbent headband.

 Cartridge type respirator approved for pesticide vapors unless label specifies another type of respirator such as a dust mask, canister type gas mask, or self-contained breathing apparatus.

TABLE 7-3.

Protective Equipment and Clothing Guide.

SUMMARIZED LABEL STATEMENT Toxicity Category:	MIXER-LOADER		APPLICATOR	
	I–II	III	I-II	III**
Precautions should be taken to prevent exposure.	A B C F G H R *	A B C F G H	B C F G H R *	C F H R *
Protective clothing or protective equipment is to be worn.	A B C F G H R *	B C F G H R *	B C F G H R *	C F H R *
Clean clothing is to be worn	C	C	C	C
Contact with clothing should be avoided.	A B C	B C	B C	C
Contact with shoes should be avoided.	B	B	B	B
Rubber boots or rubber foot coverings are to be worn.	B	B	B	B
Contact with skin should be avoided.	A B C F G H	B C F G H	B C F G H	C F G H *
A cap or hat is to be worn.	H	H	H	H
An apron is to be worn.	A	A		
Rubber gloves are to be worn.	G	G	G	G
Contact with eyes should be avoided.	F	F	F	F
Goggles or faceshield is to be worn.	F	F	F	F
Avoid inhalation.	R	R	R *	R *
A respirator is to be worn.	R	R	R *	R *

*Use this equipment when there is a likelihood of exposure to spray mist, dust, or vapors.

**If the Category III pesticide application is being made in an enclosed area such as a greenhouse, or if the application consists of a concentrate spray of 100 gallons-per-acre or less in a grove, orchard, or vineyard, then use the protective equipment guidelines for Category I-II pesticides.

is with the use of fumigants; protective clothing and gloves sometimes trap fumigants close to your skin, increasing exposure rather than protecting you. Follow label instructions explicitly for the proper, maximum protective clothing when applying fumigants.

Protective Devices

Atmosphere-Monitoring Equipment

Never enter an enclosed fumigated area, even after venting, without measuring for toxic levels of pesticide vapors (Figure 7-5). Vapors can be detected and measured with several different atmosphere-monitoring devices. When taking measurements, wear required respirators and protective clothing or use remote sensing equipment. Take measurements in several locations within the confined space, since fumigants sometimes become trapped in localized pockets. One type of atmosphere-monitoring device consists of a hand-operated air pump that draws measured amounts of air through a glass detector tube; the concentration, in parts per million, of a specific contaminant in the atmosphere is displayed. Detector tubes are available for many of the common fumigants. Tubes contain chemicals that react with pesticides in the air and produce a color change. Each tube can only be used one time; a new detector tube must be inserted into the air pump before each sample is taken.

Choose atmosphere-monitoring equipment that is suitable to the type of fumigation work you are performing. Consult with the manufacturer

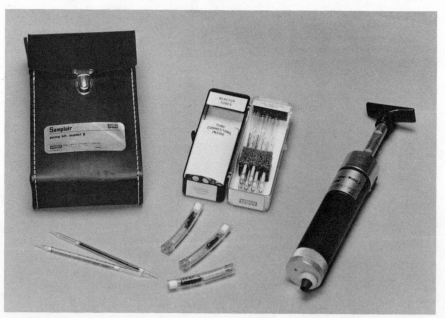

FIGURE 7-5.

Atmosphere-monitoring devices like the one illustrated here are used to determine the level of toxic material in an enclosed environment. Devices such as these should always be used before entering a treated area. Color changes in the glass tube indicate the concentration of toxicant in the atmosphere. Specific types of tubes are used for different chemicals.

or distributor to be certain that the equipment can provide accurate readings in the toxicant concentration range that you work with. Learn about the shortcomings of these devices, since other contaminants in the atmosphere can produce erroneous readings. Be sure you are properly trained in the use of the equipment and that you can reliably detect dangerous levels of pesticides.

Enclosed Cabs

Enclosed cabs are devices installed on tractors to protect the operator from exposure to pesticides (Figure 7-6). Cabs that are properly designed and maintained provide a high degree of protection, which is important to operators working in orchards and vineyards or other areas where pesticide exposure potential is high. Most models offer options of heating and cooling the air forced into the cab for added operator comfort. Cabs are often well insulated to reduce noise from the tractor and spraying equipment. Front and rear windows are usually equipped with washers and wipers.

Enclosed cabs can only protect operators if they are regularly cleaned and serviced to ensure proper operation. The air system includes pesticide filters that need to be cleaned or replaced whenever they exhaust their filtering capacity. Protection to the operator is confined to the area inside the cab. If you mix pesticides or get out of the cab during a pesticide application, be sure to remove contaminated protective clothing before getting back inside. For maximum convenience and protection to the operator, another person should be assigned the duties of mixing pesticides and refilling spray application equipment.

FIGURE 7-6.

Enclosed cabs mounted onto tractors protect the operator against pesticide exposure. Filtered, pesticide-free air is forced into the cab by hydraulic or electric blowers. Enclosed cabs are useful in orchards and vineyards where risks of pesticide exposure are high.

Select an enclosed cab carefully to assure it meets your needs and expectations. For example, consider how the blower unit is powered. Units that connect to the hydraulic system of a tractor are usually capable of moving large volumes of air, but may be noisy. Electric motor drives are quieter, but may not have as much power because of limitations of a tractor electrical system. The protection offered by an enclosed cab depends on filtered air flowing around the operator's body, creating a positive pressure within the cab and preventing entry of contaminated air or particulates. The volume of air that the unit can supply may influence the amount of protection that is provided; units that produce large volumes are usually more effective. Another important feature is the filtering system. Filtering ability is determined by the size of the unit, the volume of air being moved, number of filter stages, the type of filtering material used, and the appropriateness of the filter media to the pesticides being used. Multiple stage filters that include a pre-filter, a high efficiency particulate air filter (HEPA), and an activated carbon filter are the safest for reducing pesticide exposure. Blowers that move large volumes of air must be equipped with large capacity filters. Other features of enclosed cabs that might influence their function include the visibility afforded the operator, how the height of the cab relates to application sites where it will be used, the strength of the cab for protecting the operator from tree limbs or in case of tractor rollover, and the availability of heating and air conditioning. Some cabs are equipped with pressure gauges so the operator can monitor the positive pressure of the air inside the cab.

Personal Gear

Prevent pesticide exposure to your body by selecting adequate personal gear. This includes clothing that covers the arms, legs, torso, and head and prevents pesticide dust or liquid from contacting the skin. Gloves and boots are used to protect hands and feet, while helmets, hoods, or wide-brimmed hats prevent exposure to the head and neck. Face masks or goggles protect the eyes. Avoid breathing dusts, mists, or vapors by selecting an appropriate respirator.

Always anticipate accidents and be sure your body is adequately protected under any possible circumstance. For example, when handling pressurized fumigants, a hose from the tank might rupture, causing the hose to violently whip around while releasing large quantities of fumigant. Therefore, when using fumigants from a cylinder, wear goggles or a faceshield to protect your eyes and a self-contained respirator for safe breathing under emergency conditions.

Bodywear

Different styles of protective clothing may be selected based on personal preference and the type of work performed. Common styles include coveralls, bib overalls, jackets, and aprons. Protective clothing is made from several types of materials, providing choices in chemical resistance, weight and strength of the fabric, fabric resistance to ripping and puncturing, response to temperature extremes, comfort, ability to be cleaned, and durability. No single material provides everything, so base your selection on your most important need, protection from pesticide exposure.

FIGURE 7-7.

This person is wearing the minimal protective clothing for pesticide application. It consists of a long-sleeved shirt and long pants. The fabrics are tightly woven cotton materials.

FIGURE 7-8.

Cotton coveralls, such as those seen here, are suitable for use for many types of pesticide application; they have the advantage of being able to be easily removed if contaminated or when finished working. Coveralls should be worn over a long-sleeved shirt and long pants.

Minimal Protective Clothing. Never handle, mix, or apply pesticides without at least some minimal protective clothing to prevent chemicals from contacting your skin. Minimal protection required under low-hazard circumstances includes full-length pants and a long-sleeved shirt made from tightly woven cotton fabric (Figure 7-7). Coveralls worn over regular clothing provide additional protection from pesticides, and have the advantage of being able to be easily removed if they become contaminated or when you finish working (Figure 7-8).

Cotton fabric is popular for its comfort and coolness, but should never be worn without additional protective clothing when there is a chance of contacting wet spray or concentrated liquid pesticide, or whenever highly toxic pesticides are used. Tightly woven fabrics act as a wick and efficiently carry liquids to the inside of the garment, increasing your potential for dermal exposure (Figure 7-9).

Disposable Protective Clothing. Disposable protective clothing is manufactured from several types of materials suitable for pesticide application. Disposable fabrics made from nonwoven, bonded fiber materials are superior to woven fabrics because they do not promote wicking and are more resistant to liquid penetration. Some nonwoven fabrics are laminated or bonded to other materials to further enhance waterproofing. Disposables are usually lightweight but remarkably strong and resistant to tearing or puncturing (Figure 7-10). Available in a variety of styles to be worn over regular clothing, disposables have the major advantage of

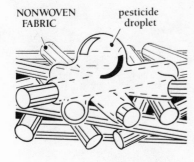

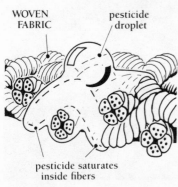

FIGURE 7-9.

Woven fabrics promote pesticide movement through the material to your skin by a process called wicking, as illustrated here. Avoid using woven materials when there is a chance of contacting spray mist or droplets.

not requiring cleaning or decontamination after use. They can be thrown away when soiled. Use this type of protection under conditions of heavy contamination in which it is impossible to effectively clean soiled protective clothing. To prevent environmental hazards, dispose of pesticide contaminated clothing in an approved disposal site, the same as any other hazardous waste.

Reuseable Protective Clothing. Most reuseable protective clothing consists of woven or nonwoven fabrics coated with or laminated to a waterproof material. The amount of protection offered against pesticide exposure depends on the type of waterproofing material. Neoprene, latex rubber, and polyvinyl chloride (PVC) are commonly used and all are very effective (Figure 7-11). Another type of reuseable clothing is made from a woven rip-stop nylon fabric laminated to nylon tricot by means of a waterproof film. This layered fabric is effective in preventing pesticide penetration, and is strong, lightweight, durable, and flexible.

Do not use protective clothing that is lined with a woven fabric. Although woven linings are more comfortable and give the garment strength, they can become wet and contaminated with pesticides, increasing the risk of exposure. Linings of nonwoven, nonabsorbent materials, such as dacron, are safer and are the only types that should be used (Figure 7-12).

Some waterproofing materials that resist pesticide penetration may, after a period of time, react with pesticides and oils to become stiff and cracked, reducing the protection they offer. Discard reuseable clothing once it loses its ability to repel water or if it becomes torn, cracked, or punctured.

Protective clothing is manufactured for many different purposes, often only for protection against rain. Select fabrics that are resistant to the chemicals you work with, such as petroleum oils, organic solvents, and abrasive dusts; they must also be resistant to tearing, snagging, and abrasion, and have a nonabsorbent lining. Look for strong and noncorrosive fasteners. Pants should be the bib overall style to provide sufficient overlap with the jacket. Select jackets with attached hoods for greater protection of the neck and head. Garments should allow movement without binding. Batwing or raglan sleeves, for example, provide more freedom than conventional set-in sleeve designs.

Aprons. Aprons protect parts of your body during handling and mixing of pesticides, but should not be used instead of other protective clothing during pesticide application. Aprons must be made of waterproof materials and be long enough to protect your clothing. Styles having a wide bib provide splash protection to the upper chest and are preferable for pesticide mixing and handling of containers (Figure 7-13). Disposable aprons, made for one-time use, are generally made of thin plastic materials that tear or puncture easily and therefore have limited use for pesticide handling. Reuseable aprons are more durable, but require regular cleaning and decontamination; they should be discarded if they develop tears or holes.

Head Protection

Hats made from fabric absorb water and can become seriously contaminated. Baseball style caps offer minimal protection to the head because they are usually made of absorbent or mesh fabrics that may allow

FIGURE 7-10.

This person is wearing disposable protective clothing which is lightweight but resists puncturing, tearing, and pesticide spray penetration; it is made from a nonwoven fabric. This type of protective clothing can be thrown away after it becomes contaminated, eliminating the need for laundering.

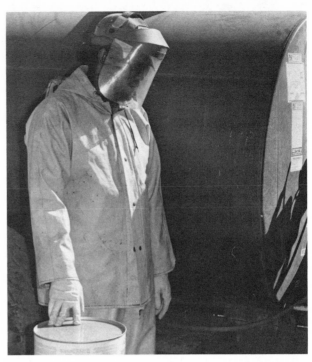

FIGURE 7-11.

Waterproof protective clothing, like the suit worn here, is made of neoprene, latex rubber, polyvinyl chloride, or blends of these materials. These materials usually provide the maximum amount of protection from pesticide exposure.

FIGURE 7-12.

The lining in this waterproof protective clothing is made of nonabsorbent material to prevent pesticide contamination.

FIGURE 7-13.

Waterproof aprons like the one here should be worn when handling or mixing pesticides or handling pesticide contaminated equipment. Select a style with a wide bib to provide added protection.

FIGURE 7-14.

A waterproof plastic helmet can protect your head from pesticide exposure; the one shown here is equipped with a plastic faceshield for protection against splashes.

FIGURE 7-15.

Protective gloves should be made of natural rubber, latex, butyl, or neoprene. Gloves should be unlined because lining material absorbs pesticide and increases risks of exposure.

spray droplets to pass through. If a hat is worn, it should be either water resistant and wide-brimmed or made out of a waterproof material such as plastic; headbands and sweatbands must also be waterproof. Most plastic helmets are impact resistant and provide added protection to the head from tree limbs and falling or flying objects. Accessories such as faceshields, goggles, and hearing protectors can be attached to several styles of helmets (Figure 7-14). All types of protective equipment, including helmets and hats, must be cleaned regularly. An alternative to protective headwear is a hooded, waterproof jacket.

Gloves

Waterproof gloves are an essential part of your safety equipment, and must always be worn when handling pesticides. Leather or fabric gloves should never be used because they absorb water and pesticide, and may actually increase exposure. Suitable gloves are made of natural rubber, latex, butyl, or neoprene (Figure 7-15). Select a material that offers resistance to the types of pesticide you are using and the amount of exposure you expect. Some materials are suitable for total immersion in a liquid toxicant for extended periods of time, while other materials only provide protection against accidental splashes or occasional immersion. The thickness of the glove material also determines the amount of protection; thicker materials are better. Choose materials that resist puncturing and abrasion.

Gloves must not be lined since fabrics used for linings may absorb pesticides, making them dangerous to use and difficult to clean. Woven removable glove liners can be used to insulate your hands from the cold.

Discard or launder contaminated glove liners. Check pesticide labels for special glove recommendations; for instance, waterproof gloves are not used with some fumigants, since they increase the chances of dermal poisoning or irritation. Fumigants are often released as cold liquids, so a severe skin burn may result if the liquid is confined by gloves. Do not use gloves that have been contaminated with a pesticide, since this increases pesticide absorption through your hands by keeping your skin in constant contact with the toxicant.

Make sure the cuffs of gloves are long enough to extend to the mid-forearm. Usually, sleeves should be worn on the outside of gloves to keep pesticides from getting in. Some special application situations, however, require holding one or both hands overhead while spraying liquids. In these cases it is best to have the protective jacket sleeve of the elevated arm(s) tucked inside the gloves to keep pesticides out. Be careful when lowering your arm to prevent pesticides from entering the glove.

FIGURE 7-16.

Protective footwear such as the boots shown here protect your feet from pesticide exposure. Wear legs of protective pants on the outside of the boots. Boots should extend to your mid-calf area. Leather or fabric shoes or boots should never be worn during pesticide mixing or application because they absorb pesticides.

Footwear

Leather or fabric shoes must never be worn while handling, applying, or mixing pesticides because they absorb most pesticides. Fabric is difficult to clean and it is impossible to remove pesticides from leather. If leather boots or shoes accidentally become contaminated with a pesticide, dispose of them. Protective footwear (Figure 7-16) should be made of rubber or synthetic materials such as PVC, nitrile, neoprene, or butyl. Select the material on its ability to resist pesticides you work with (Table 7-4). Some pesticides, 1,3-dichloropropene (Telone II) for example, are capable of penetrating through most protective waterproof substances, including rubber and synthetic materials; accordingly, these at best can only provide short-term protection.

TABLE 7-4.

Guide to Selecting Suitable Materials for Protective Footwear.

	PVC	Nitrile	Neoprene	Butyl rubber	Standard rubber	Leather	Fabric
Hydrocarbon materials, oils, and solvents (most pesticides)	•	••	•	•	–	–	–
Ketones and aldehydes	○	–	•	••	•	–	–
Alcohols	••	••	••	••	•	–	–
Organic Esters	○	–	•	••	•	–	–
Inorganic Metals	••	••	••	••	••	–	–

- ••   Highly satisfactory
- •   Often acceptable
- ○   Occasionally acceptable
- –   Do not use

FIGURE 7-17.

Protective footwear is also available in overshoe styles which can be worn over regular shoes or boots.

FIGURE 7-18.

When choosing waterproof boots, select a sole pattern that can be easily cleaned and does not collect mud.

Waterproof rubber or synthetic materials are poor conductors of electrical current and therefore offer good protection against electrocution when working in wet areas where there are electrical hazards.

Waterproof footwear is available in conventional boot and overshoe styles (Figure 7-17); some boots have internal steel toe caps for protection against falling objects. Select footwear that fits well and is comfortable to wear. Protective footwear should be calf-high, and worn with the legs of your protective pants on the *outside* to prevent spray from getting in. For increased protection, use rubber bands to seal pant legs tightly around the outside of the boots. Make sure boot seams are well sealed to prevent pesticide entry. Choose a sole design that protects against slipping on wet surfaces and is easy to clean (Figure 7-18).

Waterproof boots do not "breathe" like leather or fabric shoes, so wear clean cotton or wool socks to absorb perspiration.

Eye Protection

Eye protection must always be worn during mixing and loading, while adjusting, cleaning, or repairing contaminated equipment, and during most types of pesticide application. The only situations in which eye protection is usually not necessary are: (1) if pesticides are being injected or incorporated into the soil; (2) if pesticides are being applied through vehicle-mounted spray nozzles that are located below and behind the operator, with the nozzles directed downward; (3) if the operator is working in an enclosed cab; or (4) if the pesticides being applied are rodenticides, predacides, or avicides that are not in a liquid or gaseous form. Eye protection is available as goggles, faceshields, and faceshields combined with respirators. In some cases, safety glasses that include a brow piece and side shields may be used.

Goggles. Goggles are the most common form of eye protection. Goggles must be worn whenever the pesticide label specifically mentions their use. To prevent pesticide entry into the eyes, there must be full side shields (Figure 7-19). Nonfogging lenses are available for most styles, and solutions can be purchased to reduce or eliminate fogging of ordinary lenses. Some goggle styles are designed to be worn over eyeglasses; these styles should be the *only* type worn over eyeglasses.

Goggles are held in place either by elastic or synthetic rubber straps. Because elastic straps contain fabric, they can absorb pesticide and possibly increase contamination to the back of the head, an area that is highly sensitive to pesticide absorption. Avoid this problem by using a hood or protective headwear over the strap. Replace or thoroughly wash the elastic band if it becomes contaminated. Straps made of neoprene or other synthetic materials are safer because they are nonabsorbent and easy to clean.

Goggle lenses may become coated with spray droplets during some types of pesticide applications. If this happens, carry cleaning supplies or an extra pair of goggles. Check and clean your goggles each time you stop to refill the spray tank. Use caution, however, to prevent scratching plastic lenses. Never wipe lenses to remove dirt; clean them with soap and water. Scratched lenses make the goggles useless. Some styles of goggles are available with replaceable lenses.

Faceshields. Faceshields can be used for mixing pesticides because they protect your eyes and prevent liquids from splashing onto your face. They have limited use during application since airborne spray or dust

FIGURE 7-19.

Protective goggles like those shown here can be worn to protect the eyes during mixing and applying pesticides. Some styles are designed to allow the user to wear prescription glasses.

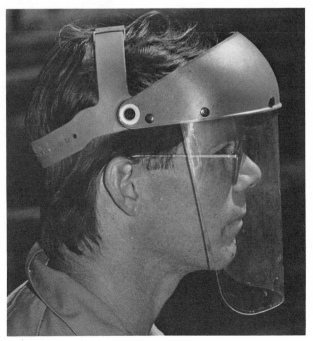

FIGURE 7-20.

A face shield will protect your eyes and face during pesticide mixing or application. Face shields can be used with eyeglasses as shown here.

can float in around the edges of the faceshield. Faceshields are comfortable to wear, allow better air circulation, and provide a greater range of vision than goggles. Faceshields attach to the visor of a plastic helmet or have a separate headband (Figure 7-20).

Faceshields are made of plastic and require the same care as goggles to protect them from scratching. When not in use, store faceshields and goggles in plastic bags to keep them clean and prevent scratching.

Respiratory Equipment

A respirator is a device that offers protection to the lungs and respiratory tract from airborne pesticides. Many types and styles of respirators are available suitable for use while mixing and applying pesticides. Select respiratory equipment based on the type and toxicity of pesticides you are using, the recommendation or requirement listed on the pesticide label, and the nature of the area where you are working. Respiratory equipment should be approved for use with pesticides by the Mine Safety and Health Administration (MSHA).

Disposable Masks. The simplest form of respiratory protection is a disposable mask made out of paper, nonwoven fabric, or foam rubber

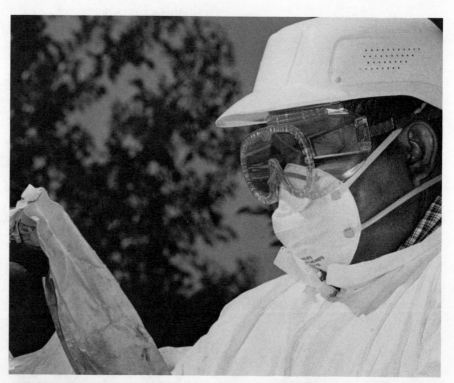

FIGURE 7-21.

A disposable face mask protects against the inhalation of irritating pesticide dusts. The type of mask shown here should be worn while mixing or applying dusts having a low toxicity or minimal respiratory hazard. This style should not be used for vapor protection or for protection from Category I liquid or dust formulations.

(Figure 7-21). These masks can be used for protection against nontoxic, nuisance dusts.

Disposable dust masks are lightweight, soft, and fairly comfortable to wear; they are held in place by one or two elastic straps. Most have a soft metal band at the top edge that can be shaped around the bridge of the wearer's nose, improving the seal. Disposable masks provide only minimal protection. Do not use them with liquid sprays or when handling toxic (Category I) materials, including dusts.

Cartridge Respirators. Cartridge respirators include a fitted rubber facepiece and replaceable filters contained in one or two screw-on cartridges (Figure 7-22). Other models have a single, larger cartridge that is worn on a waist belt and connects to the facepiece by a flexible hose; this style provides greater protection for an extended period of time because of the larger filtering capacity. Cartridge respirators have a one-

FIGURE 7-22.

A cartridge respirator must be worn when mixing or applying Category I pesticides and some other less toxic materials. Several styles are available as seen here, but each has a two stage filter for removing droplets, dusts, and vapors. Cartridge respirators usually are not suitable for use with fumigants and should never be used in atmospheres containing less than 19.5% oxygen.

FIGURE 7-23.

Cartridges suitable for use with pesticides must be approved by NIOSH/MSHA. Look for the TC approval number and the word "pesticide" on the label.

way exhalation valve; inhaled air must pass through the cartridge to be filtered, but the valve permits exhaled air to bypass the filters. Facepieces are held in place by at least two adjustable elastic headbands. Manufacturers offer different types of filters and cartridges which can be interchanged with the facepiece, providing a wide range of application including spray painting, welding, handling radioactive materials, or pesticide application. Suitable filter cartridges can be used to remove low levels of pesticide dusts and mists; they are used when mixing and applying most Category I materials. Cartridge respirators cannot be used for protection against fumigants (gases) or in atmospheres that pose an immediate threat to life or health, such as atmospheres containing carbon monoxide or having an oxygen level below 19.5%. Cartridges must have a NIOSH/MSHA "TC" ("Tested and Certified") approval number and be designated for use with pesticides (Figure 7-23). Cartridges made for pesticide use must have two stages: a particulate prefilter to mechanically trap airborne particles; and an activated carbon organic vapor cartridge to adsorb gases (Figure 7-24).

Cartridge respirators need to fit properly to be effective and safe. They should be in good working condition and be cleaned after each use. Beards and long sideburns affect the way cartridge respirators seal around the face and will prevent them from giving adequate protection (Figure 7-25). Regulations prevent pesticide applicators with beards or long sideburns from wearing cartridge respirators. A respirator that does not seal properly is more dangerous than no respirator at all because you receive inadequate protection but a false sense of security. Isoamyl acetate, a chemical with a strong bananalike odor, is used to test for proper fit of cartridge respirators (Table 7-5). If the wearer can detect this odor, the facepiece is admitting unfiltered air. Adjust the facepiece until the odor is absent or try another size, brand, or style.

Eye protection may be required when using a respirator, so be sure the respirator can be worn comfortably with goggles or a faceshield. Some cartridge respirators include a full facepiece with built-in eye protection (see Figure 7-22).

Filters and cartridges may need to be replaced during use. Increased resistance to breathing is an indication that the mechanical part of the filter is loading up with trapped particles. Although the filter becomes

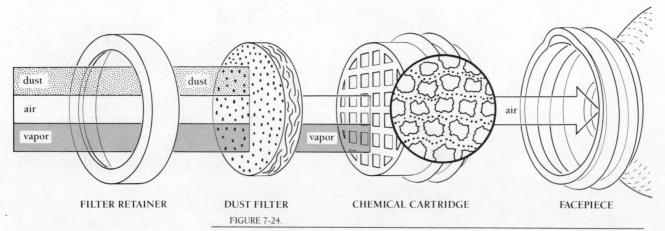

| FILTER RETAINER | DUST FILTER | CHEMICAL CARTRIDGE | FACEPIECE |

FIGURE 7-24.

Pesticide cartridges consist of two stages. This diagram illustrates the mechanical dust filter which removes dust and droplet particles and the activated charcoal chemical cartridge which removes vapors.

TABLE 7-5.

Technique for Properly Fitting Cartridge Respirators.

ISOAMYL ACETATE
(banana oil) FIT TEST*

The chemical isoamyl acetate, commonly referred to as "banana oil," is available from major chemical suppliers and is widely used to check respirator fit. Its odor is easy to detect and the chemical can be used with any pesticide respirator equipped with organic vapor cartridges or canister.

When conducting a fit test, it is important to know that some brands of respirators are available in small, medium, and large sizes. If possible, have several different sizes available during the test to ensure proper fit. Try respirators from different manufacturers since one brand may fit better than others.

If a respirator does not fit properly, the applicator will not be adequately protected. Therefore, be sure to follow the test procedures outlined below:

1. Be sure there is no banana oil odor in the test area that may influence the wearer's ability to detect its presence. Once a respirator is selected, have the wearer adjust it until there is a good face-to-mask seal.

2. Saturate a piece of cotton or cloth with banana oil. The person performing the test should wear rubber gloves and avoid skin contact with the wearer.

3. Pass the saturated material close to the respirator in a clockwise and counterclockwise motion. Have the wearer stand still and breathe normally and then deeply. If the wearer smells banana oil, readjust the respirator or select a different size or style before starting again.

4. If the odor cannot be detected while the wearer is standing still, have them perform side-to-side and up-and-down head movements. Also have the wearer talk loudly enough to be heard by someone standing nearby. Then have the person make other movements, such as bending over, that may occur during spray application.

5. If the banana oil odor cannot be detected during the above movements, it indicates a satisfactory fit. Seal the respirator in a plastic bag marked with the wearer's name. Keep a record of when the fit test was conducted, along with the the size and brand of respirator selected for each user.

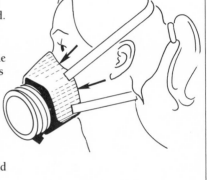

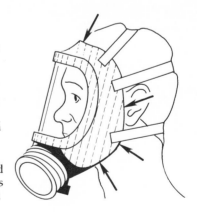

*Adapted from *A Guide to the Proper Selection and Use of Respirators*, Zoecon Corporation.

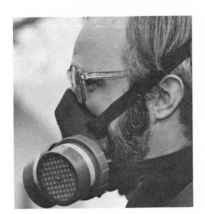

FIGURE 7-25.

Cartridge respirators must seal around the face to be effective and to prevent unfiltered air from entering. Beards or long sideburns prevent proper sealing. It is illegal for persons having a beard or long sideburns to use cartridge respirators or apply pesticides that require them.

more efficient as air passages become smaller, breathing becomes more difficult and the respirator will be uncomfortable to use. Replace the filter unit when this happens. Replace the organic vapor cartridge if you begin to smell a pesticide odor, detect any taste, or experience any irritation; these signs may indicate that the organic vapor unit cannot absorb more pesticide and has lost its effectiveness. Use extreme care when working with odorless pesticides. Change cartridges following pesticide product label instructions, equipment manufacturer's recommendations, or, lacking factual information to the contrary, at the end of each daily work period.

Powered Air Cartridge Respirator. The powered air cartridge respirator is a cartridge respirator that uses a battery-operated motor to force filtered air through a flexible tube to a hood, helmet, or face mask (Figure 7-26). The motor, pump, batteries, and filters are usually worn on a waist belt. In place of a battery pack, some units can be plugged into the electrical system of a tractor. These devices have large, efficient filters, and provide comfortable protection for lengthy application jobs. Like other types of cartridge respirators, these units can be used only when the atmosphere poses no immediate threat to life or health and when the oxygen level is 19.5% or greater. These respirators often incorporate faceshields for eye protection. They are comfortable to wear because the user does not actively force air through filters or valves. The constant supply of forced air around the user's face eliminates the need for a mask-to-face seal so this type of respirator can be worn by a person with a beard or long sideburns.

Supplied Air Respirators. Maximum respiratory protection can be obtained by using a supplied air respirator. This type of equipment must be used when working in an area being fumigated, when working with concentrated amounts of highly toxic pesticides, or when the atmosphere contains less than 19.5% oxygen. These should also be worn by emergency workers responding to toxic spills or fires or rescuing injured persons. Several styles and types are available. Supplied air respirators do not require filters or cartridges to remove toxic materials because they provide an outside source of clean, uncontaminated air.

Self-contained supplied air respirators (often called a self-contained breathing apparatus—SCBA) provide clean air from pressurized tanks that the user wears, similar to a scuba diver (Figure 7-27). External air models connect the wearer to a distant air pump or stationary tank by means of a hose; air pumps, however, must be located in an area where safe, fresh air is available.

Self-contained units are limited to the amount of air that the user can carry; once the air supply is exhausted the system cannot provide any protection, therefore these units are equipped with a bell or other warning device to alert the user when the air supply is getting low. Air tanks may be heavy and bulky, but they give unlimited mobility because no hoses are required.

Hose-connected supplied air respirators provide a large or unlimited quantity of fresh air, but the user is confined to the range of the available hose. The maximum hose length is usually 300 feet. Long hoses are cumbersome and awkward to handle; users must take precautions to avoid kinking, snagging, or hose damage.

Supplied air respirators use either half or full facepieces; full facepiece models provide eye protection. Other models attach to a hood (or helmet) that encloses the entire head and have a clear plastic faceshield.

FIGURE 7-26.

A battery-powered fan forces filtered air through a flexible hose into a helmet or hood in this type of powered air cartridge respirator. This design allows the respirator to be worn with eyeglasses, beards, and long sideburns. As seen here, the filters, motor, and battery pack are worn on a waist belt.

FIGURE 7-27.

A self-contained supplied air respirator, like the one being worn here, provides the wearer with uncontaminated air from a compressed air tank. The tank and air regulator are similar to the breathing equipment worn by a scuba diver.

Facepieces use a pressure demand regulator that admits fresh air into the mask as the wearer begins to inhale. Air flow diminishes when the user exhales. This method is used with self-contained units because of the limited air supply. Hoods provide a continuous flow of fresh air around the entire head, whether inhaling or exhaling. Hoods do not require critical sealing around the face, and can be worn with beards, long sideburns, and eyeglasses; they are used when air is supplied from an external source through a hose.

Cleaning and maintenance of supplied air respirators is critical to their safe operation. Masks must fit properly and exhalation valves have to be in good working order to prevent any outside air from entering. Keep hoods free of holes or tears. Regularly inspect air hoses, both from self-contained tanks or from air pumps, and replace them if cracked or worn. Keep air pressure regulators clean, dry, and protected from damage.

Barrier Creams

Barrier creams are skin lotions that provide limited, short-term protection against pesticide penetration. Creams containing silicone are used when working with petroleum-based liquids, emulsions, and wettable powders. Barrier creams that do not contain silicone are used for protection against water-soluble pesticides. Barrier creams should never be used in place of other types of protection such as gloves and hoods. Use these lotions on areas of your body that are otherwise difficult to protect, such as parts of the face and neck. If barrier creams are used, make sure that they are applied frequently. Never apply them over pesticide residues because they may seal in the pesticide and increase skin absorption. Barrier creams can be removed from the skin by washing with soap and water or by using a waterless skin cleanser; wash immediately after exposure.

Cleaning and Maintenance

Always keep protective safety equipment in proper working condition. Protective equipment is effective only as long as it is free from pesticide contamination and works properly, therefore frequent cleaning and inspection is required. Replace or repair equipment as soon as you spot a problem.

Respirators

Extend the life of respirators through proper care, regular cleaning, and safe storage. The ability of a respirator to protect you from harmful pesticide dusts, mists, and vapors depends in part on how well you maintain it.

Inspection. Before cleaning your respirator at the end of each day, inspect it for wear and damage. Check the headbands for fraying, tears, or loss of elasticity, and replace them if necessary. Remove filters and, if filter holders are equipped with gaskets, replace them if they are defective. (Never use these types of cartridge respirators without gaskets since gaskets prevent contaminated air from bypassing the filter cartridge). Valve assemblies are essential parts of a cartridge respirator and must be in good working order. Disassemble and inspect valve flaps for wear, deformities, or punctures. Replace parts if you suspect they might leak. Check the threads of all valves and cartridge parts to make sure they are in good condition and not cracked or scratched.

Examine the facepiece for cracks, cuts, scratches, and any signs of aging. If damage is found, defective parts must be replaced.

When replacing items on a respirator, use only approved replacement parts for that specific brand and model. If unapproved parts are used, the respirator will not be in compliance with the law and the respirator may be dangerous to use.

Cleaning. After removing filters and cartridges, soak the respirator, gaskets, and valve parts in a solution of warm water and mild liquid detergent. Do not use abrasives or cleaning compounds containing alcohol or other organic solvents. Anti-germicidal cleaners must be used if the same respirator is worn by more than one worker. Use a soft brush or cloth to remove any pesticide residue (Figure 7-28). Rinse the respirator and valve parts in clean water. Air dry rather than using applied heat.

After it is completely dry, reassemble the respirator and store it in a clean plastic bag to protect it from dirt and environmental deterioration.

Boots and Gloves

Rubber boots and gloves should be rinsed of pesticide residues under running water *before you take them off*. Use a detergent solution and soft brush for washing, then rinse with clean water (Figure 7-29). Do not get the insides of the boots wet. At the end of each day, wash rubber gloves with soap and warm water. Inspect them for holes while washing and discard the gloves if any are found. Gloves may be washed in a washing machine by placing them into a cloth net bag. Use warm water and wash according to the instructions given below for protective clothing. Turn gloves inside out for drying. Store dry boots and gloves in plastic bags to keep them clean and prevent deterioration.

Faceshields and Goggles

Use care when washing faceshields and goggles to prevent scratching the lenses. Submerge them in warm, soapy water and, if necessary, remove pesticide residue with a soft, wet cloth or soft brush (Figure 7-30). Lenses that are treated with anti-fogging materials should not be rubbed, since this reduces their effectiveness. Rinse well with clear water and air dry or blot with a soft cotton cloth; rubbing increases chances of scratching. Inspect goggles and faceshields for excessive scratches and for cracks and loss of elasticity in headbands. Scratched lenses can be replaced on many styles without replacing the entire goggle. Store goggles and faceshields in paper or plastic bags to keep them clean.

FIGURE 7-28.

Remove cartridges and wash respirators in warm, soapy water. As shown here, use a soft brush or cloth to remove pesticide residue.

FIGURE 7-29.

Wash boots before removing them. Use a brush and soapy water, then rinse with water. Do not get the boots wet inside. Let boots air out after washing; store them in a clean plastic bag once they dry.

Protective Clothing

Contaminated protective clothing must not be reworn until it has been washed. Wash contaminated garments at the end of each work day if possible since immediate washing reduces the chances of you or others being exposed to any residues. Discard clothing that has had large quantities of pesticides spilled on it and do so in a site approved for pesticide residues. Burning may be allowed in some locations, but first check with the local agricultural commissioner. Moderately or lightly contaminated clothing can be cleaned by washing.

Change out of contaminated clothing at your work site if possible. Empty pockets and cuffs of garments to remove excess pesticide residue. Place contaminated clothing into a clean plastic bag until it can be laundered; never reuse plastic bags since they may build up pesticide residues. Do not combine contaminated clothing with any other laundry before, during, or after washing.

Minimal Protective Clothing. Minimal protective clothing, such as long-sleeved shirts, full length pants, coveralls, socks, and underwear, should first be soaked in hot, soapy water for at least ½ hour. Use any type of prewash product, such as a solvent soak, prewash spray, or liquid laundry detergent, to improve pesticide removal; add extra amounts to heavily soiled spots. Launder in a standard washing machine, using hot water and liquid laundry detergent (liquid detergent removes oil-based pesticides better than powdered detergent); use the maximum amount recommended in the detergent instructions. Set the washing machine to its longest cycle (at least 12 minutes) and use the highest water level,

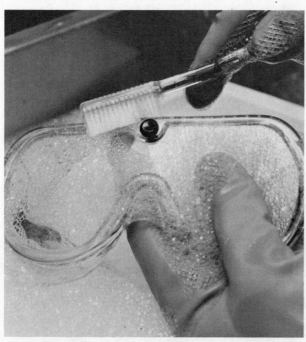

FIGURE 7-30.

Goggles and faceshields should be washed in warm, soapy water. Use a soft brush to remove pesticide residue, but do not rub or wipe the clear plastic lenses. Blot dry and store in a clean plastic bag. Inspect the headband to make sure it is in good condition.

FIGURE 7-31.

Wear gloves when handling pesticide-contaminated clothing. Wash soiled protective clothing in a washing machine, using hot water and the highest water level. Do not combine clothing used for pesticide application with any household or other laundry.

even if you are not washing a full load. Household bleach may be used if necessary, but does not contribute to the decontamination process. See Table 7-6.

Wash pesticide-contaminated clothing separate from all other laundry to prevent transferring residues. Clothing contaminated with different types of pesticide should be washed separately, not combined. Wear gloves or use some other means to prevent skin contact with contaminated clothing while putting it into the washing machine (Figure 7-31). If the garments have a pesticide odor or visible pesticide spots or stains

TABLE 7-6.

Techniques for Washing Pesticide-Contaminated Clothing.

1. Keep pesticide-contaminated clothing separate from all other laundry.

2. Do not handle contaminated clothing with bare hands; wear rubber gloves or shake clothing from plastic bag into washer.

3. Wash only small amounts of clothing at a time. Do not combine clothing contaminated with *different* pesticides—wash these in separate loads.

4. Before washing, presoak clothing:
   a. Soak in tub, automatic washer, or spray garments out of doors with a garden hose.
   b. Use a commercial solvent soak product, or apply prewash spray or liquid laundry detergent to soiled spots.

5. Wash garments in washing machine, using hottest water temperature, full water level, and normal (12 minute) wash cycle. Use maximum recommended amount of *liquid* laundry detergent. Neither bleach or ammonia seem to affect the removal of most pesticides. Never use both.

6. If garments have pesticide odor, visible spots, or stains before washing, rewash one or two more times as in step 5.

7. Clean washing machine before using for other laundry by repeating step 5, using full amount of hot water, normal wash cycle, laundry detergent, *but no clothing*.

8. Hang laundry outdoors on clothesline to avoid contaminating automatic dryer.

Do not attempt to wash heavily contaminated clothing; destroy it by burning or by transporting to an approved disposal site. Follow these suggestions for reducing chances of contaminating the family laundry and endangering family members:

1. Whenever possible, wear disposable protective clothing which can be destroyed after use.

2. Always wear all required protective clothing when working with pesticides.

3. Wear clean protective clothing daily when working with pesticides. Wash contaminated clothing *daily*.

4. Remove contaminated clothing at work site and empty pockets and cuffs. Place clothing in clean plastic bag until it can be laundered. Keep contaminated clothing separated from all other laundry.

5. Remove clothing immediately if it has had a pesticide concentrate spilled on it.

before washing, they should be rewashed 1 or 2 more times in the same manner.

After washing is completed, run the washer through another complete cycle, using hot water and detergent, but without any laundry. This step helps to remove pesticide residues left in the washer and prevents contaminating other loads of laundry.

Whenever possible, hang washed clothing outdoors for drying since the ultraviolet light in sunlight breaks down many pesticides and air drying avoids contaminating the automatic dryer. Clothing can be dried in a clothes dryer if necessary, but it should never be combined with other laundry.

Waterproof Protective Clothing. Remove as much pesticide residue as possible from waterproof clothing by washing with a hose and scrub brush out of doors in an area where the runoff will not cause contamination. Do this before removing these garments, if possible. After removal, store the protective clothing in a clean plastic bag until it can be laundered. To decontaminate the protective clothing, first soak garments in warm, soapy water for ½ hour, then wash in a washing machine, using warm (not hot) water and liquid laundry detergent. Keep these garments separate from all other clothing to prevent contamination. Hang up waterproof clothing to dry; do not put it in a clothes dryer because the heat of the dryer may damage the waterproofing material. If the clothing is being hung in the direct sunlight, turn it inside out to prevent deterioration of the waterproofing material by the sun and to help deactivate any pesticide material remaining on the inside lining.

Storage of Safety Equipment

Never use personal safety equipment for any other purpose. When not in use, keep it stored in a clean, dry place, and protected from temperature extremes and bright light. If possible, store these items in sealable plastic bags. Light, heat, dirt, and air pollutants all contribute to the deterioration of rubber, plastic, and synthetic rubber products. *Never store protective clothing or equipment in an area where pesticides are kept.*

Problems Associated with Protective Equipment

You may experience problems or occasional frustrations with personal safety equipment. Sometimes these problems cause applicators to become careless and stop wearing the required protective equipment. Fortunately, most problems can be overcome by selecting the right type of equipment for the job and weather conditions and by making sure equipment fits properly and is in good working order.

Fitting

Personal safety equipment needs to be sized and fitted to the individual user. Accurate sizing helps to improve comfort by eliminating binding or slipping. Properly fitted respiratory equipment prevents unsafe air leaks. When selecting waterproof pants and jackets, wear the same weight of regular clothing as you would during an actual pesticide application.

If weather conditions are cold, wear a long-sleeved shirt and sweater or coat under the waterproof jacket. Be sure you are comfortable and can move freely, without binding. During hot weather it will be more comfortable to wear lightweight cotton clothing under the protective equipment to provide an absorbent layer and assist in cooling your body.

Discomfort

Discomfort is probably the main reason that some applicators dislike wearing protective equipment. Eye protectors fog up, become covered with spray, and restrict the wearer's range of vision. Protective clothing can be cold during cold weather, and very hot otherwise. Some types are made of heavy, stiff materials that restrict movement. Gloves may impair feeling in the hands and are cold or hot, depending on the weather. Respirators can be uncomfortable to wear if they restrict breathing.

Required protective equipment must be worn at all times during mixing and application. If you become uncomfortable, stop and make adjustments or replace the equipment with another style. Whenever possible, plan pesticide applications during a time of day when the temperature is moderate. Or, anticipating uncomfortable temperatures, dress accordingly. Give yourself an opportunity to take a break during lengthy jobs so that you can get out of the protective equipment for a few minutes. If conditions are too extreme, trade off jobs with a co-worker.

Limits to Protection

Personal safety equipment has limitations to the amount of protection provided so you are never completely protected, even when wearing recommended safety equipment. You still must use caution to prevent pesticides from being spilled, splashed, or sprayed onto your body. The equipment helps to reduce exposure, but you must do everything possible to prevent the exposure from happening. Some pesticides can penetrate through protective materials; pesticide solvents and adjuvants may enhance penetration.

Pesticides confined next to your skin cannot dissipate through air movement or volatilization. Therefore, if you get pesticides on your skin or clothing before putting on protective equipment, the equipment may increase the amount of pesticide absorbed. You will also contaminate the inside of the protective garment. Always wear *clean* protective equipment over *clean* clothing.

FIELD-WORKER SAFETY

In agriculture, field-workers may be working in the vicinity of where a pesticide application needs to be made; field-workers must be protected from pesticide exposure. Workers are not allowed to be in a field, orchard, vineyard, or other area that is being treated with a pesticide. Field-worker safety can be further enhanced if a spray application is made while workers are not present in surrounding areas. Early morning, late afternoon, or during the night are times when it might be possible to apply pesticides without risking field-worker exposure. Preventing drift will also avoid exposure of field-workers in adjacent areas.

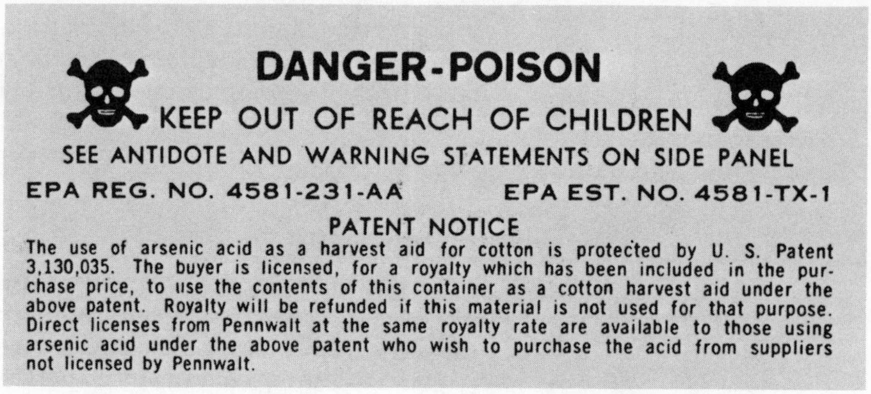

DANGER-POISON
KEEP OUT OF REACH OF CHILDREN
SEE ANTIDOTE AND WARNING STATEMENTS ON SIDE PANEL
EPA REG. NO. 4581-231-AA   EPA EST. NO. 4581-TX-1
PATENT NOTICE
The use of arsenic acid as a harvest aid for cotton is protected by U. S. Patent 3,130,035. The buyer is licensed, for a royalty which has been included in the purchase price, to use the contents of this container as a cotton harvest aid under the above patent. Royalty will be refunded if this material is not used for that purpose. Direct licenses from Pennwalt at the same royalty rate are available to those using arsenic acid under the above patent who wish to purchase the acid from suppliers not licensed by Pennwalt.

FIGURE 7-32.

In California, all Category I pesticides—those having the signal word "Danger" on the label—have a minimum reentry interval of 24 hours. If the label states a longer reentry interval, the longer period must be used.

Notification

Before applying pesticides, contact field-workers or their supervisors to inform them when the application will be made so they will not enter the treated area. Tell them what pesticides will be used, and describe the hazards if they should become exposed. Inform them when the area can be safely reentered.

Reentry Interval

A reentry interval is the period of time that must elapse after a pesticide application before anyone can go back into the treated area. Reentry intervals are given on pesticide labels; however, California regulations are often more restrictive than the label. Pesticide use recommendations written by licensed pest control advisers must indicate the appropriate reentry interval. The local agricultural commissioner is also able to provide this information. In all situations where the reentry interval on the label differs from California requirements, the *longest* reentry interval shall apply. For example, in California all Category I pesticides (signal word "Danger") have a minimum reentry interval of 24 hours (Figure 7-32). A longer period may be indicated on the label, and in these cases, the longer period must be used. For Category II and III pesticides, if no reentry interval is specified, all persons should be kept out of the treated area at least until the spray dries or dust settles.

When two or more pesticides are applied together, a new reentry interval must be calculated. Start by finding out the reentry intervals of each pesticide when applied alone. Next, compute the new reentry interval by adding to the longest period one-half of the next longest period. For example, the reentry interval is 14 days for Pesticide A and 10 days for Pesticide B. When combined, the new reentry interval will be 19 days (14 days plus one-half of 10 days). Two compounds with the same interval of 10 days would have a combined reentry interval of 15 days (10 days plus one-half of 10 days). A compound with a reentry interval of 10 days, combined with another not having a reentry interval, would have a combined interval of 10 days. Table 7-7 can be used to look up reentry intervals for combinations of two (or more) pesticides. If more than two pesticides are combined, use the two *longest* reentry intervals.

TABLE 7-7.

How to Determine the Reentry Interval
When Two (or More) Pesticides Are Combined.

PESTICIDE A

PESTICIDE B	1	2	3	4	5	6	7	8	9	10	11	12	13	14	15	16	17	18	19	20	21
0	1	2	3	4	5	6	7	8	9	10	11	12	13	14	15	16	17	18	19	20	21
1	2	3	4	5	6	7	8	9	10	11	12	13	14	15	16	17	18	19	20	21	22
2		3	4	5	6	7	8	9	10	11	12	13	14	15	16	17	18	19	20	21	22
3			5	6	7	8	9	10	11	12	13	14	15	16	17	18	19	20	21	22	23
4				6	7	8	9	10	11	12	13	14	15	16	17	18	19	20	21	22	23
5					8	9	10	11	12	13	14	15	16	17	18	19	20	21	22	23	24
6						9	10	11	12	13	14	15	16	17	18	19	20	21	22	23	24
7							11	12	13	14	15	16	17	18	19	20	21	22	23	24	25
8								12	13	14	15	16	17	18	19	20	21	22	23	24	25
9									14	15	16	17	18	19	20	21	22	23	24	25	26
10										15	16	17	18	19	20	21	22	23	24	25	26
11											17	18	19	20	21	22	23	24	25	26	27
12												18	19	20	21	22	23	24	25	26	27
13													19	21	22	23	24	25	26	27	28
14														21	22	23	24	25	26	27	28
15															23	24	25	26	27	28	29
16																24	25	26	27	28	29
17																	26	27	28	29	30
18																		27	28	29	30
19																			29	30	31
20																				30	31
21																					32

Locate reentry interval for first pesticide along top row. Locate reentry interval for second pesticide along left column. The new reentry interval in days is found where these intersect.

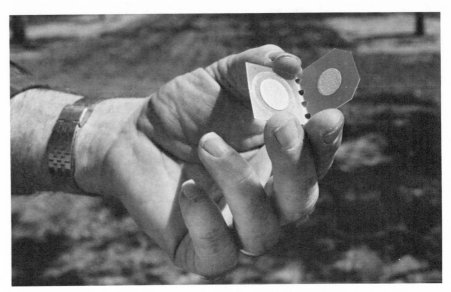

FIGURE 7-33.

Simple test kits like the one being used here are available to detect some types of pesticide residues. These can be used to monitor treated areas after reentry intervals have expired to determine if workers can be allowed to enter the area.

FIGURE 7-34.

Whenever required by the pesticide label, or by state or local regulations, post a treated area with signs to warn the public of the danger. In agricultural and other open areas, signs should be no more than 600 feet apart along usual points of entry and along unfenced areas next to roads and other public rights-of-way.

Simple, quick-acting test kits are available to detect residues of some types of pesticides in a treated area. In especially hazardous situations, consider using these before entering an area *after* the expiration of the reentry interval to be certain toxic levels have dissipated (Figure 7-33).

Posting

Areas treated with pesticides having a required reentry interval must sometimes be posted with warning signs (Figure 7-34). Regulations require these signs to be made of a durable material and printed in English and Spanish. They must contain the word "Danger" in letters large enough to be read from a distance of 25 feet. If the reentry interval is greater than seven days, the sign must also name the pesticide that was used and the date that the area may be reentered. Check pesticide labels and current federal, state, and local laws to determine requirements for posting. Local offices of county agricultural commissioners have this information.

To post a treated area, place signs at usual points of entry and along unfenced areas next to roads and other public rights-of-way. Signs should be no more than 600 feet apart. They should be posted before an application is made (but no sooner than 24 hours before the application). They have to remain in place throughout the reentry interval, and must be removed within 5 days after the end of the reentry interval and *before* workers are allowed to enter the field.

PUBLIC AND ENVIRONMENTAL SAFETY

Prevent the public from accidentally becoming exposed to pesticides during an application or from coming in contact with treated areas after the application is made. Notify people in the area where pesticides are to be used; explain the potential hazards and possible poisoning symptoms, what you are doing to reduce these hazards, and what they should do to avoid exposure. Try to make pesticide applications at times when people are not present. Prevent drift of pesticide materials out of the treatment area, and keep people and animals away.

Check pesticide labels and material safety data sheets for environmental hazards such as dangers to wildlife, endangered species, honey bees, or groundwater. Become familiar with wildlife in the area; if necessary, contact the California Department of Fish and Game for information and help (Chapter 6). Obtain specific information on groundwater conditions at the application site by consulting with the pesticide coordinator at the local agricultural commissioner's office. Use selective pesticides and make pesticide applications at times when honey bees and other beneficial insects are not present in the treated area. Do not allow sprays to drift into waterways because many pesticides are toxic to fish or other aquatic life and sprays may contaminate surface and groundwater resources.

THE SAFE HANDLING OF PESTICIDES

Undiluted pesticides are a greater risk to people and the environment than diluted spray mixtures. The safe handling and transporting of undiluted pesticides can prevent many environmental and human health hazards. In addition to toxic hazards, pesticides have a high dollar value and may be subject to theft; security measures must always be taken to prevent such problems.

Manufacturer's Packaging

Pesticides are packaged by manufacturers in several ways, depending on the formulation. The type of packaging and formulation affects the hazards of handling, transporting, and storage. Pesticides are available in paper bags, plastic bags, water-soluble bags, plastic bottles, glass bottles, and metal containers (Figure 7-35). Most packages are in convenient units of size for ease in measuring and mixing; quantities of these units are usually packed in larger, cardboard boxes for shipping and handling. Shipping containers must be approved by the U.S. Department of Transportation.

Paper and plastic bags are common packages for powder and granule formulations. Water-soluble plastic bags are used for some highly toxic or otherwise hazardous powders. However, paper or plastic containers may tear or puncture easily if they are not properly handled. Opened paper and plastic bags may be difficult to reseal, presenting possible future problems with leaking. Spilled powders can be easily scattered—they are also difficult to clean up. To prevent tearing and reduce the danger

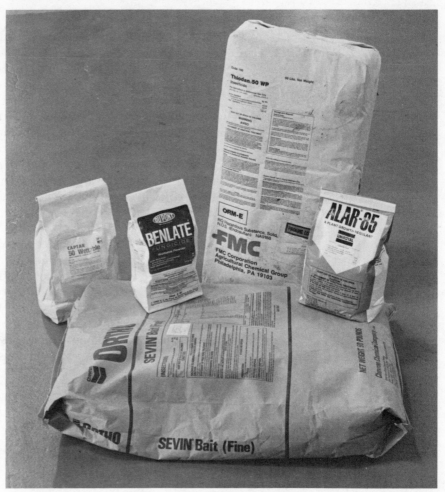

FIGURE 7-35.

Pesticides are packaged in a number of different types of containers. Often the container type is dictated by the chemical characteristics or formulation of the pesticide. Most packaging is in convenient units of size for ease in measuring and mixing.

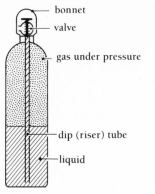

FIGURE 7-36.

Gas fumigants are often supplied in large steel cylinders like the one illustrated here. These are capable of withstanding high pressures of the gases they contain. The cylinder is equipped with a valve which dispenses gas through a hose. During transporting and storage, the valve is protected by a steel screw on cap, known as a bonnet.

FIGURE 7-37.

Pesticides should only be transported in the back of a truck. Containers must be secured in the cargo area and protected from moisture and damage. Do not allow children, adults, or animals to ride in the cargo area where pesticides are carried. Never transport food, animal feed, or clothing in the same compartment.

of spilling, cut bags open with a scissors or sharp knife; bags will also be easier to close after use.

Plastic bottles and pails are used for both liquid and granular formulations. If punctures occur from improper handling, there will be a pesticide spill. Uncapped opened containers are also subject to spilling if they are tipped. Spilled liquids are difficult to contain or clean up because liquids soak into wood, cloth, paper, and nearly everything else they contact. Granular formulations in plastic containers cause less problems if spilled because granules can be easily contained and will not soak into porous surfaces.

Glass bottles are used for some liquid formulations. Colored glass provides protection to liquids that are degraded by light. Glass resists puncturing and bottles can be resealed, but these containers may break if not properly handled.

Metal containers are used for liquid and dust aerosols and many liquid formulations, although some corrosive chemicals cannot be packaged in metal. Reuseable steel cylinders (Figure 7-36) are used for fumigants, usually liquids under high pressure. Metal containers probably are the safest packaging materials for pesticides because they resist puncturing and do not break. They are resealable and empty containers can either be recycled or disposed of in approved disposal sites. However, special precautions should be taken with metal aerosol containers, which may explode or cause injury if punctured, overheated, or burned.

Service Containers

Service containers are any container, other than the original labeled packaging, that is used to hold pesticides. They are usually specially constructed metal or plastic containers designed to apply, hold, store, or transport pesticide concentrates or use-diluted preparations; when being transported, service containers must be labeled with the common name of the pesticide they contain, the signal word, and the name and address of the person responsible for the container.

Transporting

Many pesticides are subject to state and federal hazardous materials transportation requirements including: (1) incident reporting; (2) packaging—including container maintenance and retesting; (3) labeling; (4) marking; (5) placarding; (6) emergency actions in the event of an accident; (7) loading; (8) vehicle safety equipment; and (9) routing. There is always a risk of an accident while transporting undiluted or diluted pesticides in a vehicle; serious exposure and damage may occur if these materials are spilled. Some pesticides are flammable, adding dangers of fire and toxic fumes. Once pesticides are spilled on public roads, they can be blown or splashed onto people, animals, residential areas, or nearby crops, and can be scattered by passing vehicles. Spilled chemicals may wash into ditches, streams, and rivers during rainstorms, creating the potential for serious damage, including groundwater contamination. Spilled pesticides may also contaminate the vehicle, its occupants, or other cargo; it may be impossible to completely remove all residue.

Never carry pesticides in the passenger compartment of any vehicle; pesticides are most safely transported in the back of a truck (Figure 7-37). All pesticide containers should be secured in the cargo area and protected from rain and damage. Pesticide containers must never be *stacked* higher than the sides of the transporting vehicle. Fumigant cylin-

FIGURE 7-38.

In many instances, when transporting pesticides that are classified as hazardous materials, specific placards such as these must be displayed on all four sides of the vehicle.

ders have to be secured in the vehicle in an upright position and be equipped with a screw-on steel bonnet to protect the valve mechanism. Never allow children, adults, or animals to ride in the area where pesticides are carried, and never transport food, animal feed, or clothing in the same compartment. Do not leave pesticides unattended in a vehicle unless they are inside a locked compartment. Be certain of what pesticides you are transporting so this information can be given to emergency workers in case of an accident.

Whenever pesticides are to be transported on public highways, contact the California Highway Patrol, the California Department of Transportation (CalTrans), and the California Public Utilities Commission for current regulations, to determine if you need special licenses or permits, and to determine which materials are subject to hazardous materials transportation requirements. Your vehicle may require placards signifying the type of hazardous material being carried (Figure 7-38). If placards are required, they must be on all four sides of the vehicle. Always remove placards from vehicles when the hazardous material is not being transported. Table 7-8 lists where to get information and regulations on transporting pesticides. Some vehicles are required to be equipped with devices to protect the pesticide cargo in the case of a rear-end collision. Fumigation cylinders must be carried in such a manner to prevent their rupture and to keep them contained within the vehicle in the event of an accident.

Should you have an accident involving spilled pesticides, alert the highway patrol, county sheriff, city police, or local fire department at once. Keep people and vehicles away. Never leave the scene of a spill until responsible help arrives. For advice on cleaning up spills, you or one of the emergency workers should contact CHEMTREC (Chemical Transportation Emergency Center) at 800-424-9300. *Call this number only in the case of an actual emergency*. CalTrans (on state highways) and designated city or county agencies, including some fire departments, provide or contract for clean-up of highways within their jurisdiction when the person or company transporting the material cannot quickly and safely clean up a spill.

Handling

Bulk pesticides should be handled carefully. Do not drop or throw containers or packages because this may cause damage and leaks. Check for contamination or leaks on all packages being handled and do not let damaged packages or spilled pesticide come in contact with your skin or clothing. Wear rubber gloves and protective clothing, such as an apron, when handling pesticide packages. If a leak is present, you may also need respiratory and eye protection. Check the label for all precautions and required safety equipment. Never walk through a spilled pesticide. If a container is damaged and leaking, the pesticide should be transferred to another container designed for that pesticide or to a safe container that can be properly identified (Chapter 8).

Prevent theft or danger to children and animals by not leaving pesticide containers unattended or stored in unlocked areas (Figure 7-39). Always keep pesticides away from food and water and away from sources of heat and fire. Never allow paper containers to get wet.

Do not eat, drink, or smoke while handling pesticides and pesticide containers. Wash thoroughly when finished and before eating, drinking, smoking, or using the bathroom.

TABLE 7-8.

Where to Get Information and Regulations on Transporting Pesticides.

FOR INTERSTATE MOVEMENT OF PESTICIDES:

U.S. Department of Transportation   or   U.S. Department of Transportation
Office of Motor Carrier Safety   San Francisco Regional Office
Federal Building   (415) 974-9888
801 I Street
Sacramento, CA 95814-2507
(916) 551-1300

FOR TRANSPORTATION OF PESTICIDES
WITHIN THE STATE OF CALIFORNIA:

California Department of Transportation (CalTrans)
Hazardous Materials Division
1120 N Street, Room 3200
Sacramento, CA 95814
(916) 322-8226

California Highway Patrol
Motor Carrier Section
2555 First Avenue
P.O. Box 898
Sacramento, CA 95804
(916) 445-6211

... contact the California Highway Patrol Motor Carrier Safety Unit Supervisor
at one of the locations listed below:

Northern Area:	250l Cascade Boulevard Redding, California 96003 (916) 225-2715
Sacramento Valley:	11336 Trade Center Drive P.O. Box 8001 Rancho Cordova, CA 95670-8001 (916) 366-5185
Golden Gate:	3601 Telegraph Avenue Oakland, CA 94609 (415) 658-2928
Central Valley:	5179 North Gates Avenue Fresno, CA 93711 (209) 488-4329
Southern State:	437 North Vermont Avenue Los Angeles, CA 90004 (213) 736-2996
Border:	3703 Camino Del Rio South P.O. Box 20522 San Diego, CA 92120 (619) 281-8121
Coastal Area:	1122 Laurel Lane San Luis Obispo, CA 93401 (805) 549-3261
Inland Area:	505 N. Arrowhead, Suite 506 P.O. Box 1029 San Bernardino, CA 92402 (714) 383-4747

FOR LOCAL INFORMATION: County Agricultural Commissioner's Office

Storage

Store pesticides in their original, tightly closed containers. Whenever possible, wipe or wash pesticide residue off outsides of containers. Protect pesticides from extremes in temperature and from becoming wet. A pesticide storage area should be a separate building, away from people, living areas, food, animal feed, and animals. The area must be well ventilated, well lighted, dry, and secure, with lockable doors and windows. Post signs near all primary entrances to warn others that the building contains pesticides (Figure 7-40).

Keep a record of all pesticides being stored. This record should indicate the date of purchase and the date that each chemical was placed in the building. Keep this record, or a copy of it, separate from the building so, in case of fire, you can inform emergency workers what toxic or hazardous materials are inside, possibly preventing serious injury. Before an emergency arises, contact the local fire protection agency and provide them with material safety data sheets and an inventory of the pesticides being stored. Update these records periodically and whenever additional types of pesticides are placed in the storage area.

Check stored pesticides on a regular basis to be sure that the containers are in good condition and no leaking or spilling has occurred. Any spilled pesticide in the storage area must be cleaned up immediately (see clean-up procedures in Chapter 8).

Some pesticides do not store well for long periods of time. Such materials are said to have a short *shelf-life*. Extended storage, especially after temperature extremes, may cause chemical changes resulting in some products losing their effectiveness or others becoming more toxic. Moisture and air picked up during storage may alter the composition of some pesticides, especially those stored in unsealed containers. Solvents and petroleum-based chemicals can deteriorate some types of containers after a period of time. Keep packages on shelves or on pallets to reduce exposure to excess moisture. Most chemicals should not be stored for longer than two years. Before pesticides exceed their shelf-life, use them in an appropriate application or transport them to an approved disposal site.

Certain pesticides, such as 2,4-D and other hormonal herbicides, must never be stored with other pesticides. Vapors from these herbicides may combine with other pesticides, causing problems of contamination and resulting in potential damage to treated plants. Always check the label for special storage precautions to prevent such problems.

Liquids expand when heated or frozen. Liquid pesticides in sealed containers may expand enough to rupture the containers under extremely hot or cold conditions. Proper ventilation of the storage area is important to prevent overheating. If pesticides are stored in locations subject to winter freezing, the storage building must be well insulated.

MIXING PESTICIDES

Techniques for mixing pesticides are the same for large and small volumes—the proper amount of pesticide must be thoroughly incorporated into a measured amount of water or other solvent. Before beginning, read the mixing directions on labels of all pesticides you will be using and decide on the proper order that chemicals should be added to the

FIGURE 7-39.

Pesticides must be in a lockable area of the vehicle, as shown here, to prevent unauthorized access while the vehicle is unattended.

FIGURE 7-40.

Pesticides must be stored in a separate building such as the one shown here, away from people, living areas, food, animal feed, and animals. The storage area should be well ventilated, well lighted, dry, and secure. Doors and windows must be securely locked. Signs must be posted at all primary entrances warning that the building contains pesticides.

spray tank (Chapter 4). If adjuvants are needed, these are usually added before pesticides unless label instructions give a different order.

Determine what protective clothing will be needed for mixing and application. Check the spray equipment to be sure there are no cracked hoses or leaks, and that the filters, screens, and nozzles are clean. Arrange to have an emergency supply of fresh water nearby for washing in case of an accident.

The water used for filling a spray tank should look clean enough to drink and be free of sand, dirt, algae, or other foreign matter. Sand or dirt causes excessive wear on pumps and nozzles and clogs filters, screens, and nozzles. Algae may clog filters and nozzles and can react with some pesticides to reduce their effectiveness. Smell the water to see if you can detect any chemical odor. Chemicals may react unfavorably with some pesticides. For example, chlorine (used in domestic water supplies for control of bacteria) combines with some pesticides and reduces their effectiveness. High levels of dissolved salts causes deactivation of pesticides and may even damage treated foliage. If possible, do a simple pH test as described in Chapter 4. High pH (alkaline water) causes hydrolysis, or breakdown, of many pesticides before they are sprayed onto the target surface. Use a buffer or acidifier if the pH is too high. When you have any doubts about the water quality, locate another source.

Measure pesticides carefully, accurately, and safely. Inaccurate measuring can produce large errors in the amount of pesticide being applied. Newer herbicides, the sulfonylureas for example, are applied at rates of ⅙ to ¾ ounce of formulated material per acre; small inaccuracies in measuring can produce gross errors in application rates.

FIGURE 7-41.

This person is using a closed mixing systems for mixing a liquid Category I pesticide. Closed mixing systems enable accurate measuring of pesticides and most systems enable rinsing of empty containers.

You are required to use a closed mixing system when mixing liquid Category I pesticides. Closed mixing systems enable you to accurately and safely measure the amount of pesticide being put into the spray tank (Figure 7-41). (The closed-system requirement does not apply if you handle one gallon or less of Category I pesticide per day and the liquid pesticide is in an original container of one gallon or smaller). Dry formulations and Category II and III liquids do not require a closed mixing system unless they are being mixed by a person under 18 years of age.

Liquids and some granular pesticides are measured by volume, while dusts, powders, and most dry formulations are measured by weight. Pesticide labels use the English system of measurement; liquid volumes are in fluid ounces, pints, quarts, and gallons; dry weights are in pounds and ounces. An assortment of glass or plastic measuring utensils, from one cup to one gallon, is essential for accurately measuring liquids unless a closed mixing system is used. Some pesticides react with metal, especially aluminum and iron, so avoid using metal measuring utensils. Use an eye-dropper to measure small quantities of liquid. An accurate scale and a set of measuring cups and spoons are needed for measuring and weighing dry pesticides (Figure 7-42). Some measuring equipment can be mistaken for kitchen utensils, so identify it in a very obvious manner such as painting handles with brightly colored waterproof paint or attaching waterproof labels. When not being used, keep all measuring and weighing equipment locked in the pesticide storage area so it cannot be used for other purposes. Clean and wash utensils before they are stored to prevent contaminating future mixtures.

Pesticide packages are available in different units of weights or vol-

FIGURE 7-42.

Measuring and weighing pesticides requires a variety of calibrated utensils and an accurate scale.

umes. Whenever possible, plan a mixture that uses an even, preweighed amount of pesticide. Unit cost may be greater when pesticides are bought in smaller packages, but this disadvantage can often be minimal compared to the convenience and added safety of not having to weigh or measure. Do not open pesticides packaged in water-soluble packets since these contain highly hazardous formulations; calibrate application equipment to utilize the entire packet or a number of whole packets.

Select a mixing location that can be cleaned easily should an accident occur. When not using premeasured packets, measure and weigh chemicals in a clear, open area. If outdoors, stand upwind to reduce chances of exposure. Wear an approved dust/mist respirator or cartridge respirator while weighing and mixing dry pesticides to prevent inhaling dust. Protect your hands and clothing with appropriate outerwear. Liquids are easily spilled and splashed, so wear suitable gloves and a rubber apron, or waterproof pants and jacket. Refer to the pesticide label for specific protective clothing and equipment for mixing and loading pesticides. A faceshield, goggles, or other protective eyewear must be worn, however, even if the requirement is not on the pesticide label (Figure 7-43). Reduce chances of spills or splashes into your face and eyes by always measuring and pouring pesticides *below* eye level.

FIGURE 7-43.

Wear protective clothing while measuring pesticides to prevent skin or eye exposure in case of an accidental spill or splash. Always pour and measure chemicals below eye level. If measuring outdoors, stand upwind.

Begin mixing by filling the spray tank at least half full with clean water. To allow room for the pesticide, adjuvants, and residues from triple rinsing of containers, avoid filling the tank more than ¾ full. Check and adjust the pH of the water in the spray tank at this time. Start agitators if the equipment has them.

Open pesticide containers carefully to prevent spilling and to make resealing easier. Cut paper containers open with a sharp knife or scissors, rather than by tearing. Metal containers, glass and plastic bottles, and plastic pails all have protective seals that must be broken before use. Most of these containers can be easily resealed with screw caps.

After measuring or weighing the correct amount of pesticide, carefully pour it into the partially filled spray tank (Figure 7-44). Rinse the measuring container and pour the rinse solution into the spray tank also. Use caution while rinsing to prevent splashing. Many closed mixing systems are equipped with container rinsing devices that pump the rinse solution into the pesticide tank. Unless rinsed automatically, liquid containers should be drained into the spray tank for 30 seconds after they have been emptied, then rinsed and drained three more times (triple rinsed). After each draining, fill the container about ¼ full of water, put the cap back on, and shake for several seconds to mix the residue with water. *Pour each rinse solution into the spray tank.* Table 7-9 illustrates how much pesticide is removed from the container by triple rinsing. Containers that have been triple rinsed do not have to be transported to a Class 1 disposal site, but can be recycled or transported to a Class 2 disposal site.

Once pesticide has been added, fill the spray tank to its final volume.

FIGURE 7-44.

Carefully pour pesticides into the spray tank. Rinse measuring containers and empty and triple rinse liquid pesticide containers. Always pour the rinse solutions into the spray tank.

TABLE 7-9.

Triple Rinsing Pesticide Containers.

PROCEDURE:

1. When container is empty, let it drain into spray or mixing tank for at least 30 seconds.

2. Add correct amount of water to container as follows:

CONTAINER SIZE	RINSE SOLUTION NEEDED
Less than 1 gallon	¼ of the container volume
1 gallon	1 quart
5 gallons	1 gallon
30 to 55 gallons	5 gallons

3. Close container.

4. Shake container or roll to get solution on all interior surfaces.

5. Drain container into sprayer or mixing tank. After empty, let drain for an additional 30 seconds.

6. Repeat steps 2 through 5 *two* additional times.

AMOUNT OF ACTIVE INGREDIENT REMOVED FROM
A 5 GALLON CONTAINER BY TRIPLE RINSING:

	Amount of Active Ingredient Remaining*
Drain	14.1875 g a.i.
1st Rinse	0.2183 g a.i.
2nd Rinse	0.0034 g a.i.
3rd Rinse	0.00005 g a.i.

*After draining, a 5 gallon container is assumed to still contain 1 ounce of formulated pesticide. This would amount to 14.1875 g of a.i. if the formulation contained 4 pounds a.i. per gallon.

Do not allow the tank to overflow during filling, and do not let the hose, pipe, or other filling device come in contact with liquid in the tank. If the tank is being filled through a top opening, there must be an air space, equal to at least twice the diameter of the filling pipe, between the liquid in the spray tank and the filling device. This gap will prevent siphoning of the spray mixture back into the water supply after the water flow is stopped (Figure 7-45). Side or bottom filling systems require the use of check valves to prevent back-flow of pesticides in the spray tank into the water supply. Recheck the pH of the tank mixture and adjust if necessary.

APPLYING PESTICIDES EFFECTIVELY

To use pesticides safely and effectively, make sure that all the pesticide is applied to the treatment area and in the proper amount. Pesticide coverage often must be uniform; for example, both sides of leaf surfaces need be coated with pesticide droplets for adequate control of some plant-

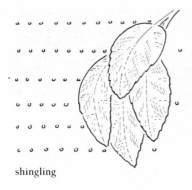

shingling

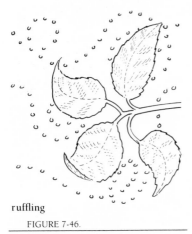

ruffling

FIGURE 7-46.

Spray that is improperly aimed at foliage may cause leaves to stick together and prevent proper coverage—a condition known as shingling. Use of an air blast sprayer or oscillating boom sprayer will cause ruffling of the foliage and improve distribution of spray droplets.

feeding pests. Sprays improperly aimed at the foliage may cause *shingling*, a condition in which leaves clump together and prevent droplets from reaching some leaf surfaces (Figure 7-46). Using an air blast sprayer or oscillating boom sprayer produces ruffling of the plant foliage and enables spray droplets to contact all surfaces.

Spills, leaks, and drift waste material and may leave pesticide in non-target areas. Improper equipment calibration results in too little or too much pesticide reaching the target site. Safe pesticide applications require using proper equipment, developing good application techniques, reducing or eliminating drift, and having an awareness for all potential hazards.

Selecting Application Equipment

The equipment you use to apply pesticides must be suited to the location and conditions of the treatment area. Equipment that is too big or too powerful may be as much of a problem as equipment that is too small. Most pesticide application equipment is designed to work efficiently only in a limited number of situations. Some conditions require that the spray be moved to target surfaces with a blast of air to improve coverage.

Choose application equipment that is easy to use and comfortable to work with. Calibration must be simple or it will never be accurate (Chapter 11 describes how to calibrate application equipment.) Equipment should be easy to repair, and parts readily available. Hand-held equipment must be lightweight so that it is convenient to use. Motor powered units should be quiet enough to prevent operator stress, yet powerful enough to do the job properly. Moving parts need shields and guards to prevent accidents and injuries. Powered equipment must have accurate gauges to enable the operator to monitor spray pressure and other functions.

Pesticide application equipment has to be durable because commercially used machinery is subjected to long hours of operation. This equipment is frequently transported on public roads and used in areas

FIGURE 7-45.

When filling a spray tank from the top, be sure there is an air gap between the filler pipe and the top level of the water in the tank. This will prevent back flow of pesticide contaminated water into the water supply.

FIGURE 7-47.

Spray tank covers must fit well to prevent pesticide mixtures from splashing out during operation or while transporting the equipment.

where there may be high risks to the public and the environment. Make sure that filler covers on spray tanks close properly, seal well, and are lockable (Figure 7-47). Hoses and fittings should be strong and durable to prevent loss of pesticide mixtures and environmental contamination. Leaks or ruptures may cause serious exposure if you are handling hoses or fittings while making an application.

Safe Application Techniques

Safe application techniques require working with the weather, controlling droplet size and deposition, having an awareness of the application site and its hazards, developing special application patterns for the site to accommodate hazards and environmental conditions, and leaving buffer zones to protect sensitive areas.

Working With the Weather. Weather influences pesticide applications in outdoor areas; its influence on pesticide applications in greenhouses, buildings, and other confined spaces is more subtle, although possible effects still must be considered. Temperature affects the phytotoxicity of certain pesticides, so label directions will usually warn against using these products when temperatures are above or below critical limits. Pesticide degradation and volatilization rates are accelerated by high temperatures. Warm temperatures are associated with clear, sunny weather, but many pesticides are broken down rapidly by ultraviolet light which is most intense during these times.

Air temperature is responsible for the inversion phenomenon which may often cause pesticide drift. Inversions occur when the air 20 to 100

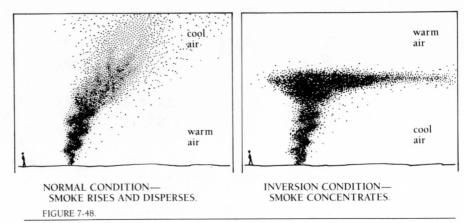

NORMAL CONDITION—
SMOKE RISES AND DISPERSES.

INVERSION CONDITION—
SMOKE CONCENTRATES.

FIGURE 7-48.

This drawing illustrates a temperature inversion. A temperature inversion is a layer of warm air above cooler air close to the ground. This warm air prevents air near the ground from rising, similar to a lid.

or more feet above the ground is warmer than the air below it. The warm air forms a cap that blocks vertical air movement. To detect a temperature inversion, observe a column of smoke rising into the air. (Black smoke from a burning tire or diesel fuel is easy to see, but check with local air quality authorities to be sure such burning is permissible.) If the smoke begins moving sideways or collects in one area a few hundred feet above the ground, an inversion condition probably exists (Figure 7-48). Inversion conditions are dangerous during a pesticide application because fine spray droplets and pesticide vapor can become trapped and concentrated, similar to the smoke column. Rather than dispersing, the pesticide often moves as a concentrated cloud away from the treatment site.

Honey bees only forage during certain temperature ranges; therefore, make applications when temperatures are not suitable for bee activity if you are using pesticides that might injure bees.

Rainfall, fog, and even heavy dew affect pesticide applications because the moisture dilutes and degrades pesticides and may wash the material off treated surfaces. Rainwater washes pesticides into the soil, producing possible groundwater and surface water contamination. Pesticides can be carried away from the application site through water movement after heavy rains.

Wind influences pesticide drift and also has an effect on volatilization. Strong air movements are responsible for uneven pesticide deposition, although some air movement has advantages in getting good coverage of treated surfaces.

Controlling Droplet Size and Deposition. Spray droplet deposition is influenced by droplet size, the pressure of the spray stream, the force and volume of the air, if any, used to distribute spray, and the speed of travel of the application equipment. Droplet size is a factor of nozzle size, style, and condition combined with spray volume, spray pressure, and weather influences. Most application equipment emits a spray with a wide range of droplet sizes, even though the best spray applications result from applying uniform-sized droplets evenly to all treated surfaces. You can increase the uniformity of spray droplets by selecting nozzles designed for the working pressure and volume of your application equipment and replacing worn or defective nozzles. The type of application equipment used must be suitable for the physical and environmental

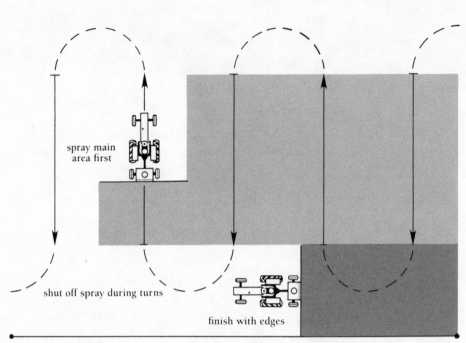

spray main
area first

shut off spray during turns

finish with edges

FIGURE 7-49.

Shut off spray nozzles during turns to avoid uneven applications. After spraying the main part of the treatment area, finish by spraying the edges.

FIGURE 7-50.

A simple bypass system like the one illustrated here allows the operator to spot clogged injectors and prevent uneven application of pesticides beneath the soil surface. When an injector clogs, the pesticide will begin filling one of the corresponding glass bottles.

features of the location. Application speed is critical and should be adjusted to the type and size area being treated. Slower speeds are usually required for large trees and shrubs because spray droplets must travel farther and more volume is required to cover larger surface areas.

Site Characteristics and Environmental Hazards. Before beginning a pesticide application, the physical characteristics of the terrain, greenhouse, building, or other application site must be observed and all potential hazards recognized. Check for organisms or structures that might be damaged by pesticides or water and other materials in the spray mixture or by the physical movement of the equipment and operator through the area. Ditches, embankments, steep slopes, electrical wires, electric fences, and, in confined areas, electrical outlets, motors, sources of sparks, and improperly ventilated spaces can also pose hazards to operators or equipment.

Application Pattern. An application pattern is the route the applicator follows while applying a pesticide. The purpose of any application pattern is to provide an even distribution of pesticide over the treated area and to avoid overlaps or gaps. Pesticide application speed usually determines the uniformity of the application pattern. At higher speeds, the equipment bounces more; with airblast sprayers, the volume of displaced air is reduced as travel speed increases. The pattern used for pesticide application must take into consideration prevailing weather conditions, what is being sprayed, and hazards in or near the application site. Patterns should also be designed to eliminate the need for the operator to travel through airborne spray or walk or drive through freshly treated areas. Operating

SENSITIVE AREA

BUFFER AREA
(DO NOT SPRAY)

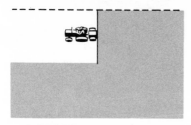

FIGURE 7-51.

Leave untreated buffer areas when an application site adjoins locations where organisms, people, or structures might be harmed by pesticide exposure. The buffer should be no less than the width of one spray swath.

equipment (such as boom applicators) during turns produces an uneven application. Figure 7-49 illustrates how to make a uniform application by shutting the sprayer off during turns. Watch for clogged nozzles that will also produce uneven applications. It is difficult to see clogging when using soil injection equipment; bypass devices similar to those illustrated in Figure 7-50 can be used to spot clogged injectors.

Buffer strips should be left untreated when a treatment area adjoins locations where organisms, people, or structures might be harmed by pesticide exposure (Figure 7-51). The size of the buffer strip depends on the type of application equipment being used, prevailing weather conditions, the nature of pesticide being applied, the type of pest problem being treated, and the sensitive nature of adjoining areas. As a general rule, the buffer should be no less than the width of one spray swath.

Pesticide Drift

Pesticide drift refers to the movement of pesticides away from the treatment site. Drift is most serious when applications are made during windy conditions, especially while using high pressure and small nozzle sizes. Spray droplets intended for a specific treatment area may be carried away through the air to other locations (Figure 7-52). Another form of drift occurs when sprayed pesticides partially evaporate before reaching the target. The resulting vapor can drift away from the treatment area through air movement, often traveling several miles.

Table 7-10 lists some of the factors that influence drift; there are many

TABLE 7-10.

Factors Influencing Pesticide Drift.

PESTICIDE:

Volatility of active ingredient.
Solvent used to dissolve or suspend active ingredient (formulation).
Solvent used to dilute pesticide in spray tank.

ADJUVANTS:

Deposition aids, thickeners, and stickers
(reduce drift by making droplets larger or less volatile).

APPLICATION EQUIPMENT:

Operating pressure of spraying system.
Nozzle size.
Distance from nozzles to target surface.
Height from which spray is released.
Speed of travel of application equipment.
Application pattern and technique.

TARGET SURFACES:

Size of target area.
Location of target area.
Nature of target surfaces.

WEATHER CONDITIONS:

Wind intensity.
Wind direction.
Air temperature.
Humidity.

FIGURE 7-52.

Spray drift is caused when pesticides are applied during windy periods. Drift is also the result of small droplet size, caused by high pressure and small nozzle orifices.

steps that you can take to reduce this problem. Eliminating very small droplets can significantly reduce drift; droplet size is increased by using larger nozzles and by lowering the output pressure of the sprayer. Adjuvants, called deposition aids, that assist in increasing spray droplet size or reducing evaporation potential can also be added to the spray tank when drift must be avoided. Do not spray during windy conditions. Usually, winds less than 5 miles per hour help provide good pesticide distribution in trees and leafy plants, especially if a blower is not used with the application equipment. Stronger winds, however, will increase drift potential. In some cases, spraying may be illegal if the wind speed is over a designated rate. Contact the local agricultural commissioner for information on pesticide application restrictions during windy conditions. Other weather conditions, such as a temperature inversion, also promote drift of small droplets and vapors. High temperatures and low humidity increase the evaporation rate which will reduce the size of droplets before they reach their target. The resulting smaller droplets may be highly subject to drift.

Special Hazards in Treatment Areas

Pesticides containing petroleum-based carriers may be flammable and must never be used in areas where open flames or other ignition sources are present. Gas-fired water heaters and electric motors may ignite flammable pesticides. When applying pesticides in areas where such hazards exist, use either a nonflammable water-based spray or shut off all ignition sources.

The spray from water-based pesticides is usually capable of conducting electricity, which may result in a potentially fatal electric shock if the spray should come in contact with an electrical source. Never direct any spray onto power transmission lines, electrical cords, outlets, or motors or appliances. Disconnect motors and appliances and shut off electricity in areas where pesticide application is taking place to prevent chances of electrocution.

CLEAN-UP AND DISPOSAL

Disposal of Surplus Diluted Pesticide. To avoid problems associated with leftover pesticide mixtures, calculate the exact size of the treatment area and mix only enough pesticide for the job. If you have some leftover spray mixture, find another appropriate location where it can be used as it was intended, otherwise it must be transported to a Class 1 disposal site. There are private companies that specialize in collecting and removing pesticide wastes, including unused spray mixtures, to Class 1 disposal sites. Excess pesticide must never be indiscriminately dumped; such dumping is a potential source of environmental and groundwater contamination and is an illegal practice. Persons convicted of dumping are subject to large fines and jail terms.

Pesticide Container Disposal. Regulations concerning the disposal of pesticide containers vary from county to county. Specific disposal information can be obtained from local Water Quality Control Boards, the Department of Health Services, and local agricultural commissioners.

Generally, containers made of paper, and some plastics, can be burned if the burning takes place in an area where people or animals will not be exposed to the smoke. In agricultural situations, if legal, burn the empty containers near where pesticides were applied. In nonagricultural areas, contact local authorities to work out arrangements for disposal of plastic and paper pesticide containers. Empty containers are considered hazardous wastes and must be disposed of according to provisions of Water Quality Control Board and Department of Health Services regulations.

Metal, glass, and plastic containers that have been triple rinsed at the time of use to remove traces of pesticide can either be offered to the pesticide manufacturer for recycling or transported to an approved Class 2 disposal site. If these containers have not been triple rinsed, they can be taken only to a Class 1 disposal site. The Water Quality Control Board or agricultural commissioner in your area can provide information on the locations of approved disposal sites.

Cleaning Application Equipment. Application equipment must be cleaned and decontaminated. Residues remaining in tanks may contaminate a subsequent pesticide mixture and might possibly alter its toxicity; there is also the problem of causing phytotoxicity to plants or other types of damage to sprayed surfaces. Pesticide residue on the outside of application equipment can be hazardous to people who must operate or service this equipment. The outside of spray equipment should be washed with water, using a small amount of detergent if necessary. Clean equipment in areas where runoff will not drain into any waterway or other sensitive area, or will not percolate into the groundwater. The inside of the tank should be rinsed with water and decontaminated by using an appropriate pesticide tank cleaning material or 1 quart of household ammonia to each 25 gallons of water. A tank washing solution can be prepared by mixing ½ pound of detergent with 30 gallons of water. Commercial pesticide tank cleaning and neutralizing compounds can be purchased through chemical suppliers and farm equipment dealers—be sure to check the pesticide label for any precautions regarding the use of cleaning and decontaminating chemicals. Follow the directions for the amount of cleaner to use for your spray tank. Be sure to run pumps and agitators, and to flush all hoses.

Personal Clean-Up

After using pesticides, you must clean your personal protective equipment, shower thoroughly, and change into clean, uncontaminated clothing. When showering, take special care to wash your hair and clean your fingernails. Clothing that was worn during the pesticide application should be immediately placed in a plastic bag until it can be laundered. Never eat, drink, smoke, or use the bathroom until you have thoroughly washed.

RECORD KEEPING

Maintain records of *every* pesticide application you make. Write down pertinent information such as the kind and amount of pesticide used, amount of water used, calibration adjustments, adjuvants added to the mixture, type of equipment used, severity of pest infestation, and, if applied to crops or animals, the stage of development of the host. Temperature and general weather information at the time of application should also be noted. Write down any other conditions that might have an influence on the effectiveness of the pesticide. Keep a record of the names of persons you spoke to regarding each pesticide application. Any follow-up information and notes of application results should also be included (see Table 9-4 on page 270 for a pesticide application follow-up checklist). Table 7-11 is an example of a pesticide application record. Copies of written recommendations should be kept with application records.

Application records will be helpful as a history of pesticide use, especially when plantback restrictions are involved. Even more important, this information is vital in case problems associated with the application should develop. Good records may also be important to your defense in any legal action.

TABLE 7-11.

Pesticide Application Record.

	Date: _____ Applicator: _____

APPLICATION SITE

Owner/Responsible Party: _____   Location: _____

Size of Treatment Area: _____   Plant Age and Condition: _____

Description (Turf, Ag Crop, etc.): _____   Soil Conditions: _____

Surrounding Sensitive Areas: _____

Previous Pesticides Used: _____

PEST PROBLEM

Primary Pest: _____   Damage Observed: _____

Other Pests Present: _____   Location of Damage: _____

Beneficials Present: _____

Severity of Pest Problem: _____

PESTICIDE(S) USED

Pesticide(s):	Formulation:	Rate:	Total Amount Used:
(1) _____	_____	_____	_____
(2) _____	_____	_____	_____
(3) _____	_____	_____	_____

Adjuvants
Type: _____   Amount: _____   Total Gallons of Diluted Spray Used: _____

APPLICATION

Date(s) of Application: _____

Weather Conditions
Temperature: _____

Equipment Used: _____

Cloud Cover: _____
Wind Speed: _____
Wind Direction: _____

Equipment Calibrated By: _____   Other: _____

Pesticides Mixed By: _____   Travel Speed: _____

Pesticides Applied By: _____   Total Hours for Application: _____

Persons Notified or Spoken to Regarding Application:
(1) _____
(2) _____
(3) _____

FOLLOW UP

Effectiveness of Application: _____   Beneficials Present: _____   Pest Resurgence Noted: _____

Injury to Nontarget Plants or Surfaces: _____

COMMENTS _____

LIABILITY

As a pesticide applicator, you assume personal responsibility for accidents and injuries that arise as a result of each pesticide application. This liability may result in fines, jail sentences, and loss of your applicator certificate or license if you are found to be negligent in your application of pesticides or have broken state or federal laws. You may also be held responsible in lawsuits for personal injury or damages. If you are working for someone else, your actions may result in lawsuits and fines to your employer. Keeping accurate records of all your pesticide applications may help you in your defense should a claim of negligence be brought against you.

Personal liability might be incurred if the pesticide you are applying drifts away from the treatment area and causes damage to plants, animals, or someone's belongings, or causes human injury. The pesticide you apply can potentially cause damage to the crop or surfaces for which it was intended. Damage might result from improper mixing, use of the wrong adjuvants, improper application, applying the wrong pesticide, poor timing, or from using a pesticide that has been contaminated with impurities. You could also be sued for destroying beneficial insects such as honey bees, and, if the bees are essential for pollinating a crop, you could be liable for the loss of the crop as well. There have been instances where applicators were sued because they applied pesticides to the wrong location. Pesticides and pesticide application equipment are attractive nuisances; children, fascinated with what you are doing, may be injured or even killed by chemicals and equipment that has been left unattended. The extent of damage (and therefore liability) from a pesticide accident or application error might be greatly reduced by prompt action once the problem has been discovered. See the following chapter for information on how to deal with pesticide emergencies.

Liability Insurance

Liability insurance is usually purchased by commercial pesticide applicators to protect themselves from claims associated with pesticide use. Clients of professional applicators may require the applicators to have liability insurance. Policies cover the costs of damages from accidents and improper use. This type of insurance may be expensive and sometimes difficult to obtain due to the nature of pesticide injury claims. Applicators who maintain complete records and are conscientious in their efforts to use pesticides responsibly are usually considered better risks by insurance companies.

Professional organizations representing pesticide applicators may have information on insurance companies and policies, including ways to participate in a group policy. Select insurance suitable to your operation and specialty. Be sure you understand the extent of coverage and policy liability limits.

8 Pesticide Emergencies

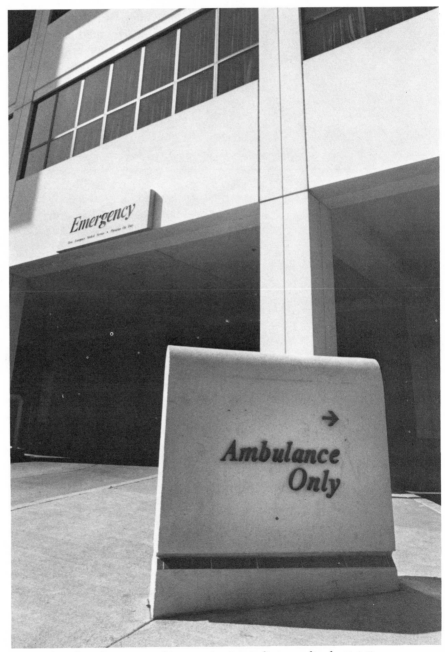

Pesticide related injuries often require immediate medical attention.

Any accident involving pesticides must be treated as an emergency because of the great potential for causing harm to people. Accidents may occur during the handling and application of pesticides even under the most careful conditions. Pesticides diluted with water may be hazardous, but undiluted pesticides can be much more dangerous. Pesticide emergencies may be the result of leaks and spills, fires, thefts, misapplication, or lack of care in storage or handling.

Whenever using pesticides, always have in your possession the names and locations of nearby medical facilities capable of treating pesticide-related injuries. If an emergency situation arises, always check first for possible injuries to applicators and other people in the area. Be prepared to offer first aid if necessary.

FIRST AID

First aid is the help you are able to give a person exposed to pesticides before professional help is obtained. *First aid is not a substitute for professional medical care.* Pesticide labels provide specific first aid information (Figure 8-1).

Poisoning or exposure can occur if pesticides are splashed onto skin or into eyes, are swallowed, or vapors, dusts, or fumes are inhaled. The

---

PRECAUTIONARY STATEMENTS
Hazards to Humans and Domestic Animals
MAY BE FATAL IF SWALLOWED • MAY BE ABSORBED THROUGH SKIN • MAY BE INJURIOUS TO EYES AND SKIN
Do Not Take Internally • Do Not Get in Eyes, on Skin or on Clothing • Avoid Breathing Vapors and Spray Mist • Wash Thoroughly After Handling
Statements of Practical Treatment
If Swallowed, do not induce vomiting. Contains aromatic petroleum solvent. Call a physician immediately.
If On Skin: In case of contact, remove contaminated clothing and immediately flush skin with soap and water. Wash contaminated clothing before reuse.
If In Eyes: Flush eyes with plenty of water for 15 minutes. Call a physician.
NOTE TO PHYSICIAN: Chlorpyrifos is a cholinesterase inhibitor. Treat symptomatically. Atropine only by injection is an antidote.
Physical and Chemical Hazards
COMBUSTIBLE.
Do Not Use or Store Near Heat or Open Flame
Do Not Cut or Weld Container
Environmental Hazards

FIGURE 8-1.

Read the pesticide label for special first aid instructions. All labels contain some type of statement of practical treatment.

type of exposure determines what first aid and subsequent medical treatment is required. Serious pesticide poisoning can arrest breathing or cause convulsions, paralysis, skin burns, or blindness. Applying the proper first aid treatment for pesticide exposure may reduce the extent of injury and even saves lives.

To be able to assist someone having breathing difficulties, it is vital that you know how to administer artificial respiration or cardio-pulmonary resuscitation (CPR). If you are unfamiliar with these techniques, contact the American Red Cross chapter in your area for training. Prepare yourself *before* an emergency arises.

Protect yourself from contamination when administering first aid to a person suffering from pesticide exposure. Avoid getting pesticides on your clothes or skin and do not inhale vapors. Do not enter a confined area to rescue a person overcome by toxic pesticide fumes unless you have the proper respiratory equipment. Remember, the pesticide that affected the injured person can also injure you.

Professional medical care must be obtained *at once* when a person is accidentally exposed to a highly toxic pesticide or shows any signs of pesticide poisoning; this requires transporting the exposed or injured person to a medical facility where equipment and trained staff are available to provide treatment. Speed in obtaining medical care often controls the extent of injury. Always provide medical personnel with information about the pesticide suspected of causing poisoning or injury. A pesticide label or clean container with a label attached should be sent to the medical center with the injured or exposed person.

Pesticides on the Skin or Clothing

Concentrated pesticides spilled on the skin or clothing can cause serious injury, either in the form of burns or rashes or through skin absorption resulting in possible internal poisoning. Follow the sequence of first aid steps listed below:

Remove the victim from the contaminated area. Get the exposed person away from fumes, spilled pesticide, and further contamination. Do this quickly.

Restore breathing. If the exposed victim has stopped breathing, begin artificial respiration or cardio-pulmonary resuscitation (CPR) at once and continue until breathing resumes or until professional help arrives.

Prevent further exposure. Once breathing has been restored, remove contaminated clothing. Thoroughly wash the affected skin and hair areas of the person's body, using soap or detergent and large amounts of water. Washing in this manner will prevent other parts of the person's body from becoming contaminated. Remember, different areas of the body do not absorb pesticides in the same way; poisoning will usually increase in severity as more skin area is involved. Remove the chemical as rapidly as possible, then dry the washed area and cover it with a blanket if available. Keep the person warm.

Chemical burns cause the skin to become red and painful like other burns. If this happens, use extreme care in washing, and use large amounts of water. Cover the burned area with a clean cloth. Do not put any ointment, spray, powder, or other medications on the injured areas.

Get medical care. Call an ambulance or transport the person to the nearest medical facility as quickly as possible. Choose the method that will provide medical care in the shortest amount of time. Be sure to send a copy of the pesticide label with the injured person so that medical personnel will know what pesticide they are dealing with. If a label is not available, write the brand name, chemical name, and manufacturer of the pesticide on a piece of paper.

Pesticides in the Eye

Pesticides can cause serious damage to eyes. Prompt first aid, followed by medical care, helps to reduce damage.

Wash the eyes. Immediately wash the victim's eyes with a gentle stream of clean, running water. Any delay, even of a few seconds, may greatly increase the possibility of permanent eye injury. Hold eyelids open to assure thorough washing. Do not use any chemicals or drugs in the wash water, since this may increase the extent of injury. Continue flushing the eyes for at least 15 minutes (Figure 8-2). If running water is not available, slowly pour clean water from a glass or other container onto the bridge of the nose, rather than directly into the eyes.

Obtain medical care. If pain or irritation persists, transport the person to the nearest medical facility as soon as the flushing has been completed. Protect the eyes with a clean, wet cloth. Be sure to send along a pesticide label, container, or written identification of the pesticide.

Inhaled Pesticides

Inhaled chemicals, such as fumigants, pesticide dusts, vapors from spilled pesticides, and fumes from burning pesticides, can cause serious injury to lungs and can be absorbed into other parts of the body through

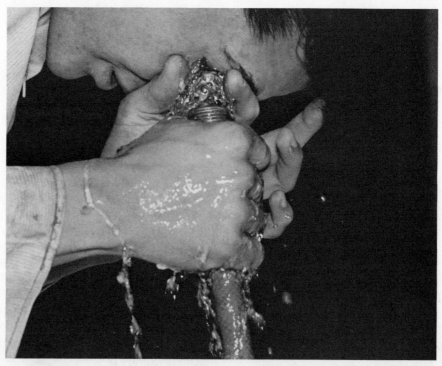

FIGURE 8-2.

If pesticides get into the eyes, they must be flushed immediately for 15 minutes with a gentle stream of clean water. Hold eyelids open while flushing. If irritation still persists, obtain medical treatment.

the lungs. Immediate first aid measures must be taken to reduce injury or prevent death.

Remove the victim from the contaminated area. A person overcome by pesticide fumes must be moved to fresh air immediately. If at all possible, carry or drag the exposure victim out of the contaminated area; physical exertion places an extra strain on the heart and lungs of a person suffering from inhalation injury and could prove fatal.

Wear a supplied air respirator when entering an enclosed area to rescue a person who has been overcome by a pesticide. Cartridge respirators are not suitable for high concentrations of pesticide fumes or deficient oxygen conditions; *use only a supplied air respirator.* If you do not have this equipment, call for emergency help. You can be of more assistance to the injured person by seeking proper emergency help than you can be if you are overcome by the pesticide yourself.

Loosen clothing. Once the victim has been brought out into fresh air, loosen all tight or restrictive clothing. This will help make breathing easier and also removes pesticide vapors trapped between clothing and the skin.

Restore breathing. If breathing has stopped, or is irregular or labored, begin artificial respiration or CPR. Continue assisting until breathing has improved or until medical help arrives.

Treat for shock. Inhalation injury often causes a person to go into shock. Keep the injured person calm and lying down; prevent chilling by wrapping the person in a blanket after removing contaminated clothing. Do not administer any type of alcoholic beverage.

Watch for convulsions. Convulsions can occur as a result of some types of pesticide poisoning, so protect the victim from falls or injury and keep air passages clear by making sure their head is tilted back.

Get immediate medical care. Call an ambulance or transport the person to the nearest medical facility. Be sure to provide information on the type of pesticide inhaled, if known.

Swallowed Pesticides

Two immediate dangers are associated with swallowed pesticides. The first is related to the toxicity of the pesticide and the poisoning effect it will have on a person's nervous system or other internal organs. The second involves physical injury that the swallowed pesticide causes to the linings of the mouth, throat, and lungs. Corrosive materials, those which are strongly acid or alkaline, can seriously burn these sensitive tissues. Petroleum-based pesticides can cause lung and respiratory system damage, especially during vomiting. *Never induce vomiting if you suspect that the swallowed pesticide is corrosive or petroleum-based.*

Dilute the swallowed pesticide. If the person is conscious and alert, give large amounts (1 quart for an adult or a large glass for a child under 7) of water or milk to dilute the swallowed pesticide. Do not give any liquids to an unconscious or convulsing person.

Induce vomiting. If you are certain that neither a corrosive or petroleum-based pesticide has been swallowed (check pesticide label), induce vomiting by placing a blunt object at the back of the victim's throat. Have the person kneel or lie face down or to their right side. Never induce vomiting while the victim is lying face up. Do not induce vomiting if the victim is unconscious or having convulsions. *Do not administer salt solutions or any other compounds to induce vomiting since these may cause further injury.* If you are in doubt regarding the type of pesticide swallowed and are unable to get this information from the pesticide label, *do not* induce vomiting. Do not spend valuable time trying to induce vomiting if this time can be used to transport the person to a hospital or other medical facility.

Obtain medical care. Call an ambulance or transport the poisoning victim to the nearest medical facility. Provide as much information as possible about the swallowed pesticide. If the person has vomited, collect some of the vomitus in a clean jar for analysis.

Regional poison information centers are located throughout the state and can be reached by telephone 24 hours a day and 7 days a week; most have toll-free telephone numbers. These centers provide quick, life-saving information on poisoning treatment. Table 8-1 lists locations and telephone numbers of regional poison centers in California; *call them*

TABLE 8-1.

Poison Control Centers Located in California and National Hotline for Animal Poison Control.

FRESNO (209) 445-1222 (800) 346-5922	Central Valley Regional Poison Control Center Fresno Community Hospital Fresno and R Streets P.O. Box 1232 Fresno, CA 93715
LOS ANGELES (213) 484-5151 (800) 777-6476	Los Angeles County Medical Association Regional Poison Information Center 1925 Wilshire Boulevard Los Angeles, CA 90057
OAKLAND (415) 547-2928 (800) 523-2222	Children's Hospital Medical Center of Northern California 51st and Grove Streets Oakland, CA 94609
ORANGE (714) 634-5988 (800) 544-4404	University of California Irvine Medical Center Regional Poison Center 101 City Drive South Orange, CA 92668
SACRAMENTO (916) 453-3692 (800) 342-9293	Sacramento Poison Control Center University of California, Davis 2315 Stockton Boulevard Sacramento, CA 95817
SAN DIEGO (714) 294-6000 (800) 876-4766	San Diego Regional Poison Center University of California, San Diego Medical Center 225 Dickinson Street San Diego, CA 92103
SAN FRANCISCO (415) 666-2845 (800) 523-2222	San Francisco Bay Area Poison Center San Francisco General Hospital 1001 Potrero Avenue, Room 1E-6 San Francisco, CA 94110
SAN JOSE (408) 279-5112 (800) 662-9886	Central Coast Counties Regional Poison Control Center Santa Clara Valley Medical Center 751 South Bascom Avenue San Jose, CA 95128

NATIONAL HOTLINE FOR ANIMAL POISON CONTROL:
(217) 333-3611

only in the case of an actual emergency. Most telephone directories list the nearest center in the white pages under "poison." Table 8-1 also lists a source for information on animal poisoning.

PESTICIDE LEAKS AND SPILLS

All pesticide leaks or spills should be treated as emergencies. Concentrated pesticide spills are much more dangerous than pesticides diluted with water, but both types should be treated seriously and immediately. Leaks or spills can occur during transporting, storing, or while using pesticides. Pesticide may be spilled indoors, in enclosed areas, or outside. When spills occur on public roadways, immediately contact the California Highway Patrol and the State of California Office of Emergency Services. These agencies will take charge of coordinating the clean-up and protecting the public. When pesticides are spilled on public roadways, a report is required to be filed with the Office of Emergency Services. If leaks or spills should occur in areas other than public roadways, follow the emergency procedures listed below. All leaks or spills of pesticides, no matter where they occur, must be reported to the local agricultural commissioner as soon as possible.

Clear the area. Immediately clear people and animals from the contaminated area and administer first aid to anyone who has been injured or contaminated. Send for medical help if needed. Spilled liquids often have toxic fumes, while dusts are easily blown around in the air, so rope off the area or use some other means to prevent anyone from getting near the contamination. .

Prevent fires. Some liquid pesticides are flammable, or are formulated in flammable carriers. Pesticide powders are potentially explosive, especially if a dust cloud is formed in an enclosed area. Do not allow any smoking near a spill. If the spill occurs in an enclosed area, shut off all electrical appliances and motors that could produce sparks and ignite a fire or explosion. Open doors and windows to provide ample ventilation.

Wear protective clothing. Before beginning any clean-up, put on rubber boots, gloves, waterproof protective clothing, goggles, and respiratory equipment. Check the pesticide label for additional precautions, but when uncertain what has been spilled, wear the maximum protection.

Contain the leak. Stop the leak by transferring the pesticide to another container or by patching the leaking container. (Paper bags and cardboard boxes can be patched with strong tape.) Use soil, sand, sawdust, or absorbent clay to contain liquid leaks. Common cat litter is a readily available material that can be used for pesticide cleanup. Powdered pesticides are subject to air movements and must be immediately contained by covering with damp sawdust, sand, or soil. Small liquid spills can be contained with special products that form a gel as they absorb the pesticide. Once collected, this gel can be added to a spray tank and diluted with water; the resulting liquid is used in a spray application, eliminating the need for disposal.

Clean up pesticide. As soon as the spill or leak has been contained, proceed to clean it up. Anything that was contaminated by the spilled material must be cleaned or disposed of. Remove uncontaminated items from the immediate area to prevent possibility of contamination.

Select a container to hold the spilled pesticide and any items that cannot be cleaned. The container must be sealable and suitable for transporting; a steel drum with lid may be used, or special containers can be purchased and kept on hand for this purpose. Containers must be labeled to indicate they contain hazardous pesticide waste; the name of the pesticide and the toxicity category must be included. Pick up liquid spills by using a special absorbent or cat litter, sand, soil, or sawdust; shovel the saturated absorbent into the holding container (Figure 8-3). Wood and paper cannot be decontaminated, so they must be placed in the holding container with the spilled pesticide. Powdered or granular pesticides should be swept up and placed into a suitable container. An industrial vacuum can be used to pick up spilled powder, but be certain the vacuum is equipped with proper filters to remove dust from discharged air; do not use an ordinary vacuum. Dispose of the vacuum bag and filters into the holding container. If the spill occurred outdoors, shovel all contaminated soil into a holding container.

Spills on cleanable surfaces, such as concrete, require thorough decontamination. Commercial decontamination preparations are available for this purpose, or prepare a solution, using 4 tablespoons of detergent and 1 pound of soda ash, dissolved in each gallon of water. (Soda ash cannot be used for detoxification of a few pesticides, so check the label or material safety data sheet before using this solution. Contact the pesticide manufacturer if you have any questions.) First, contain and collect the spill, using an absorbent material. Pick up the absorbed pesticide and put it into a holding container. Cover the affected area with sawdust or other absorbent product and thoroughly wet it with decontamination solution. After 12 hours, sweep the area and put the absorbent into a holding container. Wash the affected area with more decontamination solution, absorb it with sawdust or other material, and place this into the holding container. Thoroughly rinse the area with clean water, collecting and disposing of this water in the same manner. Do not let the rinse water or any decontaminating solution flow out of the area of the spill. Never wash a contaminated area with water under pressure from a hose

FIGURE 8-3.

Pesticide spills should be covered with an absorbent material and shoveled into a steel drum. When the clean-up is completed, seal and label the drum and transport it to an appropriate disposal site. Wear protective clothing during the clean-up.

because you will disperse the pesticide, increase potential for contamination, and make it very difficult to contain or clean up.

Because local regulations vary, contact the agricultural commissioner or Water Quality Control Board for instructions on how to dispose of the holding container and its contents. Under most circumstances, the residue from a pesticide spill must be transported to a Class 1 disposal site. Up to 10 pounds of soda ash can be added to the container before disposal to aid in pesticide detoxification. First check the pesticide label to make sure it is safe to use soda ash.

PESTICIDE FIRES

Fighting pesticide fires requires special care because smoke and fumes generated by burning pesticides cannot be contained; areas endangered by these fumes must be evacuated. Toxic fumes hamper fire fighting efforts and require the use of supplied air respirators and protective clothing. Water must be used with caution when fighting pesticide fires. Use it primarily to cool containers and prevent overheated chemicals from exploding. Do not splash or spread toxic chemicals with high-pressure water.

Once the fire has been brought under control, all hoses and equipment, including personal protective clothing, must be decontaminated (Chapter 7). Residue remaining at the fire site must be removed and disposed of.

Follow this sequence when a pesticide fire breaks out:

Call the fire department. Contact the nearest fire department as quickly as possible (call "911"). Inform them that it is a fire involving pesticides and provide them with the names of the chemicals contained in the structure or vehicle. If possible, provide material safety data sheets to the arriving fire units.

Clear the area. Get people out of the immediate area of the fire; there may be considerable risk of toxic fumes and explosion.

Evacuate and isolate the area around and downwind of the fire. Protect animals and move equipment and vehicles that could be damaged by the fire or fumes, or that would impair fire fighting efforts. Keep spectators from being exposed to smoke from the fire and runoff from fire fighting. Contact police or sheriff and have downwind residences, schools, and buildings evacuated until the danger has passed.

Do not endanger your health by attempting to fight a large pesticide fire without help. If you are involved in fighting a fire, wear protective clothing. Stay upwind and remain a safe distance from the smoke. Concentrate your efforts on cooling containers that could explode if overheated. Whenever possible, use foam or carbon dioxide extinguishers to fight the fire because they are less likely to disperse pesticides than water.

After the fire has been extinguished, contain the runoff with earthen dams and rope off the contaminated area until it has been properly cleaned up. If large amounts of pesticides were involved in the fire, contact a professional decontamination and disposal company for assistance. Small amounts of contamination can be disposed of in the same manner as pesticide spills. Consult with the county agricultural commissioner for advice and information on proper disposal methods.

PESTICIDE THEFTS

Losing pesticides through theft is a serious problem that warrants emergency action. Stolen pesticides in the hands of potentially irresponsible people can cause human poisoning and environmental damage.

Once a theft of pesticides is discovered, contact the local police or sheriff. Provide information on the type and quantity of stolen pesticide and describe containers as accurately as possible. Also notify the agricultural commissioner located in the county where the theft took place.

MISAPPLICATION OF PESTICIDES

Another form of emergency may exist when pesticides have been misapplied. *Intentional misapplication* involves intentional use of a pesticide on an unregistered site or applying pesticides in a manner inconsistent with label directions. *Accidental misapplication* involves unknowingly applying the wrong pesticide to a location, or applying the improper amount of a pesticide for control of a labeled pest. *Negligent application* involves improper calibration of application equipment as well as improper use and disposal of the pesticide; it also involves applying pesticides at the wrong time or in any other way inconsistent with label recommendations.

Making an application mistake is a serious problem; do not compound the damage by failing to take responsible, corrective action once the mistake is discovered. You or your employer may be financially responsible for damages, both physical and legal, caused by your misapplication of a pesticide. The amount of damage and liability can often be reduced by prompt action once the error has been discovered. Of primary importance is the protection of people, animals, and the environment. Responsible, quick action helps to offset penalties and legal claims.

Incorrect Amount of Pesticide Used

Insufficient quantities of pesticides usually do not give adequate control of the target pest and waste time and money, but generally present no serious problems to people or the environment. Excessive amounts of pesticide, however, can be an environmental threat as well as a danger to human health. This type of problem occurs as a result of poor calibration of the application equipment or is due to faulty mixing of chemicals in the spray tank. Residues from the pesticide may last longer than expected or a concentrated application may cause damage to the treated area, in the form of either phytotoxicity to plants, visual residues, staining, and spotting of paint, furniture, plants, or produce.

Once an improper application has been discovered, take immediate steps to notify and protect people in the area. Contact the pesticide manufacturer for help in determining what corrective measures can be taken. Investigate the possibilities of diluting the pesticide with a spray of water or some other solvent. Notify the agricultural commissioner of the problem and seek information and advice on what remedies can be taken. Remember, speed is of utmost importance when trying to reduce damage.

Application of the Wrong Pesticide

Lack of attention to the mixing operation or improper instructions given to an applicator may result in the wrong pesticide being applied to the treatment area. Besides possible damage to plants or surfaces in the treatment area, using the wrong pesticide exposes workers and the public to hidden toxic residues. Mixing and application might take place unknowingly without using necessary personal protective equipment, resulting in possible injury to the applicator.

Whenever you discover that the incorrect pesticide has been mixed or applied, contact the pesticide manufacturer and the agricultural commissioner for help. Notify people in the application area and keep them away until it can be made safe again.

Pesticides Applied to the Wrong Site

Another form of accident involves pesticides being applied to the wrong site. Owners and inhabitants of the sprayed property must be contacted immediately and told of the problem. If the pesticide was appropriate to the location, then only the normal precautions need to be taken to protect people and animals. The problem becomes more serious if the pesticide was applied to an unregistered site. Contact the pesticide manufacturer and the agricultural commissioner for assistance. Keep people and animals out of the sprayed area until it has been determined that it is safe for their return.

Effective Use of Pesticides

The uniform-sized spray droplets produced by controlled droplet applicators enhance the safety and effectiveness of some pesticide applications.

Pesticides must be used effectively as well as safely. The results of a pesticide application should usually be worth any financial or labor investment by yielding a profit or quality advantage in crops, by reducing health hazards to people and livestock or poultry, by improving health, appearance, and growth of turf and ornamental plants, or by eliminating annoying pests from buildings, workplaces, and homes. This chapter discusses several ways to improve the effectiveness of pesticide use.

PEST DETECTION AND MONITORING

Pest detection, identification, and monitoring are primary keys to effective pesticide use. Detection verifies the presence of pests and helps anticipate when and where pests will occur. Correct identification is usually necessary to obtain information about a pest's biology (life stages and habits), which will help in choosing a pesticide and determining when, where, and how to apply it. An established monitoring program allows you to detect pests, observe seasonal changes in pest populations, properly time control applications, and assess the effectiveness of control measures.

Predicting Problems

Early detection will enable you to plan a program for monitoring pest development or activity and to predict if or when treatment is necessary. Review the history of pest problems on the farm, landscaped area, building, right-of-way, or other location to know what pests to expect at different times of the year. If this information is not available for the specific location where you are working, try to get pest history information from a similar location nearby.

Look for conditions that favor pest buildup. For example, pest insects often overwinter in crop residues or field borders, so their presence in these areas may predict future crop injury. Weeds that have been allowed to produce seed provide a seed reservoir, indicating that large populations of these weeds can be expected in following seasons. Vertebrates, such as squirrels, may not be a problem while other food supplies are adequate, but if conditions change, they may move into cropped or landscaped areas for food. Cockroaches, ants, and rodents need food sources, water, and often shelter before they can seriously infest an area.

Chapter 2 describes methods used for identifying various types of pests. Also, tables in Chapter 2 provide information on how to use pest identification services and how to package pests for shipment to experts or identification laboratories.

After a while, you will probably learn to recognize the more common pests found in your work situation. When you come across pests you do not recognize, collect a sample using traps, nets, or other appropriate methods. Weeds should be dug up to include roots with the rest of the plant. Collect seedlings and/or flowering specimens if they are present. Be careful when handling birds and rodents because they may be diseased; some rodents are infested with fleas that can vector plague, while birds often carry lice, mites, or biting bugs. Rabies is prevalent in some skunks, bats, and other small mammals, so handle these animals with tongs or heavy gloves to avoid being bitten. Do not contact their urine or feces.

Do not mistake natural enemies for pests. Be sure you can recognize natural enemies of pests such as insects because these may be contributing to the control of the pest problem or other potential pests, eliminating or reducing the need for a pesticide application.

Using Life History Information. Life history information usually provides clues to nesting sites, food preferences, natural enemies, seasonal occurrences, or life cycles, that can be vital for effective control. Plan pesticide applications and other control measures that are most appropriate to the pest and apply pesticides during the most susceptible life stage. For instance, successful weed control often occurs when weed seeds are germinating or plants are seedlings. Once past the seedling stage, perennial weeds are usually most susceptible to herbicides when they are flowering. Eggs of mites and insects are often resistant to pesticides, while adult stages of scale insects are protected by a hard, waxy structure that blocks pesticide entry. When using poison baits for birds or rodents, first determine what the organism is currently feeding on, then select a bait that contains this food. When possible, time control measures around breeding seasons to prevent vertebrate pests from reproducing.

Establishing a Monitoring Program

A regular monitoring plan provides information on the day-to-day field populations of pests needed to make management decisions. Information collected may include the density, life stages, and species composition of pest populations and observation of factors controlling or favoring the pest. It is more difficult to monitor pests while their populations are low and damage minimal, but it is worth the effort because, if chemical treatment is necessary, you may be able to use less toxic pesticides or limit applications to a more restricted area.

Visual monitoring, the most common method, includes any systematic method of searching for pests, pest damage, or evidence of the presence of pests. Generally, this requires thoroughly examining a representative portion of a specific area in a uniform way, such as sampling foliage, pulling up a certain number of plants, or walking a prescribed transect. Look for patterns of distribution, damage, or activity and for evidence of natural enemies or other mortality factors. Seeds, weed remains from the previous season, or animal burrows, tracks, feeding damage, fecal droppings, webbing, and eggs may provide clues to the presence and identity of some pests. Sometimes other organisms may indicate the presence of economic pests, such as fleas in the area of rodent nests or ants climbing trees or shrubs to collect honeydew from aphids and scale insects.

Devices that are used for monitoring and observing pests include traps (such as animal traps, sticky traps, and light traps), attractants, pheromones, microscopes, binoculars, hand lenses, insect nets, beating trays, and tracking powder (Table 9-1).

TABLE 9-1.

Equipment Used for Collecting and Monitoring Pests.

HAND LENS	Used for locating, examining, and identifying insect, mite, fungal, and other pests.	Magnification range of 7× to 14× most useful. Hold lens close to eye and bring object to be examined up to lens for focus.
INSECT NET	For monitoring and collecting many types of insects. Essential piece of equipment.	Three types available: *Aerial net*—made of lightweight net material; used for flying insects. *Beating net*—made of heavier cloth such as muslin; used for sampling insects on plants and shrubs. *Aquatic net*—specially designed net to collect aquatic specimens.
BEATING TRAY	Used to monitor plant-feeding pests—especially useful on larger trees and shrubs.	Easy to make; design for own special applications.
PHEROMONE TRAPS	Excellent way of monitoring flight activity of many insects.	Several styles of traps are available for different insect species.
LIGHT TRAPS	Attract mostly night flying insects.	Only useful in indoor or enclosed areas. Some have electric grid to kill attracted insects.
ANIMAL TRAPS	Helpful in detecting small animals such as rodents and birds. Usually have to be carefully located.	Several types available: *Live traps*—animals are usually attracted by baits, captured animals are unharmed. *Spring traps*—injures or kills trapped animals. *Sticky traps*—animal gets caught in sticky substance, may die.
PITFALL TRAPS	For detecting walking insects.	Very effective in landscape and nursery areas.
OTHER TRAPS	For detecting insects and pathogens. Often attract by shape or color.	Lure traps with bait other than pheromones. Sticky yellow traps. Spherical sticky traps. Spore traps for scab and other diseases.
PYRETHRUM TEST	1–2% solution in water applied primarily to turf for insect detection.	Irritates cutworms and other subterranean pests which then come to the surface.

MICROSCOPE	Used for examining plants and other objects for pests and very helpful in identifying insects, mites, and fungi.	Low power dissecting microscope that provides magnification in the range of 10× to 50× is the best.
BINOCULARS	Especially helpful in spotting and identifying birds, but also useful for rodents. Sometimes helpful in examining crop damage and infestation.	Magnification range should be 6× to 7×, with an objective lens size between 35 mm and 50 mm. Porro prism or Dach prism binoculars are best.
TRACKING POWDER	Used to monitor movement of rodents, sometimes used to monitor insects.	Use only on floor surfaces to prevent powder from falling from upper areas.
CONTAINERS	Used for holding and transporting collected specimens: plant parts, insects, mites, nematodes, fungi, and weeds.	Glass vials, plastic bags, and paper bags.
KNIFE	Used for cutting open plants, fruits, nuts, wood, and other objects to locate pests and examine for damage.	Should be strong and sharp.
SHOVEL	Used for sampling weeds. Helpful in digging around plant roots or in removing plants for examination.	
ICE CHEST	Keeps collected specimens fresh until they can be examined. Also can be used to ship materials to identification labs.	Inexpensive foam plastic chests work very well.
IDENTIFICATION AIDS	Used to assist in pest identification.	Includes books, photos, keys, and preserved specimens. Obtain these from libraries, bookstores, and biological supply houses.

Weeds. The most important field information needed for making weed management decisions is what species are present, what is their stage of development, and whether the relative abundance of different species is changing from previous seasons. Since many weed control materials are applied preventively and most control seedling weeds much more effectively than older plants, knowing the history of weed problems in a field or managed area is especially advantageous for predicting future management needs.

Monitor in the late fall or early winter (after the first rains) to detect emerging winter annuals. Monitor in the late spring to detect emerging

ORCHARD LOCATION_____   CONTROL METHODS_____

CONTROL DATES_____

COMMENTS_____

	NOV _____		FEB _____		MAY _____	
	% of total weeds		% of total weeds		% of total weeds	
ANNUAL GRASSES	treated	untreated	treated	untreated	treated	untreated
annual bluegrass						
barnyardgrass						
crabgrass						
sprangletop						
wild barley						
wild oats						
_____						
_____						
_____						
ANNUAL BROADLEAVES						
cheeseweed (mallow)						
clovers						
groundsel						
filaree						
fiddleneck						
knotweed						
lambsquarters						
mustards						
pigweeds						
puncturevine						
purslane						
_____						
_____						
_____						
_____						
_____						
PERENNIALS						
bermudagrass						
dallisgrass						
johnsongrass						
curly dock						
field bindweed						
nutsedge						
_____						
_____						

FIGURE 9-1.

Keep records of the weed species present in an area to help select appropriate herbicides or other control measures. The form illustrated here was developed for weeds found in walnut orchards.

summer annuals. Monitor at other times as needed to detect perennial and biennial weeds. Identify all the weed species found growing in an area, preferably while they are in the seedling stage. Watch for any new species. Using a form similar to Figure 9-1, keep records of the different species and their locations. Mapping the locations of perennial weeds is especially important. Try to estimate the percentage of each weed species to the total weed population. Note areas where weeds have been allowed to produce seeds. Also, keep records of all herbicides used for weed control in the area; these will be important when plantback restrictions apply and for evaluating the effectiveness of previous control efforts. Note cultural methods that have been used for weed control.

Look for adjacent weedy areas such as roadsides or ditchbanks that may be a reservoir for seeds. Cultivation equipment may also move weed seeds or vegetative structures from one site to another. Birds and mammals also can move seeds or vegetative structures.

Nematodes. Management of nematodes must usually occur before the crop or other plants are in the ground, early enough to allow for treatment before planting. Therefore, sample in autumn for winter and spring planted crops or plants so the area can be fumigated in autumn or an alternate, nematode resistant plant species can be chosen if necessary. The simplest form of sampling is performed to determine if nematodes are present. In agricultural crops, this can be done during the previous growing season, if susceptible crops or weeds are in the ground, by digging up plants that look stressed and checking roots for galls, cysts, or swollen root tips. Sandy soils should be regularly monitored for rootknot nematode infestations in this way when susceptible plants are to be grown there. If nematodes are detected, or if you know that nematodes have been a problem in the area before, more detailed quantitative sampling should be carried out. One sample a year should be sufficient.

Take soil samples for nematodes along a transect through the area, keeping each sample separate. Sample the soil within the root zone of the plants. Also look for plants which are stunted or damaged. Include infected plants, with roots, in the sampling. Prepare and send samples to an identification laboratory as instructed in Chapter 2. Draw a map of the area which shows locations of healthy and infected plants, soil types, water drainage, and other important features. Look for ways in which nematodes might be introduced to an area such as through contaminated cultivation equipment.

Pathogens. Monitor pathogens by observing plant symptoms or damage or, with certain fungi, by looking for fruiting bodies and other structures. Damaged plant material may serve as an inoculum source for healthy plants. Look for inoculum sources before conditions favor the spread of pathogens, since plant disease pathogens are most successfully controlled by suppressing them to prevent infection. Environmental conditions such as temperature, rainfall, or heavy dew are often the controlling factors for pathogen infection or development, so these conditions should be monitored. Look for any pattern of symptoms. For example, do symptoms occur only on scattered plants, are they concentrated in certain parts of the field, or are they generally distributed? Also, monitor for the presence of insects or nematodes that are capable of transmitting certain pathogens—aphids, for example.

Identification of most plant pathogens requires laboratory analyses. Collect damaged plant material according to the instructions in Chapter 2.

Arthropods. Insects, mites, and other arthropods can be observed by visual monitoring or by collecting plant foliage and examining it with a hand lens or microscope (Figure 9-2). Use a sweep net to collect certain

FIGURE 9-2.

A hand lens is often needed for detecting, identifying, and monitoring insects, mites, and other arthropods on plant foliage. Hold the lens close to your eye and bring the object being examined up to it until it is in focus.

FIGURE 9-3.

Use a net to monitor the presence of certain pest insects on plant foliage.

pest insects found on foliage (Figure 9-3) (avoid using sweep nets on tender plants that can be damaged by this technique). Beat foliage onto a white sheet, tray, or pan for a simple way of detecting certain plant-feeding insects. These methods may also be used to estimate the size of an arthropod pest population, or to evaluate its rate of increase or decrease. Sometimes the decision to apply a pesticide is based on the number of insects or mites found on a prescribed number of leaves taken from different plants or from sweep-net samples. Control decisions often are based on studies which show that until the pest population builds up to a certain size, no economic damage will occur.

Many night-flying insects are attracted to light, especially the ultraviolet spectrum. Traps with an ultraviolet light source (black light) are used to attract and kill some nocturnal insect pests in localized areas. These traps also serve as monitoring devices for particular types of insect pests.

Use sticky traps for catching and monitoring some species of insect pests. These traps consist of a surface coated with a thick sticky paste. Several methods are used to make sticky traps attractive specifically to target insects. Placement of the traps and trap color and shape enhance specificity. For example, locate sticky traps along wall bases or other normal pathways to monitor cockroaches. Use a bright yellow color to attract whiteflies to coated cardboard. Hang sticky red or green spheres in trees to catch apple maggot adults or walnut husk flies. Attractants, such as food, sugar syrups, or chemicals with odors resembling natural food sources can be added to sticky traps or sticky surfaces to attract flies, cockroaches, and other pest insects.

Pheromones are used to lure certain types of insects into sticky traps. Pheromones are chemicals produced by insects to attract individuals of the same species. Most pheromone traps use a chemical that mimics the pheromone produced by female insects to attract males for mating. A few mimic pheromones released by males to attract females. Trap catches increase as insects emerge from their pupal stage and begin to mate. Knowing when adults are emerging can give you a good idea of when to apply insecticides for optimum control. For some agricultural pests, simple degree-day tables and guidelines are available that help to accurately predict pest insect egg hatch and insecticide application timing, based on pheromone trap catches and daily high and low temperatures.

Vertebrate Pests. Monitoring for vertebrates requires some understanding of the habits of potential pests. Many species are active only at certain times of the day or night; others cease their activities when people are around. Often the best way to monitor for them is to check for evidence of their presence, such as gopher or ground squirrel burrows, feces of rats or rabbits, or mouse trails in fields. Use animal traps and tracking powders to monitor the activities of vertebrate pests when it is too difficult or time-consuming to watch continually for the pests. Some animal traps are spring-loaded devices, such as rat traps. Live traps, which resemble cages, are used when it is important not to harm the captured animals (Figure 9-4).

Use tracking powders for monitoring rodent activity in buildings (Figure 9-5). Tracking powders are spread over an area suspected of being a rodent trail or runway. Keep tracking powders on floor surfaces to prevent contamination of counters, furnishings, or other objects in an area. Tracks left in the dust will reveal information about the population

FIGURE 9-4.

Live traps can be used to monitor the presence of small animals such as birds or rodents without injuring the animals. Domestic animals, such as this cat, may be accidentally trapped on occasion.

FIGURE 9-5.

Tracking powders are used to monitor the activity of small rodents and insects. Toxicants sometimes are added to the powder to kill the animal when it cleans itself.

size, age of individuals, and areas of activity. This technique is useful in planning the placement and timing of traps or rodenticides. Tracking powders can be combined with toxicants in areas where toxicants are safe to use; contaminated rodents ingest the poison while cleaning themselves.

MAKING PESTICIDE USE DECISIONS

How do you decide when to use a pesticide and what pesticide to use? In residential and urban situations, pests are usually controlled when the people living and working in these areas can no longer tolerate them; pesticides used in these situations are often chosen according to their safety, speed, and effectiveness. In agriculture and other commercial ventures, however, the economics of pest control is also important.

In most cases, a few individuals of a particular pest probably will not cause economic losses, but as their numbers increase, so does their damage. For some pests—mostly insects, nematodes, and mites—action or treatment thresholds have been established that indicate what population levels can be tolerated without loss and at what level a pesticide application will pay for itself. Such treatment thresholds must be flexible. If the market price for a crop fluctuates, for instance, the threshold of allowable pest damage can be changed to accommodate for this. When the price of pesticides increases or decreases, the tolerable pest injury is adjusted accordingly. Decisions like these, based on extensive sampling, pest monitoring, and life history information, prevent people from being in the position of spending more money to control a pest than the potential the pest has for causing economic loss.

Action thresholds for weeds, vertebrates, and pathogens may be very low due to the potential these pests have for building up over time and the difficulty in controlling them at later stages of infestation. For plant pathogens, it may be necessary to apply pesticides when environmental conditions favor disease outbreak. Usually once disease symptoms are seen, the damage has already taken place; control at this time may not be economical if plants are seriously injured or are no longer susceptible to further damage. Use indicators such as inoculum source, history of disease infection, and favorable conditions for development of the pathogen to determine if suppression is necessary. For weeds, herbicides are most effectively applied when weeds are in the seedling stages and are often significantly easier and cheaper to apply prior to planting or when crop plants are very young. Thus there is no "wait and see" time to allow development of higher populations. Usually, crop plants are most seriously affected when they are smaller and more susceptible to competition for light, water, and nutrients. In some cases, crop stage may be as important as weed numbers in determining treatment needs.

Factors that influence herbicide use decisions include favorable weather conditions, weed species, growth stage of problem weeds, growth stage of crop, degree of damage being caused by weeds, resistance of certain weed species to herbicides, soil type and condition, and herbicide persistence in the soil. Decisions to apply herbicides may be based on the economics of chemical control versus mechanical methods such as mowing or tilling.

SELECTING THE RIGHT PESTICIDE

Often several pesticides can be used for control of the same weed, insect, pathogen, nematode, or vertebrate pest in a particular situation; choosing the right pesticide can be a difficult task. Information about

FIGURE 9-6.

Pest management guidelines are published by the University of California for many different types of pests. These are useful for selecting the proper chemical and other control methods.

pesticides for specific uses can be obtained from pesticide label books (available by writing to the manufacturers), from farm advisors and agricultural commissioners, from licensed pest control advisers, from pesticide chemical handbooks, and from University of California publications and treatment guides (Figure 9-6). Agricultural and urban pest management guidelines, including pesticide recommendations, are included in the IMPACT (Integrated Management of Production in Agriculture using Computer Technology) computer network, part of the University of California Statewide Integrated Pest Management Program (Figure 9-7). Up-to-date pesticide use guidelines, pesticide toxicology information, and other pest management techniques are available through direct computer hook-up, or through computer terminals located in county Cooperative Extension offices. For additional information on this resource, contact your local Cooperative Extension office or the IPM Implementation Group, University of California, Davis, California 95616. Publications giving specific pest control recommendations can be obtained through the Division of Agriculture and Natural Resources Publications, University of California, Berkeley, California 94720 (a catalog of publications is available upon request). County Cooperative Extension offices can also order these publications for you.

When choosing a pesticide, consider such factors as its cost, the hazards to the user, plantback restrictions, persistence characteristics, the ease in use and ability of the pesticide to combine with other materials, its effects on natural enemies and beneficial insects, and required reentry intervals and harvest limitations.

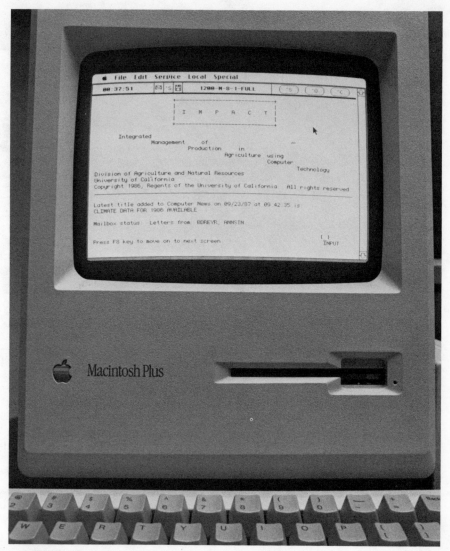

FIGURE 9-7.

The University of California Statewide Integrated Pest Management Project maintains an up-to-date listing of pest control recommendations that can be accessed through computer terminals. The system shown here, called IMPACT, also provides pesticide toxicology information.

Pest Species

Select a pesticide or pesticide combination that is suitable for the pest species or range of pest species being controlled. Be sure the materials chosen are capable of controlling the current life stage of these pests. Determine the suitability of the pesticides by reading pesticide labels; if the target site is not on the label, the materials should not be used.

Toxicity of the Pesticide To Be Used

Each pesticide has a toxicity category rating that suggests the relative hazard of the pesticide to people and organisms in the environment. Hazards are modified by such factors as formulation type (for example, microencapsulated formulations are safer for applicators to use than wettable powders), persistence in the environment, and amount of pesticide used. As a general rule, and if you have a choice, select pesticides from a lower toxicity category. These will usually be safer for you to work with, and often less harmful to the environment, beneficial insects, natural enemies, and animals.

Pesticide Persistence

Depending on the nature of the pest control problem, selection of the most effective pesticide should be based in part on persistence characteristics. Residuals are desirable in situations where reinfestations are constantly a problem, such as for termite control. Persistent pesticides may be more hazardous in areas where people live and work, however. For protection of beneficial insects such as honey bees, low persistence is often as important as low toxicity. Persistence is an important consideration when choosing herbicides because toxic residues may damage subsequent crops.

Certain types of pesticides, such as some chlorinated hydrocarbons, persist in the environment for a long time. Other pesticides, many organophosphates, for example, break down rapidly under normal environmental conditions. The persistence of a pesticide is often referred to as its half-life—the measure of how long it takes for the material to be reduced to half of the amount originally applied. Besides the type of pesticide, there are other factors that influence persistence. For instance, the amount of pesticide applied to a location dictates how much of the active ingredient will remain after a period of time.

Pesticide formulation types affect persistence. Microencapsulated and granular formulations tend to release the active ingredient over a longer period of time, therefore only part of the material begins to break down at the moment of application. Pesticides dissolved in oils or petroleum solvents may volatilize more slowly than water-soluble materials, and therefore persist longer. Wettable powders have a longer persistence to insects than do emulsifiable solutions.

The pH of the water used for mixing pesticides affects the speed of degradation (breakdown); the pH of the soil or plant or animal tissues may have a similar influence. Tissue or soil that is highly alkaline tends to cause more rapid breakdown of some pesticides than neutral or acidic tissue or soil.

Aspects of the physical nature of the surface being treated influence pesticide persistence. Porous surfaces or soil with high organic matter absorbs pesticide, reducing the amount of active ingredient available for pest control. Oily surfaces and waxy coatings on leaves and insect body coverings prevent uptake of the pesticide and may even combine with the active ingredient, reducing toxicity and persistence.

Soil microorganisms (bacteria, fungi, protozoans, algae, etc.) break down many pesticides and influence the persistence of pesticides in the soil environment. Water-soluble pesticides that percolate deeper into the soil are degraded more slowly than those that remain near the surface because fewer microorganisms are present at greater soil depths. Soils that have diverse populations of microorganisms break down many pesticides faster. High levels of organic matter in the soil will often slow down degradation because the organic matter binds to the pesticide, making it unavailable to the microorganisms. Repeated use of the same pesticide in the soil can increase the breakdown rate; this increase is apparently the result of either a greater soil microorganism population or an enzyme change in the microorganism population that makes the microorganisms more efficient in decomposing the specific pesticide.

Weather affects persistence. Wind and rain remove or dilute pesticides from target surfaces, lessening their effectiveness. High temperatures and humidity cause chemical changes in some compounds, accelerating breakdown. Sunlight produces photochemical reactions that decompose many pesticides. Cooler soil temperatures usually increase pesticide persistence.

Cost and Efficacy of Pesticide Materials

The cost of a pesticide is an important factor, but be careful not to base selection on cost alone. Check labels to see what rates of active ingredient are required, then convert the cost per pound of active ingredient into cost per unit of area treated. Cost must be balanced with the degree of effectiveness that can be expected. A pesticide that costs 30% more but gives 60% better control is often the better bargain unless you need a less effective pesticide to protect natural enemies. Unfortunately, the effectiveness of a pesticide is hard to measure and unbiased opinions are difficult to obtain. Local environmental conditions and methods of application also influence efficacy. Often you must make a value judgment based on personal experience. Keep a notebook and evaluate the results after each application to increase your knowledge of pesticide efficacy.

Weather conditions influence the quality of an application and efficacy of pesticides. Rainfall shortly after an application can wash off or dilute sprays, while windy conditions produce drift. Some pesticides are more effective in controlling target pests when temperatures are within an optimum range. High temperatures cause some pesticides to be phytotoxic to plants. Effective pesticide use, therefore, involves timing to coincide with optimal weather conditions as well as pest susceptibility and protection of natural enemies. Ideal conditions are not always possible during pesticide applications and often some compromises that affect efficacy must be made.

Ease of Use and Compatibility with Other Materials

Pesticides that are simple to use and are compatible with other pesticides have an advantage. Compatibility and ease of use also depend on how the pesticide is being used, what it is mixed with, and the nature of the treatment area.

Effect on Beneficial Insects and Natural Enemies

Conservation of beneficial insects and natural enemies must be considered in certain applications. If an integrated pest management program is being used, the ability of the selected pesticide to work within the goals of the program must be considered. This sometimes means compromising for less immediate control of the pest in order to achieve greater long-term control or better control of other pests.

Reentry Intervals and Harvest Limitations

The pesticide selected must work within constraints of legally established reentry intervals and the allowable days before harvest. These limitations have been established to protect workers, consumers, and the public from excessive residues.

SELECTIVE PESTICIDES

Selectivity refers to the range of organisms affected by a pesticide. A *broad spectrum* pesticide kills a large range of pests as well as nontarget species. A *selective* material controls a smaller group of more closely related organisms, often leaving beneficials and nontarget organisms unharmed. However, selectivity is not always desirable; there are some definite advantages to controlling multiple pests with a single broad spectrum pesticide, including cheaper pesticides (due to a larger market) and reduced application time and costs.

Pesticide selectivity is controlled by factors such as the penetration rate of the toxicant through an organism's outer body covering (or the cuticle of plant tissue), the speed at which the toxicant is excreted by organisms, and how the toxicant binds to tissues of different organisms. Some organisms have metabolic ways of altering or detoxifying pesticides and therefore are not harmed.

THE SELECTIVE USE OF PESTICIDES

Selectivity is also achieved through the way the material is mixed and applied. Application timing, application techniques, dosage level, types of formulations, use of adjuvants, and the ability to keep the pesticide on target help make broad spectrum pesticides more selective.

Timing of Application

The timing of an application is important for good control of target pests as well as protection of natural enemies and beneficial insects. Because some pesticides are more effective than others at different life stages of the target pest, pesticide application should be timed to coincide with the most susceptible stage. Understanding the biology of the pest will help you determine its susceptible life stages and decide if a pesticide application will be effective.

The life stage of nontarget plants in the treatment area is another important consideration. Some herbicides, for example, may be toxic to crop plants as well as weeds once the crop plants have reached a certain growth stage. Check pesticide labels for precautions on using pesticides during inappropriate life stages of nontarget plants.

Whenever possible, avoid injury to nontarget organisms by timing applications to periods when they are not present in the treatment area. This technique works well for honey bees because they forage only during warm, daylight hours; pesticides applied during the early morning, late afternoon, or on cold and cloudy days usually reduce hazards to honey bees. Insecticides and miticides applied while perennial plants are dormant may provide protection to nontarget and beneficial organisms; this technique is often an effective control method for some species of plant-feeding mites, aphids, scales, and other insects that overwinter in dormant plants.

Pesticide Application Techniques

Specific pesticide application techniques should be used to improve coverage, reduce drift, and achieve better control of pests. The amount of pesticide can often be lowered without sacrificing the quality of pest control. The right choice of application techniques can also reduce human and environmental hazards.

Equipment Operation. Learn how to properly operate pesticide application equipment. For example, the ground speed must remain constant at all times to assure even pesticide deposition. Check the nozzles frequently to make sure none have become clogged. Shut off all nozzles during turns to prevent an uneven spray pattern. When injecting pesticides into the soil, shut off nozzles and raise the boom before making a turn; leave enough room after the turn to bring equipment up to specified ground speed before restarting the pesticide flow.

Irregular ground may cause booms on tractors or other equipment to bounce unless they are well supported and ground speed is controlled. Make sure the boom is parallel to the ground at all times during application. If it is tilted, the application will be uneven from one side of the boom to the other. Adjust boom height to the recommended range specified by the type of nozzles being used (Figure 9-8).

When using handguns with high-pressure sprayers, keep the application uniform on all parts of the surfaces being treated. Make sure the spray reaches upper foliage or branches of trees or shrubs, and direct spray to all sides of the plants, but avoid excessive runoff.

The ground speed of orchard sprayers and other air blast sprayers must be regulated to keep them in balance with the volume of air being moved by the fans in order to provide even dispersion of pesticide droplets. Traveling too fast will result in poor spray droplet distribution because the sprayer blower cannot displace the air surrounding the trees, vines, or other plants fast enough.

When operating backpack sprayers, walk at a regular pace and avoid uneven steps. Hold the nozzle steady and keep it a constant distance from the target surface.

Preventing Gaps or Overlaps. Pesticide swaths have to be uniform, without overlaps or gaps, to be most effective and to make the best economic use of spray materials. In some agricultural settings, there are established furrows or rows for the operator to follow which assist in

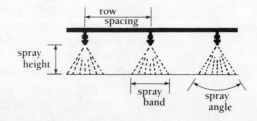

FIGURE 9-8.

Boom height must be adjusted to correspond to the type of nozzle being used.

achieving uniform application. In open fields, landscape, right-of-way, forest, and other types of settings, the operator needs to depend on some other method to prevent overlap or gaps. Foam markers can be used to mark sprayed areas to prevent overlap or gaps in the application pattern. These markers, attached to the ends of a spray boom, intermittently leave a deposit of long-lasting foam. The operator aligns the application equipment to the foam trail left from a previous pass. In some situations, colored dyes can be added to the spray mixture to visualize where spraying has taken place.

Leaving the application site to refill spray tanks presents the problem of relocating the exact point where spraying stopped; unless this location is found, there will be an uneven application. Marking devices can help avoid such problems. In some situations, use colored surveyors tape tied to plants to locate where spraying was stopped. Foam marking devices should be used when hazardous pesticides are being applied to eliminate the chance of the operator contacting treated surfaces.

Electronic positioning devices are an accurate way of guiding pesticide application equipment. These devices function on the same principle used to navigate aircraft by monitoring radio signals from a pair of nearby, fixed transmitters. Swath width and direction of travel is entered into a control unit mounted on the tractor near the operator. The positioning device monitors the location of the sprayer and, through readout instruments, guides the operator. When the application equipment is moved from a treatment area for refilling, maintenance, or repairs, the positioning unit electronically records its last location so that it can be returned to the exact spot.

Spot Treatments. Selectivity is enhanced if pesticides are used as spot treatments rather than being applied to an entire area. In addition, reduction of 70 to 90% of the pesticide used can be achieved through spot treatment methods. For example, some perennial weeds may grow in clumps scattered throughout a field, usually after all the other weeds have been controlled by cultivation or with herbicides; spot treatment involves treating just these clumps or patches rather than the whole field. Periodically, insect and mite pests congregate in localized areas before dispersing more generally, especially if the infestation is just beginning; control these pests by treating only the infested plants. In landscaped areas, pests may occur only on certain plant species, so avoid applying pesticides to uninfested plants. Frequently, just the edges of fields or landscape areas need to be treated for invading pests.

Some special types of application equipment and components can be used to assist in and improve the effectiveness of spot treatments. Field crop sprayers equipped with chemical injection pumps provide the operator with the option of mixing several different pesticides for the same application; concentrated, liquid formulations can be automatically metered into the boom and diluted with appropriate amounts of water from the spray tank before being emitted through nozzles. Rope wick applicators can be adjusted to wipe herbicides onto target weeds growing above the crop and can be used as selective, spot treatment devices. Compact, hand-held sprayers allow operators to efficiently apply pesticides as spot treatments to small areas. All-terrain cycles (ATCs) eliminate tedious walking and speed up spot treatment applications.

Band Treatments. In orchards and vineyards, herbicides can be applied as bands within the tree or vine row, leaving an area between the

FIGURE 9-9.

Strip spraying is a way of controlling weeds in tree or vine rows while reducing the use of herbicides. Only about ¼ of the actual acreage needs to be treated. The area between rows is usually mowed and maintained as a groundcover.

rows with a ground cover of weeds which may be mowed or cultivated. Only about one fourth as much herbicide is used per acre with this method (Figure 9-9) as opposed to total orchard or vineyard floor weed control with herbicides. Mowed weeds between trees or vines may reduce soil compaction, prevent erosion, reduce dust, and lower orchard or vineyard floor temperatures. However, weeds in untreated strips may compete with the trees or vines for water and nutrients under certain circumstances.

Treating Alternate Rows or Blocks. Spraying alternate rows or blocks in orchards, vineyards, or field crops is an application technique used to reduce the amount of insecticide, miticide, or fungicide applied and provide protection to natural enemies in untreated sections. This technique is often used when frequent treatments are required for the same pest, such as disease organisms, insects, or mites that occur over an extended period of time. Rows or blocks left untreated one time are sprayed during the following application (Figure 9-10). Only half as much material is used per application, although the concentration of spray material is not changed.

Low-Volume Applications. Techniques using low-volume and ultra-low-volume sprays improve the efficiency of pesticide application in certain situations. Low-volume applications use only about one fourth as much water carrier, therefore spray mixtures are more concentrated. Ul-

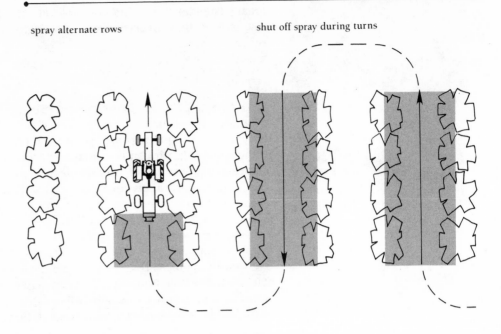

spray alternate rows · shut off spray during turns

FIGURE 9-10.

Applying pesticides to alternate rows or alternate blocks is a method that is sometimes used when several sprays are required for control of the same pest. This technique reduces the amount of pesticide used each time by 50%; it also provides locations in the treatment area for the protection of natural enemies.

tra-low-volume applications involve highly concentrated mixtures of pesticide combined with a carrier such as vegetable oil; these require the use of special application equipment. In some cases, pesticide amounts can be reduced by about one third with low-volume applications, and by as much as one half with ultra-low-volume applications as long as the application is consistent with label instructions or current UC recommendations. Savings of time, fuel, labor, and water can also be realized. Pesticide applications must be more accurate and calibration more precise when making low-volume applications. Operators using ultra-low-volume equipment work with more concentrated mixtures with potentially greater hazards, so precautions for wearing adequate protective equipment must be emphasized.

Dosage Level of the Pesticide

Reducing the amount of active ingredient applied to the treatment area for control of certain insects or mites may often be a way to reduce damage to natural enemies. This technique must only be used if the lower dosage can still be effective against target pests; the rate must be consistent with pesticide label instructions or current University of California recommendations. The percent of target pests controlled might be reduced, but by protecting natural enemies, overall control can be improved. Protecting natural enemies helps in long-term management of some insect and mite pest problems, further reducing the need for addi-

FIGURE 9-11.

Use of a bait station, like this one being used for control of ground squirrels, is a method of selectively using pesticides. The bait station excludes most nontarget organisms.

tional pesticides. Lowering dosage levels in this manner only works when adequate numbers of certain species of natural enemies are present before treatment, however.

Type of Formulation Used

The way a pesticide is formulated influences its selectivity. Granular formulations, for example, do not stick to foliage, increasing their selectivity to soil or aquatic pests. Applying a granular systemic formulation to the soil so that it can be taken up through plant roots is a selective application practice because pests feeding on plant tissue are killed, although most natural enemies and beneficial insects will not come in contact with the pesticide. (Nectar feeders like honey bees and many parasitic wasps may be harmed if the pesticide is translocated to nectar.) Liquid sprays and baits containing attractants improve the chances of target pests finding the pesticide and may make the pesticide more selective. Additional selectivity can be obtained by placing baits in feeding stations to exclude nontarget animals. Squirrels, for example, can be controlled with poisonous bait, while dogs, livestock, and children are excluded due to the design of the bait station (Figure 9-11).

Use of Adjuvants

Adjuvants are used to alter the selectivity as well as the effectiveness of pesticides. Use stickers, spreaders, and drift control agents to keep spray mixtures on target. Use surfactants to enhance uptake by target pests, and attractants to make pesticides attractive specifically to target organisms. These techniques—reducing drift, enhancing effectiveness, improving uptake by pests, and making the material more attractive to specific pests—may allow you to apply less pesticide, which will result in greater safety to natural enemies, beneficial insects, people, and the environment.

Keep the Pesticide On Target

Preventing drift improves the effectiveness of a pesticide application because more active ingredient is deposited on target surfaces. Drift factors include the skill of the operator, type of application equipment used, droplet size of the spray being applied, operating pressure of the sprayer, physical properties of the pesticide formulation, and general weather conditions as well as the unique local weather conditions (microclimate).

Operator Skill. Learn to use application equipment in ways that reduce or prevent drift problems. Use the lowest amount of pressure that will still produce an adequate spray pattern. Be sure nozzles are in good condition and properly aligned. Thickeners sometimes may be used to reduce the quantity of fine droplets. Apply pesticides only during times when there is positive air movement away from sensitive locations, or leave a buffer strip between the treated area and sensitive downwind areas. Shut off nozzles during turns.

Application Equipment. Your choice of application equipment can have an impact on how much pesticide drifts away from the treatment area. Sprayers that produce extremely small droplets increase drift. Lower the working pressure of the sprayer to reduce the quantity of small-sized droplets; use low-pressure nozzles to maintain a suitable spray pattern. Herbicides should be applied with low-pressure nozzles whenever possible. Use shields around nozzles to confine the spray and reduce drift problems. Rope wick applicators, used for weed control, wipe herbicides onto leaves of weed plants, thus eliminating drift. Electrostatic spray applicators electrically charge spray droplets as they leave the sprayer, causing droplets to be attracted to the oppositely charged treatment surface; because of the attraction between the spray particle and the treatment surface, fewer droplets drift away. Controlled droplet applicators (CDAs) produce more droplets of a uniform size, reducing the quantity of very small and very large droplets, and in this way reduce some of the drift potential.

Droplet Size. Droplet size, a factor in drift, is determined by spray pressure, nozzle size, and spray solution characteristics. Most conventional sprayers produce a wide range of droplet sizes, averaging between 40 to 500 microns (μm) in diameter. Table 9-2 compares the relative sizes, in microns, of different types of droplets. In a typical sprayer, about 70% of the droplets will be in the 100 to 250 μm diameter range. An additional 20% will be larger than this, while about 10% are smaller than 100 μm. Larger droplets drift less because they are heavier and fall quickly, but they are usually less effective for many types of pest control because they are inclined to bounce or run off target surfaces. Although a 400 μm droplet is only 10 times larger in diameter than a 40 μm droplet, it contains 1000 times more pesticide. Therefore, when large droplets bounce off or miss the target surface, a considerable amount of active ingredient is lost.

Smaller droplets, those below 100 μm, are the most efficient size for controlling many types of pests; most of the droplets that actually reach

TABLE 9-2.

Droplet Size Comparison. Most sprayers produce droplets ranging in size between 40 and 500 microns; droplets smaller than 100 microns are most effective for the control of insects and mites but are highly subject to drift.

DROPLET SIZE COMPARISON
in microns

microns	type	
1000		
	moderate rain	
500		
	light rain	average range of
300		droplets produced by
	drizzle	pesticide sprayers
200		
	mist	
100		
50		
	cloud	
30		
20		
	coarse aerosol	
10		
	sea fog	
5		
	smoke	
0		

insects and mites are smaller than 50 μm. Droplets of 100 μm or less in diameter are also best for effective penetration of dense foliage; however, these smaller droplets are the ones most likely to drift.

Physical Properties of the Pesticide Formulation. Factors such as the viscosity and volatility of pesticide formulations influence drift potential. Viscosity relates to the thickness of the liquid; the thicker or more viscous a liquid is, the more difficult it is to break it up into smaller droplets. Deposition aids include adjuvants used to increase the viscosity of a spray mixture, reducing drift by increasing droplet size. Volatilization is the process by which the pesticide is transformed from a liquid into a vapor; it takes place during application, or occurs within several hours after the spray has been deposited. Volatilization increases with higher temperatures, and volatilized spray material may drift many miles on slight air currents.

Evaporation of the carrier (usually water) in the spray droplet causes concentration of the pesticide and reduces droplet size. This happens while droplets travel between the sprayer and target surfaces. Hot, dry weather favors droplet evaporation, while high humidity and cool temperatures retard this process.

Weather Conditions. Wind is the most important component of weather contributing to pesticide drift. Slight winds of about 3 to 5 miles per hour help to distribute pesticide droplets and increase coverage because they contribute to the mixing of pesticide spray with surrounding air. When air movement is below 2 miles per hour, spray may not mix evenly with the air, resulting in an uneven application. Winds greater than 10 miles per hour are too strong for safe pesticide application be-

TABLE 9-3.

Method of Calculating Wind Speed.

For a practical means of estimating low wind velocities near the ground, toss a handful of dust into the air and walk downwind with the cloud of dust. If you can keep up with it in a slow walk, the wind is approximately 2 mph. If you can just barely keep up with it in a fast walk, the wind is approximately 4 mph. A lively run is approximately 10 mph, and only a well-trained athlete can dash 20 mph.

WIND VELOCITY *Miles Per Hour*	OBSERVATIONS
Less than 1	Smoke rises almost vertically.
1 to 3	Direction of wind shown by smoke drift, but not by wind vanes.
4 to 7	Wind felt on face; leaves rustle; ordinary wind vane moved by wind.
8 to 12	Leaves and small twigs in constant motion, wind extends light flag.
13 to 18	Raises dust and loose paper, small branches are moved.
19 to 24	Small trees in leaf begin to sway; crested wavelets form on inland water.

cause they promote drift and increase evaporation. Table 9-3 describes ways to judge wind speed. Small, hand-held devices known as *anemometers* can be used to measure wind speed more accurately.

In addition to wind, which is mostly horizontal air movement, vertical air movement contributes to mixing and dispersion of pesticide droplets. Air temperature influences vertical movement since warm air rises. Both the horizontal and vertical air movement and wind direction can be detected by burning a pot of oil or an old tire, although this practice may be illegal in some areas due to air quality standards.

Temperature inversion is a weather condition that may promote pesticide drift. It occurs when the air from 20 to 100 or more feet above the treatment location is warmer than the air at ground level. This warm air forms a cap, preventing vertical air movement and pesticide dispersion (see Figure 7-48 on page 225). Fine droplets and spray vapor are trapped and become concentrated as an invisible cloud under these conditions. A mild wind moves this pesticide cloud away from the treatment location, where it may return to the ground or surfaces of nontarget plants or other objects. This condition may be dangerous because the concentration of small droplets moves together downwind instead of dispersing widely into the atmosphere.

Treatment areas may have their own microclimate, different from surrounding areas. An irrigated field has higher humidity and lower temperatures than dry areas nearby, causing a localized, low-level temperature inversion which influences the concentration and movement of airborne pesticides.

FOLLOW-UP

Follow up after every pesticide application to learn if the application was successful. Table 9-4 is a follow-up checklist. Begin by comparing the amount of pesticide actually used with the anticipated amount; this should vary by no more than 10%. If more or less pesticide was applied, determine the cause. Check sprayer calibration, check tank mixing procedures, and recalculate the size of the target area. Look for clogged or worn nozzles and wear or blockage in the sprayer pumping system.

Inspect the application site to make sure coverage was adequate and uniform. (Wear protective clothing if necessary.) Look for signs of pesticide runoff, lack of penetration into dense foliage, or uneven coverage from the top to bottom of large plants.

When insecticides or acaricides have been applied, a second follow-up visit should be made to the treatment area a day or so after the application was made. (If the reentry interval has not lapsed, be sure to wear protective clothing and avoid unnecessary contact with plant foliage.) At this time look for signs indicating control of the target pest. Check for damage to natural enemies in the area and look for other problems, such as phytotoxicity to plants or spotting of painted surfaces. Watch for pest resurgence and secondary pest outbreaks.

Follow up fungicide applications with an inspection to verify that the pathogen has been suppressed.

When herbicides have been applied, follow up to see which weed species were controlled and which were not controlled or only partially controlled. Also look for damage to nontarget plants. Record this infor-

TABLE 9-4.

Pesticide Application Follow-up Checklist.

Amount of Pesticide Used:
(a) Calculated amount required for job: _____
(b) Actual amount used: _____
(c) Variation—divide (a) by (b) then multiply by 100.
  Subtract answer from 100. (This should be between +10 and −10.)
  _____

Coverage:
(a) Uniform _____ or uneven _____
(b) Runoff? _____
(c) Penetration into all areas? _____

Effectiveness:
(a) Target pests controlled or reduced below economic injury level? _____
(b) Condition of natural enemies: _____
(c) Secondary pest outbreak? _____

Problems:
(a) Spotting or staining of surfaces? _____
(b) Injury to plants? _____
(c) Other: _____

Comments: _____
_____
_____

mation on the treatment record and use it to determine if an additional herbicide treatment is needed.

Record your follow-up observations in the same notebook used to record other aspects of the pesticide application. This information will be useful when planning future applications for the same pest or in similar target locations.

10 Pesticide Application Equipment

Air blast sprayers have powerful fans to propel spray droplets to target surfaces.

Pesticide application equipment ranges from simple devices attached to a garden hose to elaborate machines mounted on self-propelled ground applicators, fixed wing aircraft, and helicopters. Some equipment is designed especially for application of dusts, other units are used only for granules, and many different types of equipment are used to apply liquids. Special equipment is available for weed control, for applying pesticides to orchards, vineyards, row crops, and field crops, for injecting pesticides into the soil, for applying pesticides to animals, for controlling aquatic pests, for controlling pests in buildings and homes, for applying aquatic pesticides, and for many other purposes.

In this chapter, the different types of pesticide application devices, except those used with aircraft, are described. Features of components, such as tanks, pumps, and nozzles, are explained and illustrated. Maintenance procedures are also discussed.

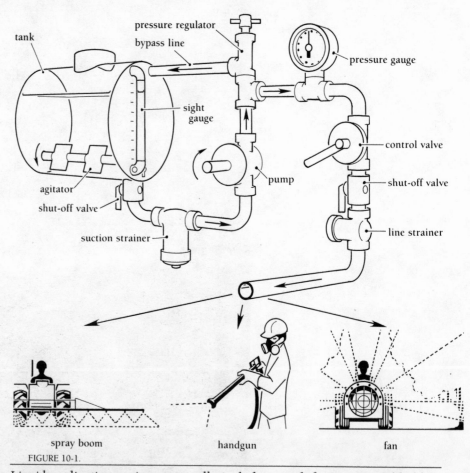

FIGURE 10-1.

Liquid application equipment usually includes a tank for mixing and holding pesticides (often equipped with an agitator), a pump for creating hydraulic pressure, and may include a pressure regulator, pressure gauge, control valve, and several types of strainers. Spray is emitted through nozzles on a spray boom, manifold, or handgun, and may be dispersed by a fan.

LIQUID APPLICATION DEVICES

Most of the application devices used for liquid pesticides employ hydraulic pressure or air to propel pesticide droplets. This equipment is either hand-operated or powered by a mechanical source such as a tractor power take-off (PTO) or an electric, gasoline, or diesel motor. Liquid application devices consist of several components, including: (1) a tank for mixing and holding the pesticide; (2) a pump or other device for creating pressure to move the liquid; (3) one or more nozzles for breaking the spray up into small droplets and directing it toward the target area; and, on some, (4) fans, pressure regulators, filter screens, control valves, agitators, booms, hoses, and fittings to improve pesticide handling, mixing, and application (Figure 10-1).

Components of Liquid Application Equipment

Before considering a pesticide application machine as a whole unit, look at all the individual components to make sure they meet your application needs. Some components, such as nozzles, are easy to replace when worn or damaged or if application needs change. Other parts, such as tanks, are expensive to replace, and may be difficult to repair if they become damaged, so should be selected with care to be sure they are appropriate to your requirements.

Pesticide Tanks

Tanks used for mixing and holding liquid pesticides are commonly made from either metal, fiberglass, or thermoplastic materials such as polyethylene and polypropylene. Tanks need to be corrosion and rust resistant to protect them from reacting with corrosive pesticides. They must also be nonabsorptive so pesticide residues can be easily removed. Tanks should have a large opening for easy filling and cleaning (Figure 10-2), and a cover to prevent pesticides from spilling or splashing out.

FIGURE 10-2.

Pesticide tanks must have a large top opening for ease in filling. The opening should be fitted with a splashproof cover.

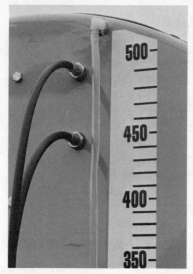

FIGURE 10-3.

A sight gauge enables the operator to tell how much pesticide is contained in the tank at all times. External tubes must be equipped with a shut-off valve to prevent leaks if the gauge becomes damaged.

Many counties require that covers be lockable. Although not always required by law, a lockable cover can be a worthwhile safety feature that prevents unauthorized or accidental exposure to the contents of the tank. Larger tanks need a bottom drain so they can be completely emptied. All pesticide tanks with a capacity of 50 gallons or more are required to have a sight gauge or other accurate means of determining the amount of liquid contained in the tank. An external sight gauge must be equipped with a shut-off valve to prevent loss of tank contents if it becomes damaged (Figure 10-3).

Tanks are commonly available in capacities from 3 to 1600 gallons. Table 10-1 is a guide for selecting tanks based on pesticide application needs.

Metal Tanks. Metal tanks are either made from one of several different grades of stainless steel or are galvanized or coated with epoxy or a similar substance to prevent them from rusting or corroding. Stainless steel has superior qualities for rust and corrosion resistance, so most pesticides can be used in them. Stainless steel cleans easily and is strong and durable. If damaged, stainless steel tanks can be repaired, although repairs may require considerable skill. Stainless steel tanks are more expensive than tanks made from other materials but generally last longer. If you select a stainless steel tank, be sure it is a grade that is compatible with the types of pesticides you use.

TABLE 10-1.

Pesticide Tank Selection Guide.

	Coated metal	Stainless steel	Fiberglass	Polypropylene	Polyethylene
Rust and corrosion resistant	Fair to good	Excellent	Excellent	Excellent	Excellent
Easily cleaned	Good	Excellent	Fair	Excellent	Excellent
Easily repaired	Yes	Yes	Yes	No	No
Cost	Moderate	High	Moderate	Low	Low
Acid resistance	Fair to good	Depends on grade	Fair to good	Good	Good
Alkali resistance	Excellent	Excellent	Fair	Good	Good
Organic solvent resistance	Excellent	Excellent	Poor to fair	Good	Fair
Strength and durability	Good to excellent	Excellent	Good	Good	Good
Weight	Heavy	Heavy	Medium	Light	Light
Absorb pesticides	If scratched	No	If scratched	No	No
Requires external reinforcement	No	No	No	Yes	Yes

Galvanized or coated metal tanks require more care and attention. Scratches or chips in the protective coating expose bare metal and may cause serious corrosion problems. In addition, some pesticides cannot be used in coated metal tanks (glyphosate, for example, reacts with metal to produce hydrogen gas, resulting in an explosion hazard). Epoxy coatings have fair to good resistance to acids, but excellent resistance to alkaline materials and organic solvents. Metal tanks with an epoxy coating must be inspected on a regular basis and blemishes touched up to prevent rust or corrosion.

Fiberglass Tanks. Fiberglass tanks are strong and durable and can be repaired easily if damaged. They are lighter than metal tanks, are highly resistant to organic solvents, have good to fair resistance to acids, but have only fair resistance to alkali materials. Scratches or abrasions to interior walls of fiberglass tanks allow some pesticides to be absorbed into the fiberglass material, which may possibly cause contamination of future tank mixes. Scratched areas must be recoated with resin to negate this problem.

Thermoplastic Tanks. Most thermoplastic materials have good resistance to acids and alkalis. Polyethylene is a common material used for spray tanks, although low-density polyethylene does not resist organic solvents well. Most plastic sprayer tanks are constructed from either high-density polyethylene or polypropylene. These plastics are lightweight and durable; when warmed, however, they become flexible and will deform unless they are reinforced or supported. Minor scratches or abrasions do not cause absorption problems, but polyethylene and polypropylene are difficult to repair if they become punctured or cracked.

Pumps

Most liquid sprayers use a pump to move the pesticide from the tank to the nozzles and to create adequate pressure to propel spray droplets to the target (Figure 10-4). Small, hand-operated sprayers use simple bellows or piston pumps that are operated by squeezing a trigger or moving a lever. Some compressed air sprayers employ a piston-type air pump (similar to a bicycle tire pump) to compress air in a sealed pesticide tank; when a valve is opened, liquid is forced out of the tank by this air pressure. Other types use cartridges of compressed carbon dioxide gas to propel the spray. Powered sprayers are equipped with more complex mechanical pumps designed for use with liquid pesticides. Selection of the appropriate pump for a particular application depends on the formulations of pesticides being used and on the volume and pressure required to apply them. Water-soluble and emulsifiable concentrate pesticide formulations are less abrasive to pumps than wettable powder, flowable, or dry flowable formulations. Construction materials and type of pump also affect how well pumps perform and wear. When choosing a sprayer pump, the following features should be considered:

Capacity. A pump must be able to supply enough volume for all nozzles under every use condition. If the sprayer is equipped with hydraulic agitation, the pump must have additional output for recirculating material in the tank. Output capacity of a pump is given in gallons per minute (gpm).

Pressure. Pumps must produce the desired capacity of spray material at a pressure suitable for the type of work being performed. Some high-capacity pumps are only able to produce low pressures. High-pressure

FIGURE 10-4.

Pumps for powered sprayers are available in many different styles and capacities. Be sure to select a pump that is suitable to the type of pesticide being used and one that can supply the pressure and volume requirements of your application needs.

pumps are often able to be regulated for low-pressure work as well. Pressure is measured in pounds per square inch (psi).

Resistance to corrosion and wear. The type of materials that the pump is constructed from, as well as the design of the pump, dictate its ability to resist corrosion and wear. Pumps with the fewest parts coming in contact with spray chemicals are the most suitable for corrosive pesticides. Pump design is also important in reducing the amount of wear due to abrasion by wettable powders.

Ease of repair. An important feature of any pump is the ease with which it can be repaired. Parts must be readily available.

Type of drive. Pumps require different operating speeds depending on their design. Most tractor power take-off shafts rotate at 540 and/or 1000 rpm. Gasoline and diesel engines and electric motors all have specific operational rpm ranges. Because each pump also has particular horsepower requirements to operate efficiently, take into consideration the speed and horsepower of the drive unit. When higher speeds are required by a pump, you will need a transmission or gear unit to increase the necessary rotational speed.

Table 10-2 is a guide for selecting common pump designs most suitable for the pesticide formulations and applications you may encounter. Other types of pumps not included in this table may also be available and suitable for certain applications.

Diaphragm Pumps (Figure 10-5). Diaphragm pumps are a popular style used on several types of spray equipment. These pumps are made of durable materials, including aluminum, steel, and high-impact plastic. They are moderately priced, and can be used for low- and high-pressure applications. Diaphragm pumps handle abrasive and corrosive chemicals well because only the chemical-resistant diaphragm is exposed to pumped liquids. They are also simple to maintain and repair. Diaphragm pumps have a low- to medium-volume capacity of between 5 and 40 gallons per minute and can produce pressures in the range of 200 to 700 psi. These pumps operate in the range of 500 to 800 rpm.

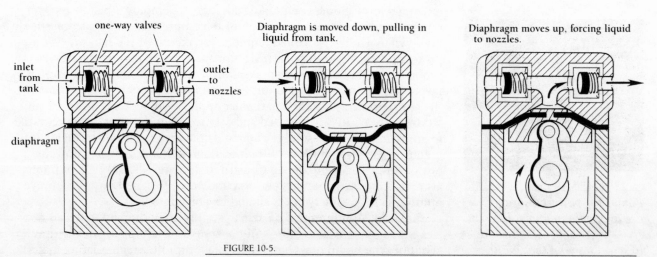

FIGURE 10-5.

In a diaphragm pump, a flexible diaphragm is moved up and down by a cam mechanism. This oscillation moves liquid through one-way valves. Some diaphragm pumps incorporate two or three diaphragms moved by the same cam.

TABLE 10-2.

Guide to Selecting the Proper Pump for a Pesticide Sprayer.

	PRESSURE RANGE (psi)	OUTPUT VOLUME (gpm)	OPERATING SPEED (rpm)	SUITABLE PESTICIDE FORMULATIONS	COMMENTS
Centrifugal	5 to 200	>200	1000 to 5000	All.	Used on large, heavy duty sprayers. Common on air blast sprayers. Best for high-volume uses.
Diaphragm	20 to 700	5 to 40	500 to 800	All (organic solvents may deteriorate some parts).	Often used on low-volume weed sprayers. Also used with some high-pressure equipment.
Gear	20 to 100	5 to 65	500 to 2000	Nonabrasive.	Limited uses. Good for low volume and low pressure.
Piston	20 to 1000	2 to 60	500 to 800	Nonabrasive unless equipped with wear-resistant cups.	Excellent for high-pressure applications. Very versatile.
Roller	10 to 300	8 to 40	300 to 2000	Nonabrasive only; may be damaged by organic solvents.	Limited low-volume uses. Can produce moderate pressure.

Pumping action is accomplished by a cam that oscillates one to three diaphragms. Movement of a diaphragm in one direction creates a negative pressure inside a pumping chamber; this negative pressure forces open a one-way valve, pulling liquid pesticide from the tank into the chamber. As the diaphragm reverses direction, the one-way valve seals shut. Positive pressure in the chamber forces open a second one-way valve, expelling the liquid pesticide out another orifice. Pressure in the system may pulsate due to this action. More expensive pumps are equipped with two or three diaphragms working opposite each other to minimize pulsating pressure. Some designs incorporate surge chambers to reduce pulsating pressure.

Diaphragm pumps have only a few moving parts. Diaphragms usually wear out after a period of time and must be replaced when they begin to leak. Valves, usually made of neoprene rubber, should be replaced when they fail to seal properly. Rubber materials may be deteriorated by petroleum-based solvents found in emulsifiable concentrate formulations.

Low-pressure diaphragm pumps are suitable for most herbicide applications. High-pressure styles can be used in hydraulic and air blast sprayers.

Roller Pumps (Figure 10-6). Roller pumps are among the least expensive pump types. They are capable of producing moderate volumes, between 8 and 40 gpm. Low to moderate pressures, in the range of 10 to 300 psi are possible with roller pumps, while operating speed is in the range of 300 to 2000 rpm.

A series of rollers fit into slots around the circumference of a rotating disc or impeller. The impeller spins off-center to its housing, allowing the rollers to move further in or further out of their slots. Liquid is picked up at the point where rollers are furthest out; impeller rotation forces the rollers back into their slots and pressurizes the liquid.

Roller pumps are subject to considerable wear, especially from abrasive materials like wettable powders. Rollers made of rubber last longer, but nylon or teflon rollers must be used to pump petroleum-based pesticides such as oils or emulsions, which deteriorate rubber. Rollers can usually be replaced easily when they become worn. Roller pumps are most suitable for herbicide applications, especially if flowables, emulsifiable concentrates, soluble powders, or other nonabrasive formulations are used.

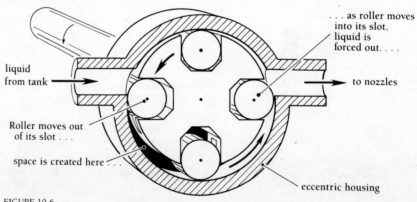

FIGURE 10-6.

Roller pumps consist of cylindrical rollers that move in or out of slots in a spinning rotor. This action creates space for liquid during half of the rotor rotation and discharges the liquid out of the pumping chamber during the remainder of the rotor rotation.

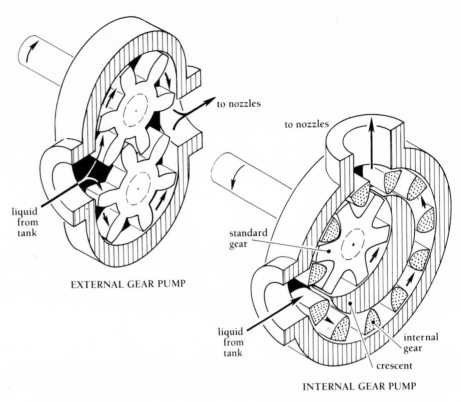

to nozzles

to nozzles

standard
gear

liquid
from
tank

liquid
from
tank

internal
gear

crescent

EXTERNAL GEAR PUMP

INTERNAL GEAR PUMP

FIGURE 10-7.

The external gear pump moves liquids by a meshing action of two identical gears. The internal gear pump consists of a standard gear that meshes with and drives an internal gear to move liquids. The close meshing of the gears in both of these designs forces fluids to move in only one direction through the pumping chamber.

Gear Pumps (Figure 10-7). Gear pumps are used for low-pressure applications between 20 and 100 psi. They are capable of producing spray in a range of 5 to 65 gpm, and operate in the range of 500 to 2000 rpm.

Two types of gear pump designs are manufactured. The *external* gear design employs two identical gears that mesh with each other and move fluid through the pumping chamber. An *internal* gear design utilizes a smaller gear meshing inside a larger gear, producing the pumping action.

Gear pumps are made of brass, bronze, or alloy steel. These are usually molded parts, making them difficult to repair. Gear pumps are suitable primarily for lubricating liquids such as oil sprays or emulsifiable concentrates. Wettable powders and similar formulations are not recommended for use in gear pumps because of their abrasiveness.

Centrifugal Pumps (Figure 10-8). Centrifugal pumps are made of high-impact plastic, aluminum, cast iron, or bronze, and are heavy duty and adaptable to a wide variety of spray applications. They produce volumes in excess of 200 gpm at pressures ranging between 5 and 200 psi. They require operating speeds between 1000 and 5000 rpm. Pumping action is created by a high-speed impeller forcing liquids out of the pump. Increased pressures are obtained by adding one or more stages of impellers; fluids pass from one impeller to the next.

Centrifugal pumps have a wide range of applications. They are well adapted for spraying abrasive materials because there is no close contact

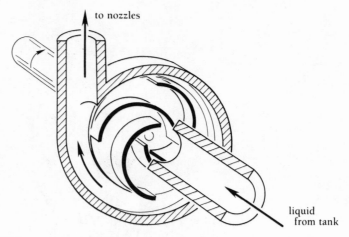

FIGURE 10-8.

In a centrifugal pump, liquid enters near the center of a vaned rotor. As the rotor spins, the liquid is moved away from the center by centrifugal force. Rotors must turn at a high rpm in order to build up sufficient pressure for most spray applications.

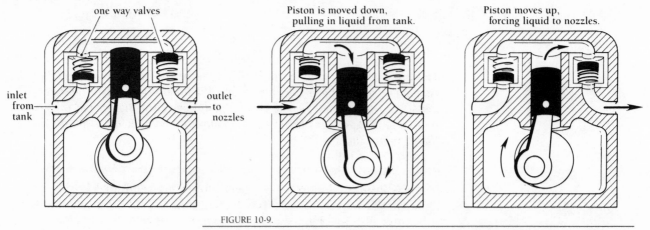

FIGURE 10-9.

This sequence shows how a piston pump functions. The downward movement of a piston draws liquid through a one-way valve into the cylinder. When the piston moves up, liquid is forced out through another one-way valve. Some pumps consist of several pistons working opposite each other.

between moving parts. They are often easy to repair and are well suited for high-volume air blast sprayers.

Piston Pumps (Figure 10-9). Piston pumps are capable of producing pressures in the range of 20 to 1000 psi at volumes between 2 and 60 gpm; they operate between 500 and 800 rpm. These are often the most expensive pumps but are usually required where high pressures are needed or where both high and low pressures are used.

One or more pistons travel inside cylinders, forcing fluids through one-way valves. This action is similar to that of diaphragm pumps, although the displacement of the piston is usually greater than diaphragm displacement. Pulsating pressure may be a problem with piston pumps, however. Piston pumps are usually subject to wear from abrasive chemicals, although most have easily replaceable cylinder liners and piston cups. More expensive piston pumps have stainless steel or ceramic cylinder liners to resist wear.

Agitators

Agitators are used both for initial mixing of pesticides and to keep insoluble mixtures from settling inside spray tanks. Equipment with agitators is recommended whenever wettable powders, water dispersible granules, or emulsions are used. Agitators are of two types, hydraulic and mechanical.

Hydraulic Agitators (Figure 10-10). Hydraulic agitators circulate spray material back through jets located in the bottom of the spray tank. In some designs, this fluid comes from a bypass on the pressure side of the pump; other sprayers have a separate pump to circulate fluid for tank agitation. Jets located in the bottom of the tank must be at least 1 foot from tank walls to prevent the walls from being weakened or developing holes.

The main disadvantage with hydraulic agitators is that they are not able to break up settled spray material when the pump has been shut down for a period of time. Severe settling requires mechanical agitation to suspend insoluble particles.

Mechanical Agitators. Mechanical agitators are propellers or paddles mounted on a shaft that is positioned near the bottom of a spray tank (Figure 10-11). The shaft passes through the tank wall and connects to the drive line by belts or chains. Mechanical agitators provide constant mixing in the tank as long as the sprayer is running. They are usually effective in suspending settled formulations. Some maintenance is required, especially where shafts pass through tank walls. Packings and grease fittings prevent leaks, but need periodic tightening and servicing. Be sure to use a marine-grade grease on bearings and seals exposed to liquids. Belts or chains also require periodic tightening and servicing.

Filter Screens and Strainers

Filter screens and strainers are used to protect pumps and prevent clogged nozzles. They remove undissolved clumps of pesticide formulation, sand, soil, and other debris from the spraying system. Filter screens and strainers help prevent problems of clogged nozzles when using water from agricultural wells that may contain small quantities of sand.

Strainers. Strainers (Figure 10-12) are devices containing filter screens to remove foreign particles that would otherwise clog nozzles or damage pumps. Strainers are usually positioned between the tank and pump (*suction strainer*), between the pump and nozzles (*pressure strainer*), and at the nozzles (*nozzle strainer*).

A simple suction strainer may connect to the end of the intake hose and is usually positioned near the bottom of the spray tank. This type of strainer may be used in low-capacity systems, often with roller pumps. Low-capacity suction strainers, sometimes called *line strainers*, are used in other systems, and are inserted into a section of the hose connecting the tank to the pump. Both the suction and low-capacity line strainers have an effective straining area of about 3 to 5 square inches. Larger

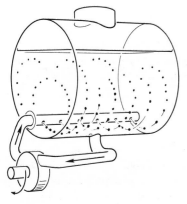

FIGURE 10-10.

Hydraulic agitators recirculate spray material back into the spray tank, providing continual mixing of the solution.

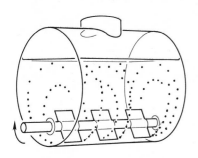

FIGURE 10-11.

Mechanical agitators consist of paddles or propellers that continually stir the liquid in the spray tank.

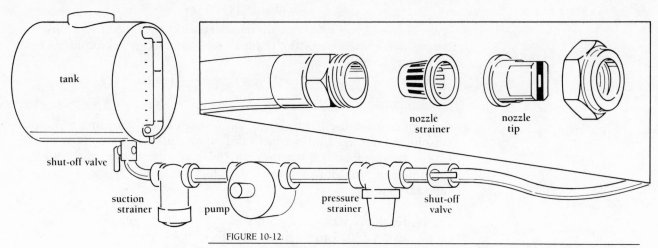

FIGURE 10-12.

Strainers hold filter screens and are located in different parts of the system. The suction strainer is positioned between the tank and pump. The pressure strainer is located between the pump and nozzles. Nozzle strainers are located adjacent to nozzles.

capacity sprayers are equipped with a *Y* or *T line strainer*, located between the tank and pump. They contain a screen providing from 7 to 30 square inches of filter surface. A capped opening on the line strainer allows removing and cleaning the filter screen without having to disassemble any plumbing or disconnect hoses. Shut-off valves placed between the strainer and the tank prevent loss of spray material when the filter is removed for servicing.

The pressure strainer is similar to the suction strainer, but is positioned between the pump and nozzles. It also contains a capped opening so the filter screen can be removed for cleaning. The sprayer should have a shut-off valve between the pressure strainer and the pump to prevent the loss of spray material while cleaning filters.

Nozzle strainers are used to protect nozzle orifices from smaller particles missed by the suction and pressure strainers. Nozzle strainers are placed between the nozzle and nozzle retainer. Several types of nozzle strainers can be selected to accommodate a range of applications. Some nozzle strainers are equipped with spring-loaded check valves to prevent nozzles from dripping when the spray is turned off; check valves usually cause a drop in pressure at the nozzle, however, so the system may require extra capacity to accommodate them.

Filter Screens. Filter screens range in size from 10 to 200 mesh. A 10 mesh size denotes 10 openings per inch, therefore the larger the mesh number the finer the screen. For most spraying equipment, the suction strainer should have a coarse screen in the range of 10 to 20 mesh, because any smaller size mesh restricts liquid flow going to the pump and plugs easily, resulting in a pressure drop within the system and increased strain on the pump. The pressure strainer should have a finer screen in the range of 40 to 50 mesh. This screen collects particles missed by the suction strainer. Screen sizes as small as 80 mesh should be used at the nozzles where small orifices are being used. Screens of nozzle strainers most commonly range from 50 to 200 mesh. Smaller mesh screens are used when using smaller nozzle orifices. Filter screens must be matched with nozzle orifice dimensions to prevent clogging; they should never be much smaller than the orifice. Nozzle strainers are not required if the nozzle orifice is larger than the pressure strainer screen mesh.

Nozzles

Spray nozzles control the application rate, drop size, spray pattern, thoroughness, and safety of the pesticide application (Figure 10-13). Nozzles may be used singly on a boom or hand-held spray wand or spray gun; in most cases, multiple nozzles are arranged on a boom or manifold to get the desired spray pattern. Several different nozzle types are available, depending on the type of application. Nozzles may well be the most important part of the sprayer. If nozzles are not properly selected and maintained, all efforts for effective pest control can be wasted. Nozzle selection is based on several criteria, including the type of material the nozzle is made of, the style of nozzle, and the nozzle orifice size.

Nozzle Construction and Wear. Nozzles are made out of several different materials, all of which are subject to wear. The design of the nozzle, the kinds of materials being sprayed, and the spray pressure influence nozzle wear. Flat-spray styles with sharp-edged orifices wear much faster initially than, for example, a flooding tip with a circular orifice. Also, as the spray pattern angle is increased, the wear on the nozzle is increased. Further, the size of the orifice affects wear; larger orifices wear more slowly than smaller ones.

Spray materials influence wear differently based on whether solids are dissolved or suspended in the liquid; true solutions cause the least amount of wear, while suspended solids cause different amounts of wear depending on particle size, size distribution, shape, hardness, and concentration. The solids that influence wear may be the pesticide or may be an inert carrier in the formulation. Rate of nozzle wear, even when using the same type of pesticide over a period of time, may be variable because chemical companies may make small changes in inert ingredients

FIGURE 10-13.

Spray nozzles control the application rate, drop size, uniformity, thoroughness, and safety of the pesticide application.

in their formulations from time to time that have no effect on the performance of the pesticide. Also, formulations of the same pesticide can vary from one manufacturer to another. Some pesticide chemicals may form crystals in the suspension depending on factors such as water pH, water temperature, and the presence of other chemicals; these crystals often increase wear on nozzles. Higher liquid pressure increases the rate of nozzle wear.

As a nozzle wears, the volume and pattern of spray changes, affecting the quality of application. Nozzles must be replaced when they fail to deliver an accurate amount and desired spray pattern. The output from nozzles of the same size, used together such as on a boom, should not vary from each other by more than 10%. To ensure uniform wear, all nozzles used together must be made of the same material. Nozzles are constructed of the following materials:

Brass. Brass nozzles are inexpensive but wear quickly from abrasion. Brass is an acceptable material when abrasive sprays are not used, or if nozzles are replaced frequently.

Stainless steel. Stainless steel nozzles do not corrode and they resist abrasion. Although hardened stainless steel wears exceptionally well, nozzles made from it are more expensive than most other materials.

Aluminum and monel. Aluminum and monel nozzles resist corrosion but are highly susceptible to abrasion because they are such soft metals. Aluminum and monel nozzles are not generally recommended unless specific corrosion resistance is required.

Plastic. Plastic nozzles are inexpensive and resist corrosion, but when made totally of plastic they may swell when exposed to organic solvents; plastics also have low abrasion resistance. Solid plastic nozzles can be used only with selected pesticides. Some plastic nozzles have stainless steel orifice inserts, making them much more resistant to wear; the inserts also reduce problems of swelling.

Tungsten carbide and ceramic. Tungsten carbide and ceramic nozzles are highly resistant to abrasion and corrosion. To reduce costs, manufacturers use tungsten carbide or ceramic inserts with brass or plastic nozzle bodies; these types of nozzles are recommended for high-pressure and abrasive sprays.

Nozzle Types. Different applications require nozzles adapted to specific requirements. Nozzles used to apply herbicides in a field are generally unsuitable for applying insecticides or fungicides to foliage. Spraying weeds along a roadside may require different nozzles than spraying weeds in a corn field. Orchard sprayers have different nozzle requirements than row crop sprayers. Residential, industrial, and institutional applications need nozzles suitable for confined spaces.

Flat-spray nozzles (Figure 10-14). Flat-spray nozzles distribute pesticide in a flat fan shape, with fan angles ranging between 50 and 160 degrees. Flat-spray nozzles produce a pattern with more spray droplets in the center of the fan, tapering off at each end, allowing for a series of flat-spray nozzles to be evenly spaced on a boom, with a slight overlap of the spray pattern of each nozzle. If the nozzles are operated at the correct height, the spray patterns will form a smooth overlap with nearly uniform deposits between nozzles. Flat-spray nozzles may be used for herbicide, fungicide, and insecticide applications.

An *off-center* flat-spray nozzle emits a pattern of spray more to one side than the other (Figure 10-15). This nozzle style is commonly used at the end of a boom to increase the spray swath width. It is commonly

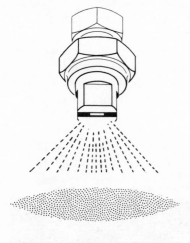

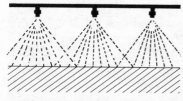

FIGURE 10-14.

Flat-spray nozzles produce a fan-shaped pattern having more droplets in the center part of the fan than at either edge. This allows for overlap of the spray, eliminating gaps and providing an even pattern when multiple nozzles are used.

used in orchards and vineyards for applying herbicide to both sides of the plant row.

Low-pressure flat-spray nozzles provide an acceptable spray pattern at pressures as low as 10 psi. These nozzles are used for the same types of applications as conventional flat spray nozzles, but have less problems with drift because larger drops can be produced at a lower pressure.

Even flat-spray nozzles (Figure 10-16). Even flat-spray nozzles are available in 40, 80, and 95 degree fan angles. They are flat nozzles, except there is no tapering of spray volume at the pattern ends. These are used when applying one or more separate bands which do not overlap.

Cone nozzles (Figure 10-17). Cone nozzles are used for applying insecticides and fungicides to dense foliage. Spray is produced in either a *hollow* cone or *solid* cone pattern, with spray angles between 20 and 110 degrees; hollow cone spray nozzles are used for most applications. Solid cones are used where larger, heavier droplets are required to reduce drift and if greater volume is needed.

Disc-core nozzles (Figure 10-18), which are a type of cone nozzle, are used in air blast sprayers for high pressure and high flow rate applications of insecticides and fungicides. Standard disc-core nozzles produce a hollow cone spray pattern, while full cone spray patterns are available for greater volume output. The orifice is located in a disc made of brass, hardened stainless steel, ceramic, or tungsten carbide. Fitted behind the disc is a core (sometimes called a spinner plate) that produces a high rotation speed of the liquid into the whirl chamber. Cores are made of brass, aluminum, nylon, hardened stainless steel, and tungsten carbide. Discs and cores are available in many sizes. Using different combinations of discs and cores provides a wide range of volume output and droplet size.

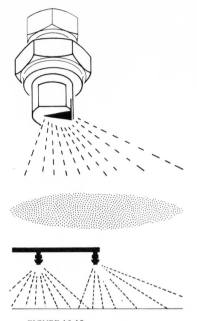

FIGURE 10-15.

Off-center flat-spray nozzles emit a full pattern of spray to one size of the nozzle. These are used on the ends of spray booms to extend the reach of the nozzle.

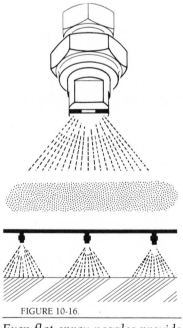

FIGURE 10-16.

Even flat-spray nozzles provide a uniform distribution of spray throughout the fan pattern. The spray from these nozzles is not overlapped; these nozzles are used to apply separate bands of pesticide without overlap.

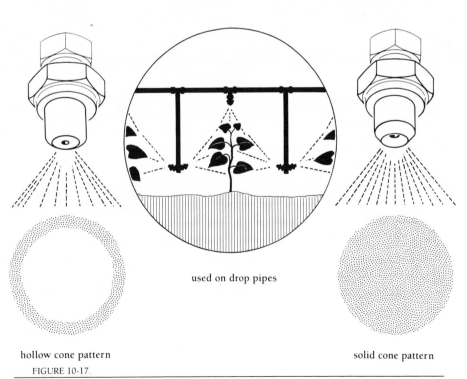

hollow cone pattern

used on drop pipes

solid cone pattern

FIGURE 10-17.

Cone nozzles are used for applying insecticides and fungicides to foliage, especially when larger volumes are required to assure complete coverage.

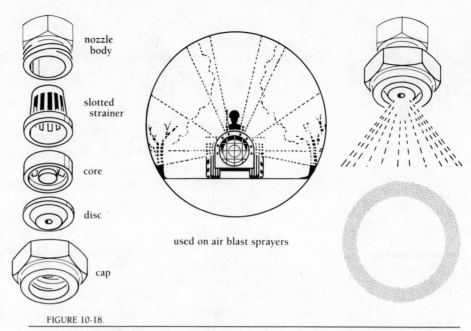

nozzle body

slotted strainer

core

disc

cap

used on air blast sprayers

FIGURE 10-18.

Disc-core nozzles are used in high-pressure applications that have a high flow rate, such as air blast sprayers. They are also used for some low-volume applications with air blast sprayers. Cores (or spinner plates) break up spray droplets and improve the deposition pattern.

Solid stream nozzles (Figure 10-19). Solid stream nozzles produce a single, solid stream of pesticide. These are used in handgun sprayers for spraying distant objects, for crack and crevice treatment in and around buildings, and for banding fluids in row crops. Different orifice sizes determine the volume of output.

Flood nozzles (Figure 10-20). Flood nozzles produce a relatively wide fan angle, up to 160 degrees, and are used to apply large volumes of liquid at low pressure. They are commonly used for liquid fertilizers; flood nozzles are rarely used for pesticide application because it is usually unnecessary to apply large volumes of liquid. They are normally widely spaced when used on booms because of the large fan angle.

Broadcast nozzles (Figure 10-21). Broadcast nozzles are used on boomless sprayers. They consist of a cluster of nozzles all attached at one point, and provide a wide side-to-side dispersal of between 30 and 60 feet. These nozzles are useful where a spray boom cannot be used but a wide swath is needed. Broadcast nozzles, like flood nozzles, are often used where large volumes of liquid need to be applied. It is more difficult, however, to achieve accurate pesticide placement with broadcast nozzles than with a series of flat spray nozzles evenly spaced on a boom.

Bifluid nozzles. Bifluid nozzles break liquids up into extremely fine droplets, such as a mist or fog, by the use of a high-velocity airstream. These are used on some types of aerosol generators for fogging enclosed areas such as greenhouses and warehouses, and for fogging confined outdoor areas.

Nozzle Tip Numbers. Most manufacturers have a method of coding nozzle tips; identification numbers are printed on the face of the nozzle. For example, for flat spray nozzles, a common nozzle is number 8004. The first two digits indicate that the nozzle produces an 80 degree fan

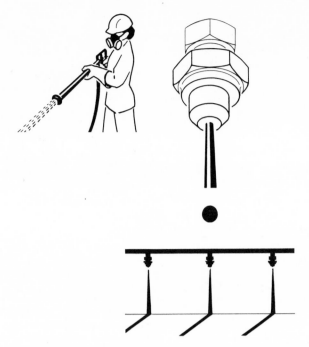

FIGURE 10-19.

Solid stream nozzles are used in high-pressure handguns and in low-pressure crack and crevice units. They are also used on boom sprayers for applying fluids in bands.

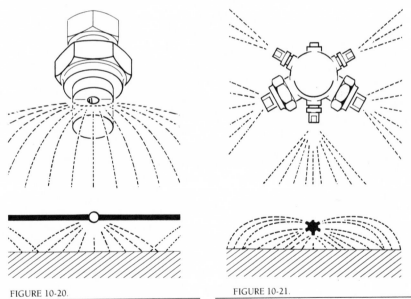

FIGURE 10-20.

Flood nozzles are used to apply large volumes of liquid under low pressure. They are occasionally used for pesticide application, but are more commonly used for liquid fertilizers.

FIGURE 10-21.

Broadcast nozzles enable a wide swath to be sprayed without using a series of nozzles on a boom. Swath widths from 30 to 60 feet can be produced.

TABLE 10-3.

Guide to Selecting Different Styles of Nozzles.

	STYLE OF NOZZLE	SUGGESTED USES	RECOMMENDED SPRAY PRESSURE	TYPE OF SPRAY PATTERN
	FLAT-SPRAY NOZZLES	Preemergence and postemergence herbicides, insecticides, and fungicides. Used on a boom.	20 to 60 psi Keep pressure as low as possible when spraying weeds.	Fanlike pattern with fewer droplets at sides than in center of fan pattern. Suitable for overlapping with other nozzles to produce wide spray swath.
	OFF-CENTER FLAT-SPRAY NOZZLES	Used on ends of spray booms to extend reach of spray pattern.	Same as flat-spray nozzles.	Fan-shaped with angle to one side.
	EVEN FLAT-SPRAY NOZZLES	Preemergence and post emergence herbicides, insecticides, fungicides. Use on boom. Do not overlap spray pattern.	20 to 40 psi Keep pressure low when used for weed control.	Fanlike pattern with even distribution of spray across width of fan.
	CONE NOZZLES	Insecticides and fungicides applied to foliage. Often used with air blast sprayers.	40 to 120 psi	Hollow or solid cone pattern. Fine spray droplets, good penetration.
	SOLID STREAM NOZZLES	All types of pesticides. Used on booms or hand guns.	5 to 200 psi	Low- or high-pressure solid stream. High pressure breaks spray into fine to medium droplets.
	FLOOD NOZZLES	Herbicides and fertilizers. High volume and low pressure to reduce drift. Used on booms.	5 to 20 psi	Wide, fanlike pattern of coarse droplets.
	BROADCAST NOZZLES	Weed and brush control in pastures and turf. Nozzles are clustered without boom.	10 to 30 psi	Wide, fan-shaped pattern ranging from fine to coarse droplets.
	BIFLUID NOZZLES	Used for developing extremely fine, airborne droplets. Flying insect control in enclosed spaces.	None. Uses air pressure to move liquids.	Fog or mist.

spray, while the last two numbers indicate the volume of spray (0.4 gallons per minute) output at 40 psi. Nozzle number 65155 is a 65 degree fan spray producing a volume of 1.55 gallons per minute at 40 psi. It is necessary to check the manufacturer's catalog to determine the rated operating pressure; some nozzle styles are designed to operate at higher or lower pressures. For example, flood nozzles are commonly rated at 10 psi.

Solid stream, flood, broadcast, bifluid, and disc type nozzles are coded in a similar way. Disc nozzles, for example, are assigned numbers such as 4, 6, 7, and 10. Sometimes this number is preceded by the letter "D" to indicate a disc nozzle. The number represents the size, in 64ths of an inch, of the orifice opening (except for the smallest sizes). A D7 nozzle has an orifice diameter of $\frac{7}{64}$ of an inch. Cores are available in several sizes which can be matched to discs to regulate the output capacity of the nozzle at different pressures. Follow the manufacturer's instructions for the proper installation of discs and cores.

When selecting nozzles, read the manufacturer's catalog to understand how nozzles are sized, their proper application, and the optimum pressure range. Manufacturers have developed charts for selecting spray volume or determining nozzle size. Table 10-3 is a guide for selecting different styles of nozzles.

Hand Spray Guns

Hand spray guns are used to apply a high-pressure stream or spray of pesticide (Figure 10-22). They are used for applying insecticides and fungicides to trees, vines, and shrubs, and are commonly used in land-

FIGURE 10-22.

Hand spray guns usually produce high-pressure streams or sprays; they are used for applying insecticides and fungicides to trees and shrubs, and are also used for spraying aquatic areas, livestock, buildings, and roadsides.

scape, nursery, aquatic, and greenhouse settings. Hand spray guns are also used for application of herbicides along roadsides, rights-of-way, and fence rows and are used to apply insecticides to livestock for control of external parasites.

Hand spray guns can be attached to many different types of spraying equipment. Often orchard sprayers or low-pressure, row crop sprayers have connectors for attaching a hand spray gun for occasional touch up or specialized spraying. In some greenhouses, pumps and tanks are located in a separate room. Pesticide is pumped through permanent plumbing to different locations within the greenhouse where a flexible hose and handgun can be connected. Hand spray guns are connected to portable sprayers also by a flexible hose. Long hoses enable the operator to spray areas distant from the pumping equipment; however, the system must accommodate for pressure drop due to the long hose. Spray guns have a handle, a valve, and a nozzle (or a small boom with several nozzles). The valve is sometimes built into a trigger mechanism or may be operated by rotating a knob at the end of the handle. Nozzles are usually interchangeable so the handgun can be used for different types of applications. In some models, the pressure and spray pattern can be adjusted by the valve mechanism.

Pressure Regulators

A pressure regulator is a spring-loaded valve that controls the pressure of liquid going to the nozzles; it is located between the pump and the nozzle, spray boom, or manifold (Figure 10-23). The amount of tension on the valve is adjusted by turning a screw. Increasing the spring tension increases the liquid pressure going to the nozzles. When the pressure in the system exceeds the pressure of the spring-loaded valve, the valve opens and some spray material is bypassed back into the tank, preventing pressure in the system from going any higher. An equilibrium is maintained by the regulator, but should the pump output pressure drop (by slowing the pump down, for example), the regulator will reduce or stop the flow of liquid into the tank, maintaining a constant pressure, if possible, to the nozzles.

Pressure regulators must be adjusted while fluid is pumped through the nozzles of the system. When the flow to the nozzles is stopped,

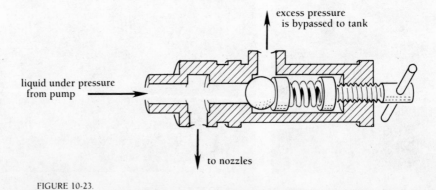

FIGURE 10-23.

A pressure regulator is a spring-loaded valve that controls the pressure of fluid going to the nozzles. If pressure increases, excess pesticide is bypassed back into the spray tank.

pressure in the system increases slightly and all the liquid is routed through the bypass.

Unloaders

An unloader is a device designed to sense a pressure increase or drop due to turning on or shutting off the flow of liquid to the nozzles. If nozzles are shut off, the unloader will bypass all the pumped liquid back into the spray tank. Once the flow to the nozzles is started, the unloader will redirect the liquid at the pressure determined by the pressure regulator. Unloaders are an important part of a high-pressure system because they protect pumps, valves, hoses, and other components from excessive and sudden pressure surges.

Pressure Gauges

FIGURE 10-24.

A pressure gauge is used to monitor the pressure of spray going to the nozzles and will alert the operator to problems in the system.

Liquid sprayers should be equipped with a pressure gauge to enable monitoring fluid pressure in the system (Figure 10-24). A change in pressure also warns the operator of potential malfunctions. The pressure gauge is usually located between the pressure regulator and the nozzles, and in this position measures pressure while spray is being emitted through the nozzles. Proper equipment calibration depends on an accurate pressure gauge; therefore recalibrate the gauge on the sprayer periodically by comparing its readings to another calibrated gauge. This second gauge should also be used to measure pressure at the nozzles from time to time to determine any pressure losses in the plumbing of the sprayer. Chapter 11 describes methods of calibrating spray equipment.

Gauges are available to measure different ranges of pressures. For example, some measure from 0 to 20 psi, while others measure from 1 to 200 psi, or 1 to 500 psi, or 1 to 1000 psi. It is important that the gauge be compatible with the sprayer pressure range. If the sprayer produces a maximum of 50 psi, a gauge with a range of 1 to 500 psi will be difficult to read and will have reduced accuracy; a gauge with a range of 1 to 100 psi is much more satisfactory. The 1 to 500 psi gauge is best suited to a sprayer that operates at maximum pressures of 300 or 400 psi (Figure 10-25). Gauges should operate at about 50% of their maximum pressure so they will be protected against damage in the case of unexpected pressure surges.

Liquid-filled gauges are commonly used on spray equipment because they last longer and can absorb the shock of rapid pressure changes. These gauges are distinguished by the liquid (glycerin) visible inside the face of the dial.

suitable for 30 psi

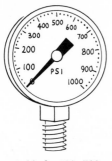

suitable for 500–700 psi

FIGURE 10-25.

Select a pressure gauge that is compatible with the pressure range of the sprayer; it should have a higher maximum pressure than the sprayer to prevent damage due to unexpected pressure surges.

Control Valves

Control valves are used by the operator to turn on and shut off liquid being pumped to the nozzles (Figure 10-26). They may be a trigger valve in a hand spray gun, or a lever valve controlling spray to nozzles on a boom. Other control valves are operated by cables, such as on air blast orchard sprayers, or by electric solenoids. Sprayers can be set up so that each nozzle is individually controlled by an electrically operated valve. On air blast sprayers, nozzles are usually attached to two manifolds, each with a separate control valve. The operator selects spray to be emitted from either side of the sprayer or from both sides at the same time. Spray

FIGURE 10-26.

Control valves are used to turn on or shut off spray to nozzles. Often multiple valves are used to regulate spray to different nozzles, giving the operator options to adjust the application according to the characteristics of the location.

booms used in field and row crop applications of pesticides often have 2 or 3 controllable sections and special valves are designed to operate these booms. A control valve for a 3-section boom has 7 spray selections available to the operator. The boom is divided into left, center, and right sections; the valve supplies spray to the right section only, left section only, center section only, right and left sections, right and center sections, left and center sections, or all three sections.

Electronic Sprayer Controllers. Electronic sprayer controllers can be connected to electric spray control valves to provide very accurate metering of pesticide sprays. Electronic sprayer controllers use microcomputers to monitor and regulate spray output and/or pressure, and are usually used to control multi-nozzle spray booms. Electronic controllers, by changing the output pressure or output volume of the pesticide, enable the operator to apply consistent and precise amounts of pesticide even if the travel speed of the equipment varies. When using a controller that regulates pressure to control output, the travel speed should not vary by more than 20% faster or slower than the optimum speed to keep the equipment operating within an acceptable pressure range. Some units warn the operator of malfunctions of nozzles or pumps.

Hoses, Couplings, and Fittings

Hoses, couplings, and fittings must be strong and durable, able to withstand the pressures produced by a spraying system and the corrosive action of spray material. Neoprene is the most common material used for sprayer hoses. Hoses must be reinforced to reduce chances of bursting under the pressure of the system because leaking or ruptured hoses expose the operator to pesticides and cause loss of unregulated amounts of

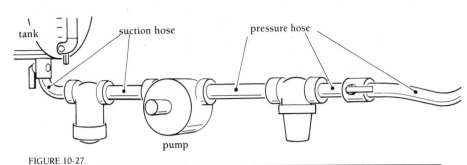

FIGURE 10-27.

Suction hoses are used to connect the pump to the tank of the sprayer. Pressure hoses are placed between the pump and the nozzles. Suction hoses must be larger in diameter than pressure hoses and should be sturdy enough to prevent them from collapsing.

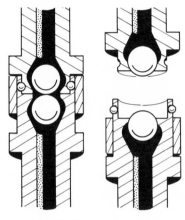

FIGURE 10-28.

A dry break coupling allows hoses containing pesticides to be disconnected from equipment without spilling any material. Spring-loaded ball bearings seal the hose openings when dry break couplings are separated.

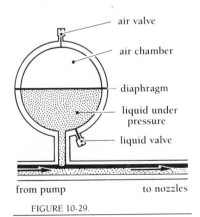

FIGURE 10-29.

A surge chamber can be installed in the pressure system of a sprayer to minimize pressure fluctuations caused by the pumping action of piston and diaphragm pumps.

pesticide into the environment. Two types—pressure hoses and suction hoses—are used to connect tanks, pumps, and booms or nozzles (Figure 10-27).

Pressure hoses should be rated to handle twice the operating pressure of the sprayer. Pressure hoses which will be used indoors for termite control or other similar purposes must be made of a material that will not leave skid marks on floors or other surfaces they contact.

Suction hoses that carry liquid from the tank to the pump must be larger in diameter than pressure hoses. Using hoses that are the same size or smaller than the pressure hose could impede liquid flow, reducing the discharge rate at the nozzles and often causing damage to the pump. Suction hoses must also be stiff enough to resist collapsing under the suction pressure of the pump.

Select couplings and fittings made of materials that are not corrosive and that can withstand the solvents used in pesticide formulations; brass, stainless steel, and high-density plastic are commonly used. All couplings and fittings should be designed for quick and easy removal in case repairs need to be made to the system. Couplings and fittings should not reduce the internal diameter of the hoses they connect, because this causes a pressure drop at nozzles and additional pressure on the pump. Couplings of hoses that need to be attached or removed during operation should be the *dry break* type to prevent pesticide leaks when they are disconnected (Figure 10-28). Dry break couplings have a spring-loaded check valve that automatically plugs the hose and fittings when they are not connected.

Surge Chambers

Surge chambers may be used in the pressure line to minimize pressure fluctuations caused by the pumping or oscillating action of piston and diaphragm pumps (Figure 10-29). One style of surge chamber consists of a hollow metal tank connected to the sprayer pressure line. Air trapped in this tank compresses or expands according to changes in pressure of the pumped liquid; this helps to minimize pressure variations. Another type of surge chamber consists of a round tank separated by a diaphragm into two hemispheres. Air is trapped on one side of the diaphragm and pumped liquid is on the other. As pressure in the system increases, the pumped liquid distorts the diaphragm, compressing the trapped air; pressure peaks in the system are moderated. When the pumping cycle reverses and pressure begins to drop, the compressed air forces liquid back into the system, slowing the rate of pressure drop. Compressed air can be pumped into some surge chambers to improve the dampening action at high pressures; some must be drained of liquid daily to prevent loss in air volume in the unit so it will function properly (return this liquid to the spray tank). If air volume in the surge chamber decreases too much, the surge chamber becomes less effective and high pressure pulsations may not be dampened.

Spray Shields

Spray shields are used with boom sprayers to confine pesticide droplets and prevent drift. They consist of metal boxes or shrouds that surround the nozzles and confine the spray to a small area of ground or to specific plants. These are attached to the spray boom and move with the tractor as a unit.

Closed-System Mixing Equipment

Most uses of Category I liquid pesticides require closed-system mixing equipment (Figure 10-30); exceptions include applications made in residential, industrial, and institutional locations, although closed-system

FIGURE 10-30.

Closed mixing systems are used to measure and transfer liquid pesticides. Liquid pesticide is pumped from its container, measured, and is then transferred into the sprayer mixing tank.

mixing equipment is available for these uses. Closed-system mixing equipment enables safe handling of toxic liquid pesticides because opened pesticide containers are not handled by operators, eliminating accidental contact. The important part of any closed-system mixing equipment is the probe which must be inserted into the pesticide container. Some probes puncture the container to draw out the liquid, while others pass through the container opening. Probes must seal well to the pesticide container to prevent any leaks. Select a unit with a probe that will fit the size containers being used. Some probes may be adapted or adjusted to fit several container sizes.

A transfer pump may be used to move the pesticide from its original container to the sprayer tank where it is combined with water, or the closed-system mixing equipment is connected to the pressure system of the sprayer. Some type of metering device is used to measure the quantity of pesticide being transferred. Some closed-system mixing units automatically rinse containers after they are empty; some have provisions for crushing containers for disposal. If the closed-system mixing equipment does not automatically rinse the container for return to the tank, you must use a separate rinsing device that will pump the rinsate into the spray tank.

Closed-system units are commonly installed on *nurse tanks*. A nurse tank is a special piece of equipment having a large tank with an agitator and often a pump; it is designed for mixing and holding diluted pesticide. One or more sprayers are filled in the field from this unit.

Nonpowered or Hand-Operated Equipment

Nonpowered or hand-operated liquid spray equipment is comparatively inexpensive, easy to use, and usually simple to repair and maintain because few moving parts are involved. These devices are used for applying pesticides to small areas, specific targets, or locations that are difficult to reach with larger equipment. Nonpowered or hand-operated equipment is lightweight; most units can be carried by an individual. Some are low-pressure sprayers with small tanks. Most are not equipped with agitators for keeping pesticides suspended in the tank and require occasional shaking if wettable powder or emulsifiable concentrate formulations are used. Table 10-4 is a guide to selecting the nonpowered and hand-operated devices suitable for different application needs.

Aerosol Cans. *Pressure spray applicators* and *aerosol foggers* are examples of aerosol cans. A fine spray of premixed pesticide is expelled when the pesticide mixture is forced through a small opening in the nozzle at the top of the can. The propellant is an inert, compressed gas. Some are designed for intermittent uses as needed, but aerosol foggers are one-time, total release units. Aerosol cans with a capacity of one quart or less are not reuseable, so the containers must be disposed of when empty. Pesticides packaged this way are commonly used around the home and are popular because of their convenience. Larger volume aerosol cans are used by pest control operators for structural and greenhouse applications; these are sometimes refillable. The aerosol cans can be carried on the applicator's waist belt and connected to a hose and spray wand (Figure 10-31). Two or more cans may be coupled to a single wand, giving the pest control operator an ability to select different pesticides during the same operation. Sprays packaged in aerosol cans offer convenience and portability to professional applicators and eliminate the need for chemical mixing.

Hose-End Sprayers (Figure 10-32). Hose-end sprayers are used for applying pesticides to lawns, flowers, and shrubs, usually in small areas. Concentrated pesticide mixtures are combined with water from a garden hose and expelled through a high-volume nozzle. A 1- or 2-quart plastic or glass container holds the concentrated pesticide. One filling of the container will produce about 20 gallons of dilute spray. Nozzles are adjustable for droplet size and to aim the spray in different directions. These sprayers generally have a valve to start and stop the flow of pesticide in the stream of water. Some contain another valve to regulate and shut off the water flow from the garden hose.

FIGURE 10-31.

Aerosol dispensers used in commercial applications are attached to a hose and spray wand enabling the operator to inject liquids into cracks and crevices.

FIGURE 10-32.

Hose-end sprayers are sometimes used to apply pesticides to lawns and shrubs. They apply high volumes of dilute pesticide.

TABLE 10-4.

Selection Guide for Nonpowered and Hand-Operated Application Equipment for Liquid Pesticides.

	TYPE	USES	SUITABLE FORMULATIONS	COMMENTS
	AEROSOL CAN	Insect control on house or patio plants, pets, small areas, cracks and crevices, and confined spaces.	Liquids must dissolve in solvent; some dusts are available.	Very convenient. High cost per unit of active ingredient.
	HOSE-END SPRAYER	Home garden and small landscaped areas. Used for insect, weed, and pathogen control.	All formulations. Wettable powders and emulsifiable concentrates require frequent shaking.	Convenient and low-cost way of applying pesticides to small outdoor areas. Cannot spray straight up.
	TRIGGER PUMP SPRAYER	Indoor plants, pets, and small home yard areas. Used for insect and pathogen control.	Liquid-soluble formulations best.	Low cost and easy to use.
	COMPRESSED AIR SPRAYERS	Many commercial and homeowner applications. Can develop fairly high pressures. Used for insect, weed, and pathogen control. Often used indoors for household pest control.	All formulations. Wettable powders and emulsifiable concentrates require frequent shaking.	Good overall sprayer for many types of applications. Needs thorough cleaning and regular servicing to keep sprayer in good working condition and prevent corrosion of parts.
	BACKPACK SPRAYERS	Same uses as compressed air sprayers.	All formulations. Wettable powders and emulsifiable concentrates require frequent shaking.	Durable and easy to use. Requires periodic maintenance.
	WICK APPLICATORS	Used for applying contact herbicides to emerged weeds. Agricultural and landscape uses.	Only water-soluble herbicides.	Simple and easy to use. Clean frequently.

Trigger Pump Sprayers (Figure 10-33). The trigger pump sprayer is a simple liquid applicator. A small pump, activated by squeezing a trigger, forces pesticide mixtures through a nozzle, producing a fine spray. Some styles have an adjustable nozzle for controlling droplet size. Diluted pesticide is contained in a plastic jar, ranging in capacity from 1 pint to 1 gallon. These applicators are used for applying pesticides to small areas, such as houseplants or pets, and in confined areas.

Compressed Air Sprayers (Figure 10-34).Compressed air sprayers are designed to hold diluted pesticide in a small, airtight tank. Air inside

FIGURE 10-33.

A trigger pump sprayer can be used to apply small quantities of diluted pesticide to surfaces such as houseplants or pets, and for applying some types of pesticides to confined areas.

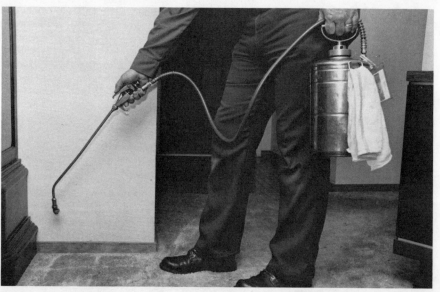

FIGURE 10-34.

Compressed air sprayers usually hold from ½ to 5 gallons of spray mixture. Air inside the tank is compressed with a self-contained pump or a carbon dioxide cartridge. Pesticide under pressure is forced through a hose attached to an adjustable nozzle at the end of a hand-held wand.

the tank is compressed with a hand pump. The compressed air forces the liquid through a hose and nozzle whenever a valve is opened by the operator. Some models use compressed carbon dioxide cartridges as the propellant, eliminating the need for hand pumping. Tanks are made of metal or durable thermoplastic and have a capacity of less than 5 gallons. Conventional tank sizes include ½, 1, 2, and 3 gallon capacities. These sprayers are convenient for treating small areas and for applying liquid pesticides indoors. Most have adjustable nozzles to control droplet size and spray pattern. For indoor use, there are adaptors to enable liquid spray to be injected into small cracks and crevices. Some compressed air sprayers are equipped with harnesses for backpack use.

Larger compressed air sprayers utilize separate air tanks that are charged from an air compressor or from a portable hand pump. These tanks are connected by hoses or pipes to an airtight chamber containing the diluted pesticide mixture.

Backpack Sprayers (Figure 10-35). Backpack sprayers have a hand-operated hydraulic pump that forces liquid pesticides through one or

FIGURE 10-35.

This type of hand-operated backpack sprayer usually requires continuous pumping action to maintain pressure for spraying. These units hold from 3 to 5 gallons of spray mixture; most have adjustable nozzles.

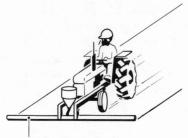

pesticide saturated pad or rope

FIGURE 10-36.

A wick applicator is used for application of contact herbicides. They can be used in areas where weeds are taller than the crop plant. Wick applicators reduce problems of drift and waste of herbicides.

more nozzles; this pump is activated by moving a lever up and down. Pressures in excess of 100 pounds per square inch (psi) can be obtained with some models. Tanks, usually made of plastic, have a capacity of around 5 gallons. These sprayers are popular with commercial applicators working in small areas and in locations where there is no access for larger equipment. Backpack sprayers are also known as *knapsack* sprayers.

Wick Applicators (Figure 10-36). Wick applicators, also called *rope wick applicators*, are used exclusively for the application of contact or systemic herbicides. Their basic design consists of a cloth or carpeting pad, rope, or wick, saturated with herbicide, which is wiped onto the leaves of target weeds. A simple hand-held model holds liquid herbicide in its hollow handle; liquid is fed to the pad through a series of small holes. Other types of wick applicators incorporate a boom and herbicide reservoir attached to a tractor. Where weeds are taller than the crop, the wick applicator is adjusted to contact just the weeds. Little or no pesticide is wasted with wick applicators, and no environmental contamination takes place as a result of the application process. More elaborate versions of this design consist of units with ropes that are mechanically wound through an herbicide reservoir, ensuring a constant and more uniform herbicide distribution.

Powered Application Equipment

Many pesticide applications require powered equipment capable of applying pesticides to large areas. Some units have self-contained motors, while others are powered by tractors or other external power sources. These machines are usually equipped with hydraulic or mechanical agitators, pressure regulators, and a variety of spray booms, hand-held spray guns, or other nozzle arrangements. Powered equipment contains many more components than hand-operated equipment, and often requires considerable maintenance and servicing to keep it operating properly. Table 10-5 is a guide to help you select the powered liquid application equipment suitable to your needs.

Powered Backpack Sprayers. The smallest powered sprayer consists of a backpack unit with a compact gasoline engine. The engine drives a pump that forces diluted or concentrated liquid pesticide from the tank out through one or more nozzles. Air blowers, also driven by the engine, help propel spray droplets. Because of the limited tank size and ability to produce high pressures, backpack sprayers are best suited for applications requiring low volume.

Controlled Droplet Applicators (Figure 10-37). Controlled droplet applicators (CDAs) are used to apply low volumes of specific types of pesticides (usually herbicides). Rather than passing through a nozzle, the liquid pesticide mixture is dropped onto a spinning disc or cup; the serrated edges of the disc or cup distribute the spray by centrifugal force and may, under certain conditions, produce a narrow drop size spectrum. Droplet size depends on the rotation speed of the unit and the nature of pesticide being used. More viscous liquids produce larger droplets. Rotation speeds are adjustable between 1000 and 6000 rpm; higher speeds produce smaller droplets. The size range of droplets produced by CDAs is 100 to 400 microns. In most units, the pesticide mixture flows from a reservoir tank to the spinning disc or cup by gravitational force, eliminating the need for a pump. Flow rate is controlled by the size of an orifice in the pesticide hose and the height of the reservoir.

CDAs are powered by variable speed, low-voltage DC electric motors or hydraulic motors. Speed is controlled in some units by changing drive belts on pulleys, while other styles have electronic speed controllers; hydraulic units accomplish speed control by adjusting the flow rate of the hydraulic fluid. Some CDAs are inexpensive, self-contained, hand-held units. Others are connected to a backpack tank by a flexible hose. One or more controlled droplet applicators can be mounted to a boom and attached to a tractor or all-terrain cycle (ATC). Controlled droplet applicators are also used in place of nozzles in air blast sprayers.

Controlled droplet applicators are used for applying contact, pre-emergent, and postemergent herbicides. They are also used for some insecticide and fungicide applications. Because CDAs can control the droplet size by the rotational speed, the units can make very low volume applications without excessive drift. Volumes between 1 quart and 3 gallons per acre are usually applied with CDAs, so accurate calibration is essential for effective use. Low-volume applications must be consistent with pesticide label directions or current University of California recommendations.

FIGURE 10-37.

Controlled droplet applicators are designed to apply small volumes of uniform sized droplets. These are available as hand-held units or can be mounted in groups on spray booms or on air blast sprayers.

TABLE 10-5.

Selection Guide for Powered Liquid Pesticide Application Equipment.

	TYPE	USES	SUITABLE FORMULATIONS	COMMENTS
	POWERED BACKPACK SPRAYER	Aquatic, landscape right-of-way, forest, and agricultural applications.	All. Some may require agitation.	May be heavy for long periods of use. Requires frequent maintenance.
	CONTROLLED DROPLET APPLICATORS	Used for application of contact herbicides and some insecticides. Some are hand-held while others are mounted on spray boom. May also be used with air blast sprayers. Produces uniform droplet sizes.	Usually water-soluble formulations.	Plastic parts may break if handled carelessly.
	LOW-PRESSURE SPRAYER	Very common type of sprayer used in commercial applications for weed, insect, and pathogen control. Used with spray booms or hand-held equipment.	All. Equipment may include agitator.	Frequent cleaning and servicing is required. Powered by own motor or external power source.
	HIGH-PRESSURE HYDRAULIC SPRAYER	Landscape, right-of-way, and agricultural applications. Use on dense foliage and large trees or shrubs.	All. Equipment may include agitator.	Important to clean and service equipment frequently. Requires own motor or external power source. Abrasive pesticides may cause rapid wear of pumps and nozzles.
	AIR BLAST SPRAYER	For application of insecticides, fungicides, and growth regulators to trees, vines, and shrubs. Sometimes used on row crops and with livestock. Also used in aquatic areas.	All. Equipment usually equipped with agitator.	Frequent service and maintenance is required. High horsepower motor is often needed to power pump and fan. Volutes may be used to direct spray.
	ULTRA-LOW-VOLUME APPLICATORS	Primarily used in agricultural and aquatic situations. Used with insecticides, fungicides, growth regulators.	Usually only pesticides that dissolve in water or organic solvents.	Requires extreme care in calibration. Applies highly concentrated pesticides. Often used with a blower.

TYPE	USES	SUITABLE FORMULATIONS	COMMENTS
ELECTRO-STATIC SPRAYER	Agricultural uses for applying insecticides, fungicides, and growth regulators to trees, vines, and row crops.	All. Requires agitation of some formulations.	Usually equipped with blower. Maintenance and frequent cleaning is required. Electrically charged spray droplets are attracted to target surfaces.
AEROSOL GENERATORS AND FOGGERS	Mainly used to apply insecticides in confined areas. Used in aquatic areas for airborne insects.	Requires water- or solvent-soluble formulations.	Emitted spray is highly subject to drift. Keep equipment clean.
INJECTION PUMPS	Injects concentrated pesticides into spray boom or irrigation water. May be used with all classes of pesticides.	Requires liquid formulations.	Accurate metering is necessary to assure proper calibration. With irrigation systems, must have backflow protection to prevent contaminating water supply.

FIGURE 10-38.

A common sprayer used for agricultural, right-of-way, forest, and landscape applications of pesticides is the low-pressure applicator like the one being used here to apply insecticides to lettuce. These apply sprays through a series of nozzles attached to a boom.

Low-Pressure Sprayers (Figure 10-38). Low-pressure sprayers are used for pesticide application in many different agricultural crops, turf, aquatic, and right-of-way settings. This equipment operates in the pressure range of 10 to 20 psi. Diluted pesticide is emitted through nozzles attached either to a boom or attached directly to the sprayer. Pesticides can also be injected into the soil with low-pressure applicators by using a series of specially designed soil shanks or chisels attached to a tractor tool bar. As shanks rip the soil, pesticide (or liquid fertilizer) is injected below the surface. This is a common way of applying soil fumigants.

Low-pressure applicators may be mounted on a trailer (tag-along models) or attached to a tractor or truck; some are self-propelled. Some units have engines to pump liquids through nozzles, while others are powered by a PTO shaft from a tractor or through hydraulic power provided by the tractor or truck. Occasionally, pumps are driven by the wheels of the trailer (ground-wheel-driven) so pesticide is metered as the unit is being towed; this method provides a constant rate of application per unit of area even if the travel speed of the unit varies.

Most low-pressure sprayers have tanks with a capacity of 100 gallons or more. Often two or more tanks can be attached to a tractor to increase total capacity. Low pressure is not suitable when thorough coverage of dense foliage is required or where the spray must travel any distance, unless a blower is used to propel the spray droplets.

The *recirculating sprayer* is a modification of low-pressure equipment used to apply herbicides in row crops. Concentrated herbicide is applied through nozzles to target weeds, and excess spray is caught in a trough or sump located below the nozzles and recirculated back through the system. Less material is used by recapturing pesticide that would ordinarily be wasted. Reclaimed spray must be filtered to remove seeds, leaves, or other foreign particles.

High-Pressure Hydraulic Sprayers (Figure 10-39). High-pressure hydraulic sprayers are used for orchards and field crops as well as for turf and landscape and right-of-way trees and shrubs. They are also used in aquatic areas and for treating livestock. High volumes of dilute pesticide are forced either through a single nozzle on a handgun sprayer, or through a series of nozzles mounted on a boom. Special boom designs improve spray coverage to all sides of crop plants. An oscillating boom may be used on densely foliated trees, such as citrus, and on large vines; several high-pressure spray nozzles are rotated from side to side and up and down as the sprayer moves along. This oscillating action, coupled with high pressure and high volume, provides thorough spray coverage to all parts of target plants.

FIGURE 10-39.

Oscillating boom sprayers, like the one being used here in citrus, are high-pressure units that direct large volumes of spray through dense foliage. They are used where thorough coverage is essential.

Pressures between 100 and 400 psi or more are common with high-pressure sprayers. Most have large capacity tanks, between 100 and 1000 gallons. High-pressure hydraulic sprayers are mounted on trailers, trucks, or tractors and may have a self-contained engine for powering the pump. Some are powered by a tractor PTO. The operator must be careful to maintain a constant tractor engine speed on PTO-driven units to assure a uniform application rate. Many models are equipped with pressure regulators and bypass mechanisms so they can be used as low-pressure sprayers as well.

Air Blast Sprayers (Figure 10-40). Air blast sprayers use fans or blowers to propel spray mixtures into dense foliage, to tops of trees, and across fields or aquatic areas. Proper use of air blast sprayers eliminates shingling (see Figure 7-46 on page 223) and improves pesticide coverage. Because spray droplets are propelled to target surfaces by air, air blast sprayers do not require extremely high pressure, and usually operate in a range between 80 and 150 psi. Most air blast sprayers have tanks ranging in capacity from 100 to 1000 gallons. Low-volume or concentrate sprayers produce between 30 and 100 gallons of finished spray per acre, while high-volume sprayers (dilute sprayers) have an output of 400 to 1000 gallons per acre. A high-volume sprayer can usually be converted to low-volume by changing nozzle size, adjusting the pump output pressure, and controlling ground speed. Nozzles of air blast sprayers are positioned in the air stream to allow the air to break up the spray droplet sizes and force them into dense foliage and tree tops. Orchard and vineyard sprayers often use different sized nozzles at different locations on the manifold so that varying amounts of spray can be applied to different parts of the

FIGURE 10-40.

An air blast sprayer has a powerful fan to move pesticide droplets through dense foliage or high into the upper parts of target trees.

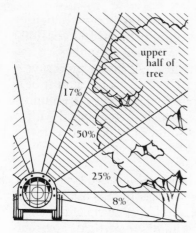

FIGURE 10-41.

An orchard air blast sprayer may be adjusted to apply greater amounts of pesticides to some parts of the tree. This is done by using several different nozzle sizes or by using more nozzles in some locations. In most situations, two thirds of the spray output is directed to the upper half of the tree. This drawing shows an ideal pattern for most mature orchards.

trees or vines (Figure 10-41). Controlled droplet applicators are sometimes used in place of nozzles on orchard, vineyard, row crop, and livestock air blast sprayers.

Special attachments for air blast sprayers, called *volutes*, are used to direct the spray into tall trees, around vines, across fields, and into dense foliage of vegetable crops. Different designs of volutes are available depending on the needs of the application.

Another type of air blast sprayer has been developed for use in row crops. Spray booms carry high-velocity air from a blower through a large diameter rigid or inflatable tube. This tube has openings adjacent to each nozzle, so that spray emitted from nozzles is propelled to target plants. Air turbulence circulates the spray droplets through dense foliage, covering the undersides of leaves as well as top surfaces. Low volumes, in the range of 3 to 15 gallons per acre, are applied with these types of sprayers, which are available as tag-along styles or as truck- or tractor-mounted units (Figure 10-42).

Air blast sprayers are either self-powered by a gasoline or diesel engine, or externally powered by a tractor PTO; units with large fans require 50 to 100 horsepower to drive them. Some smaller sprayers can be mounted to a truck or tractor, but larger ones are wheeled and designed for pulling behind a tractor. Other variations include a style having the fan, pump, and spray nozzles attached to a tractor, connected by hoses to a wheeled tank pulled behind as a separate unit. Self-propelled spraying machines, including an enclosed operator cab, are also available.

Ultra-Low-Volume Sprayers (Figure 10-43). Ultra-low-volume (ULV) sprayers apply from 1 quart to a few gallons of spray per acre. Low-volume nozzles—or sometimes controlled droplet applicators—are used to break up the spray into small sizes; air from a fan or blower propels droplets to the treatment surfaces. Highly concentrated pesticides, sometimes mixed with vegetable oil carriers to reduce droplet evaporation, are applied with ULV sprayers. Vegetable oil also improves the spreading ability of droplets once they have contacted the target surface. ULV sprayers usually have small tanks and are powered by lightweight gasoline engines. These sprayers are smaller and much lighter than higher volume machines. Because of the smaller drop size and smaller blowers, the concentrate sprayers generally are limited to applications during low winds to minimize drift and obtain satisfactory penetration and coverage.

Accurate calibration of ULV sprayers is critical because of the high concentration of pesticide being applied. There may be greater hazards to the operator with ULV sprayers because of the concentrated pesticides used.

Electrostatic Sprayers (Figure 10-44). Electrostatic sprayers apply 10 to 50 gallons per acre of pesticide in the form of small, electrically charged droplets. Droplets average about 50 microns in diameter and are given a negative electrostatic charge as they leave the sprayer volute. Because plant material has a positive electrostatic charge, spray droplets are attracted and become attached to these surfaces. The negatively charged spray droplets repel each other, so they do not clump together to form larger sized droplets. Electrostatic sprayers are usually powered by a tractor PTO, and an electrical charge of about 15,000 to 20,000 volts is produced by a transformer energized by the electrical system of the tractor. Volutes of different configurations may be used to direct the spray droplets toward surfaces being treated.

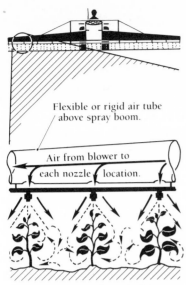

Flexible or rigid air tube above spray boom.

Air from blower to each nozzle location.

Spray contacts lower leaf surfaces due to turbulence.

FIGURE 10-42.

Air blast sprayers may also be used in row crops. As illustrated, air from a blower is carried to nozzles through a flexible or rigid tube. Pesticide droplets are distributed around the plant by the air turbulence, giving coverage to upper and lower plant surfaces.

FIGURE 10-43.

An ultra-low-volume sprayer produces low volumes of very small droplets; these are usually propelled by a fan or blower. Pesticide mixtures are much more concentrated than in higher volume sprayers.

FIGURE 10-44.

An electrostatic sprayer emits electrically charged spray droplets. Spray droplets are attracted to oppositely charged surfaces, increasing the pesticide deposition and target coverage.

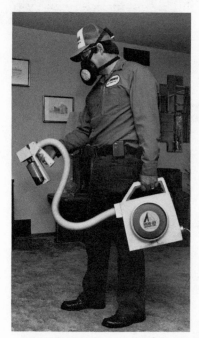

FIGURE 10-45.

Aerosol generators and foggers produce a fine insecticide mist that remains suspended in the air for a long period of time and penetrates cracks and inaccessible areas, killing insects on contact.

Aerosol Generators and Foggers (Figure 10-45). Aerosol generators and foggers are used to produce small airborne particles of pesticide. These units are commonly used for control of insects in confined spaces, such as residences, greenhouses, and warehouses. The fog-laden insecticide produced by aerosol generators remains suspended in the air for long periods of time and penetrates small cracks and inaccessible areas, killing insects on contact. Some fogging is done outdoors for mosquito control and in localized recreational areas to kill biting or irritating insects. Effective insect control in outdoor areas depends mainly on proper weather conditions to keep the fog confined and airborne within an area long enough to contact target insects.

Thermal fog applicators use heat to generate pesticide aerosols. Other types use bifluid nozzles with high-velocity air to produce extremely fine droplets.

Chemical Injection Pumps. Chemical injection pumps inject undiluted liquid pesticides directly into the nozzle line where it is mixed with water simultaneously pumped from a water tank. Some injection pumps draw liquid pesticides directly from original containers and therefore eliminate the need for mixing chemicals or cleaning tanks. When used with sprayers, chemical injection pumps enable one or more chemicals to be applied during the same operation and also allow the concentration of pesticide mixture to be varied during application. They are sometimes controlled by electronic systems that regulate the concentration of pesticide applied. Many future possibilities exist using chemical injection systems. For example, current development is under way for an orchard sprayer that uses sensors and a computer system to apply pesticides proportional to tree volume, with no spray applied between trees.

Injection pumps are also used for applying pesticides through irrigation systems, known as *chemigation* (Figure 10-46). To prevent groundwater contamination through the well, injection pumps must be equipped with a shut-off device to stop pesticides from being injected into the system when irrigation stops. Safety valves must also be installed in the irrigation system to prevent possible contamination of the water supply by back flow of irrigation water.

FIGURE 10-46.

Small injection pumps are used to meter pesticides into irrigation water, a technique known as chemigation.

Chemical injection pumps used on spray applicators and with irrigation systems are low-volume and low-pressure units. They are designed to meter small quantities of concentrated pesticide into pressurized water and do not replace the higher volume and higher capacity pumps used on sprayers or in irrigation systems.

DRY APPLICATION EQUIPMENT

Dry pesticides are applied either as dusts or granules depending on the formulation. Dust formulations are susceptible to drift over great distances if applied during windy conditions. Thus, outdoor use of dusts is restricted and their greatest use is for pest control in buildings and other confined areas, and for control of external parasites on livestock, poultry, and pets. Granules, which are usually incorporated into the soil or applied to bodies of water, are the dry formulations used most for pest control in landscape, agricultural, and aquatic situations. Table 10-6 is a guide for selecting dust and granule application equipment.

Dust Applicators

The function of a dust applicator is to combine the dust with air so that it can be evenly distributed over a large area. Various types of dust application equipment are available; selection should be made based on the intended use.

Bulb Applicators (Figure 10-47). Hand-held bulb applicators are used for applying dusts to small confined areas, cracks, and crevices. A bulb made of rubber or similar flexible material is squeezed to expel dust-bearing air through a small tube. Some bulb dusters have attachments to extend the reach of the tube and direct the dust.

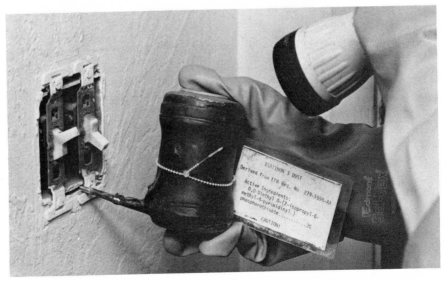

FIGURE 10-47.

Bulb applicators can be used to apply dusts to small, confined areas such as electrical outlets and cracks and crevices.

TABLE 10-6.

Selection Guide for Dust and Granule Application Equipment.

	TYPE	USES	SUITABLE FORMULATIONS	COMMENTS
	DUST APPLICATORS			
	Bulb Applicator	For forcing dusts into small cracks and crevices.	Dusts	Simple, easy to use.
	Compressed Air Duster	Used to apply dusts in confined spaces such as wall voids.	Dusts.	Avoid breathing dusts.
	Mechanical Duster	For landscape and small agricultural areas.	Dusts.	Avoid drift. Do not breathe dust. May have bellows to disperse dust.
	Power Duster	Vine crops and some special applications. Also used in buildings.	Dusts.	Equipped with blower to disperse dusts. Considerable hazard of drift.
	GRANULE APPLICATORS			
	Hand-Operated Granule Applicator	Landscape, aquatic and some agricultural areas.	Granules or pellets.	Suitable for small areas. Easy to use.
	Mechanically Driven Granule Applicator	Turf and other landscape areas. Also commonly used in agricultural areas.	Granules or pellets.	Requires accurate calibration.
	Powered Granule Applicator	Agricultural areas—usually row crops. Some large landscape applications.	Granules or pellets.	Frequent servicing and cleaning is required. Some units may have blowers to disperse granules. Others may distribute granules along a boom.

FIGURE 10-48.

A compressed air duster may be used to apply pesticide dusts in enclosed areas, wall voids, and crawl spaces and attics.

Compressed Air Dusters (Figure 10-48). Compressed air dusters propel dusts through a nozzle or hose. High-velocity escaping air picks up some pesticide from an airtight chamber and expels it as a fine powder. Dust formulations of some pesticides are available in aerosol cans and are applied in the same way as liquid aerosols; however, a common problem associated with dusts is their tendency to cake inside the aerosol applicator. Moisture or high humidity enhance caking, although this can be overcome by using formulations containing anti-caking materials.

Mechanical Dusters. Mechanical dusters utilize either a crank-operated fan and agitator or a lever-operated bellows to force dust-laden air out of a hopper. Dust is emitted through an orifice in the applicator or through a hose aimed by the operator. Most mechanical dusters can be strapped to the operator's back or chest. Smaller units are hand-carried.

Power Dusters. Power dusters use fans powered by electric or gasoline motors (Figure 10-49). Small, hand-held units are usually used in structural settings. These units are either battery powered or plug into a standard electrical outlet, while larger units are designed to be carried on a backpack and are powered by a lightweight gasoline engine. Pesticide dust is blown out of a hand-held flexible hose and can be directed at targets up to 15 feet away.

Large power dusters for agricultural crops have only limited use due to environmental problems resulting from drift. These units attach to tractors and are used for application of dusts in date palm gardens and to row or vine crops. Several nozzles may be positioned on a boom, and each is connected by large-diameter flexible tubing to a central blower where the pesticide and air are mixed. Extension tubes or fan-shaped air volutes are often used to direct dusts onto target plants.

FIGURE 10-49.

Power dusters are used to broadcast dust into confined areas. They have a blower to carry dust to treatment surfaces.

FIGURE 10-50.

A mechanically driven granule applicator may be used to apply pesticide granules to agricultural crops. Smaller versions are used for application of granular pesticides to turf.

Granule Applicators

Granule applicators may be hand-operated, mechanically driven, or engine powered, depending on the needs of the application site. Because granules are of varying sizes and shapes, equipment must accommodate size differences. Some formulations, known as pellets, are granules of identical size and shape; when applied through specially designed applicators, pellets provide more accurate calibration and more uniform application rates than other granule formulations. For some aquatic applications, granule applicators may be mounted in a boat. Granules applied to the soil are usually incorporated by tillage equipment pulled behind applicators.

Hand-Operated Granule Applicators. Hand-operated granule applicators are usually worn on the operator's chest. An adjustable opening at the bottom of a cloth, metal, or plastic hopper feeds granules onto a spinning plate operated by a hand crank. The operator walks at a steady pace while turning the crank to achieve an even distribution of granules.

Mechanically Driven Granule Applicators (Figure 10-50). Several types of mechanically driven granule applicators are available. Some consist of ground-wheel-driven metering devices attached to a hopper; often several of these units are attached to a tractor tool bar and spaced according to the crop row spacing. Larger units, employing the same principle, may have only one large hopper but several drop pipes aligned to the row spacing.

Another design of granule applicator consists of a large hopper with a chain conveyor or auger at the bottom. The chain or auger move granules onto a spinning disc that disperses them evenly over a wide area. These applicators are ground-wheel-driven; rate of application is controlled by adjusting the openings that granules pass through. The width of dispersal is controlled by ground speed and adjustable deflectors located alongside the spinning disc. Some units provide options for dispersing granules to the right, left, or both sides as needed.

Powered Granule Applicators (Figure 10-51). Backpack powered granule applicators are driven by small gasoline engines, similar to those used to apply liquids or dusts. These units have a blower to aid in the

FIGURE 10-51.

Granules can also be applied with a backpack granule applicator. A small gasoline engine powers a blower that disperses the granules.

dispersal of granules through a flexible hose. The operator aims the tube at the target area while walking slowly.

Another type of powered granule applicator consists of a long boom attached to a tractor or truck. Granules are augered down the boom and metered out at preset spacings. This provides for accurate calibration and even distribution of granular pesticides.

LIVESTOCK AND POULTRY APPLICATION EQUIPMENT

Several methods are used to apply pesticides to livestock and poultry in addition to some of the liquid and dust applicators described above. Those described here are used for the external application of liquids or dusts. In addition, systemic pesticides are used under the supervision of veterinarians to protect animals from internal and external parasites. Systemic pesticides are administered through feed, by mouth as a paste, capsule, or tablet, and by subcutaneous injection. Several methods are used to apply pesticides to the animal's skin.

Livestock Face and Back Rubbers and Dust Bags

Dry or liquid pesticide formulations are packaged into bags or other containers that can be hung or mounted in areas frequented by livestock (Figure 10-52). As the animals contact these packages, usually to scratch themselves, a small amount of pesticide is transferred to their bodies, effectively controlling many different species of external parasites.

FIGURE 10-52.

Certain types of pesticides are applied to livestock by means of dust bags or face and back rubbers. Small amounts of dust or liquid are deposited onto the animal each time it rubs against the device.

Poultry Dust Boxes

Poultry dust boxes are used in raised wire battery-type cages for laying hens. These boxes contain an insecticide dust used for control of mites. Hens instinctively wallow in the boxes and pick up the dust on their feathers and skin.

Dipping Vats

Dipping vats are large, permanent or portable containers filled with a dilute pesticide solution. For control of external parasites, livestock are forced into the vats so that they become totally covered with liquid pesticide. Ramps are used to assist animals entering and leaving a vat.

BAIT APPLICATION EQUIPMENT

Pesticide bait formulations usually require special application because a major problem associated with baits is exposing nontarget organisms to toxic pesticides. Bait stations or bait applicators can help prevent such exposure.

Bait Stations

Bait stations hold supplies of poisoned food and attract target pests; they should be designed to prevent children, pets, and nontarget animals from having access to baits (Figure 10-53). Bait stations are most often

FIGURE 10-53.

Bait stations are helpful in preventing nontarget organisms from being exposed to the pesticide.

FIGURE 10-54.

This device is used to apply poisoned bait for control of gophers; it forms an artificial burrow that intersects with the natural burrows made by gophers. Poisoned grain is deposited in the artificial burrow.

used for flies around poultry and livestock, squirrels in agricultural and right-of-way locations, and for rodents in warehouses and residential areas. They are also used for cockroaches, ants, snails, other invertebrates, and occasionally for birds.

Bait stations must be located out of the reach of nontarget organisms, pets, and children. Fly bait stations should be hung above poultry or livestock in poultry houses, loafing sheds, or barns. V-shaped troughs are often used to hold baited seeds or grains for pest bird control, and should be secured high up in trees. For rodents, locate bait stations in crawl spaces, attics, and other out-of-way places.

Bait Applicators

Bait applicators are used to apply poisons to control gophers and moles. Hand-operated models can inject a small quantity of poisoned bait directly into an underground burrow made by the gopher or mole. *Mechanical bait applicators* are tractor-mounted machines that form baited, artificial burrows which intersect natural gopher burrows (Figure 10-54). Gophers explore the artificial burrows and feed on the bait.

SPRAYER EQUIPMENT MAINTENANCE

Effective pesticide application depends on properly maintained and adjusted application equipment. Accidents or spills caused by ruptured hoses, faulty fittings, damaged tanks, or other problems can often be avoided through a periodic maintenance program. Spraying equipment problems can be reduced or eliminated by regular inspection, servicing, and maintenance. Application equipment should be inspected for wear or damage before each use and faulty components replaced or repaired. Equipment should also be thoroughly cleaned after each application. When not in use, sprayers must be properly stored to prevent deterioration or damage. Wear protective clothing, rubber gloves, and eye protection when cleaning, servicing, or repairing spray application equipment.

Preventing Problems

Several preventive steps should be taken to reduce problems of sprayer malfunction or breakdown and to maintain uniform and accurate application:

Use clean water. Water that contains sand or silt causes rapid pump wear and may clog screens and nozzles. Whenever possible, use water pumped directly from a well and make sure all filling hoses and pipes are clean. If water is taken from ponds or irrigation canals, it should be filtered before being put into the sprayer tank. Also, measure the pH of the water to be sure it is adequate for the intended pesticide use. Chapter 4 describes how to check and adjust pH.

Keep screens in place. Filter screens are designed to remove foreign particles from the spray liquid. It is a nuisance to remove collected debris from the screens, but debris accumulation indicates that the screens are doing the job they were designed for. Removing screens because they keep plugging only increases wear on the pumps and nozzles. Make sure screens are the proper size for the type of material being applied and, if excessive plugging does occur, try to eliminate the cause—change water sources, for example.

Use chemicals that the sprayer and pump were designed for. Spray chemicals are corrosive to some metals and can also cause deterioration of rubber and plastic components. It is important to recognize limitations in existing spray equipment and avoid problems by modifying the equipment or by using it only for chemicals that are not corrosive. Sometimes it is possible to replace parts of a sprayer with parts made of corrosion-resistant materials.

Properly clean nozzles. Spray nozzles are precision made. Never use any metal object to clean or remove debris, because the orifice may be damaged, adversely changing the spray pattern and spray volume. Nozzles are best cleaned by flushing with clean water or a detergent solution; remove stuck particles with a soft brush (such as an old toothbrush) or a round wooden toothpick. Always wear rubber gloves when handling or cleaning spray nozzles, and never blow through them with your mouth because nozzles may contain residue of toxic pesticides. Compressed air can be used to clean nozzles, but protect your eyes and skin (Figure 10-55).

Flush sprayers before use. New sprayers and sprayers that have been stored for any period of time must be flushed with clean water to remove foreign particles, dirt, and other debris before being used. The manufacturing process may leave metallic chips, dirt, or other residue in the tank or pump. Storage always subjects spraying equipment to the possibility of being contaminated with dirt, leaves, rodent debris, and rust.

Clean sprayer after use. To avoid possible crop damage or injury of treated surfaces, thoroughly clean a sprayer to remove pesticide residues at the completion of each job. At the end of each day, spraying equipment should be flushed and thoroughly cleaned. Avoid leaving pesticide mixtures in a sprayer overnight or for longer periods of time. Prolonged contact increases chances of corrosion or deterioration of sprayer components. Some pesticides settle out and may be difficult to get back into suspension after being left in an idle sprayer. Certain pesticides lose their effectiveness within short periods of time once they have been mixed with water. Finally, pesticides left in an unattended sprayer may present a hazard to people, wildlife, or the environment.

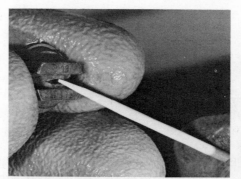 

FIGURE 10-55.

To clean a clogged nozzle, use compressed air or water for flushing the orifice. Never put your mouth to a nozzle. Use a wooden toothpick or soft brush (such as an old toothbrush) to remove stuck objects. Do not use any type of metal device to remove debris because you may damage the orifice.

Leftover spray material should be used on an appropriate target site, otherwise it must be considered a hazardous waste and will need to be disposed of in a Class 1 dump site. Water used for rinsing spray tanks can either be used for mixing other pesticides of the same type or taken to a Class 1 dump site as a hazardous waste. Rinse water cannot be drained onto the ground.

Inspection and Maintenance

Spraying equipment should receive regular inspections and have periodic maintenance to keep it in good operating condition. Inspections should focus on detecting weakened hoses, leaking fittings, damage to the tank or tank protective coating, broken regulators and gauges, and worn nozzles. Inspection may also be the best time to look for worn bearings, damaged tires (if equipped), and any other mechanical defect or wear. Simple maintenance, such as greasing bearings and drive lines, can be accomplished while inspecting the equipment. Equipment used regularly should be inspected and lubricated every day. Spraying equipment should always be thoroughly checked before each use.

Equipment with self-contained engines requires additional maintenance. Oil and water levels must be checked regularly. Air filters, oil filters, and motor oil need to be changed according to the manufacturer's recommendations. Batteries need cleaning and servicing.

By spending a few minutes each day inspecting and servicing spray equipment, you will increase its useful life and will help to avoid costly breakdowns and possible dangerous leaks. Develop a checklist for servicing and inspecting each piece of equipment to help remember what needs to be done. The checklist can also serve as a service record.

Sprayer Troubleshooting

A sprayer may not exhibit any external signs of problems, but still not function properly. Problems such as lack of pressure, too much pressure, or inadequate output at the nozzles require troubleshooting to locate and correct the cause. Table 10-7 is a guide to troubleshooting problems associated with poor sprayer performance.

Sprayer Storage

Improper storage of spraying equipment can shorten its useful life. Before storing a sprayer, decontaminate and clean it thoroughly. Wear rubber gloves and appropriate protective clothing to avoid contact with pesticide residues. Remove, clean, and reinstall all filters. Partially fill the tank with clean water and add a commercial neutralizing cleaner (or ½ pound of detergent to 30 gallons of water). Circulate this solution through the system for at least 30 minutes and flush it out through the nozzles. Refill the sprayer about half full. Add more commercial cleaner according to directions or add a quart of household ammonia to each 25 gallons of water. Circulate this solution for about 5 minutes and flush a small amount through the nozzles. Shut off the sprayer and let the solution remain in the tank for 12 to 24 hours.

While the cleaning solution is soaking in the tank, thoroughly wash all external parts of the sprayer. Use detergent or ammonia solution, or use a commercial cleaner. Scrub residue off all surfaces using a bristle brush. Rinse external parts with clean water.

TABLE 10-7.

Troubleshooting Problems Associated with Poor Sprayer Performance.

PROBLEM	POSSIBLE CAUSE
Uneven spray pattern	Clogged nozzles. Worn nozzles. Mismatched nozzle sizes. Nozzle screens not uniform. Boom not level. Hoses to nozzles or boom sections not uniform in size. Pressure not adjusted to operating range of nozzles. Foam in spray tank.
Clogged nozzles	Rust, sand, or other contaminants in spray tank. Improper or missing filter screens. Incompatible spray mixture. Poorly mixed spray components. Agitator not working properly. Failure to use marine grease on mechanical agitator fittings.
Pressure too low	Worn pump. Nozzles too large. Nozzles excessively worn. Air in pressure system. Pesticide mixture foaming. Broken or maladjusted pressure regulator. Needed pressure is beyond the capacity of the pump. Pump speed too slow. Drive belts slipping. Restricted or defective suction hose. Clogged suction strainer.
Pressure too high	Pump speed too fast. Pressure regulator not working. Bypass system blocked, restricted, or undersized. Nozzles too small. Filter screens clogged.
Pump not primed	Air trapped in the system. Suction line not completely full of liquid. Worn pump. Drive belts slipping. Shear pin broken on drive line. Foam in tank. Suction line blocked. Leak in suction hose.
Pesticide mixture settles out in tank	Agitator insufficient or not working properly. Incompatible mixture. Tank and hoses not properly cleaned before use. pH too high or too low.
Pulsating pressure	Worn piston pump seals. Highly foaming tank mix.

PROBLEM	POSSIBLE CAUSE
Excessive drift	Nozzles too small.
	Pressure too high.
	Application made during windy weather.
	Spray boom too high.
	Nozzles improperly aligned.
	Temperature too high while applying volatile materials.
	Treated surfaces not receptive.
	Failure to use a drift control agent.

To prevent rusting, touch up scratched areas on all painted surfaces of the trailer, boom, tank, and accessories. Lubricate bearings to prevent them from rusting during storage.

Remove and clean nozzles and nozzle strainers. Store these in a clean plastic bag to keep them free of dirt.

After the tank has finished soaking, flush the solution out and rinse with clean water. Seal nozzle outlets with corks or plastic bags to prevent insects or dirt from getting into the lines. Remove and clean all remaining filter screens and store these in a clean plastic bag. Remove "O" rings from filters and strainers and store them in a plastic bag to prevent them from becoming brittle. Cover the tank loosely to prevent dirt, insects, and rodents from entering during storage; do not close the tank cover tightly as this may permanently distort its rubber seal. Store the sprayer inside a building, preferably covered with a tarp for additional protection. Block up equipment having rubber tires to remove weight from the tires and bearings. Small pumps can be removed and stored in a can of new, lightweight motor oil to prevent rusting. Pumps having rubber or neoprene parts, however, should not be exposed to oil.

Hoses used on hand-held nozzles should be removed, coiled, and hung around a pail, basket, or other large round object to prevent sharp bends that might cause cracks in the rubber. Never hang hoses over a nail, rack, or board. Store hoses in an area away from direct sunlight.

Release the tension from the pressure regulator and remove the "O" ring seal. Lubricate the internal cylinder of the regulator and reassemble without the "O" ring. Place the "O" ring in a plastic bag and tie it to the regulator.

11 Calibration of Pesticide Application Equipment

All pesticide application equipment must be carefully calibrated to assure that the proper amount of pesticide is applied.

The term *calibration* refers to all the operations that ensure the correct amount of pesticide can be applied to the treated area. Failure to calibrate equipment properly is a frequent cause of ineffective pesticide applications and always carries the potential for excessive or illegal residues remaining on sprayed surfaces.

To calibrate your equipment, you must first determine the amount of pesticide and water to be applied and the appropriate rate of application. You may have to adjust ground speed or sprayer pressure, change nozzle sizes, or modify application patterns to achieve the desired rate of application. It is important to check and test equipment periodically, especially nozzles. Nozzles can wear out over a short period of time, resulting in inaccurate liquid sprayer output and uneven spray patterns. Because operators fail to understand how rapidly equipment becomes maladjusted and worn, most sprayers are not calibrated often enough. Failure to understand the effects of altering speed, pressure, or nozzle size also is a source of application and calibration errors.

This chapter discusses the *basic* principles involved in calibration of any type of pesticide application equipment. Shortcuts and quick calculations applicable only in specific situations or for specific types of application equipment are not presented here but may be available in equipment instruction manuals or trade journals and publications. Learn the principles of proper calibration first; then, if appropriate, adopt a quick calibration technique that applies to your own equipment and special needs.

Table 11-1 is a list of conversion factors that may be helpful when calibrating pesticide application equipment.

WHY CALIBRATION IS ESSENTIAL

The main reason for calibrating pesticide application equipment is to determine how much pesticide must be put into the spray tank to assure that the correct amount of chemical gets applied. This is necessary for: (1) effective pest control; (2) protecting human health, the environment, and treated surfaces; (3) preventing waste of resources; and (4) to comply with the law. Also, it is often important to have control over the volume of water being applied to a given area.

Studies conducted to evaluate how well applicators calibrate equipment have shown that many consistently either underestimate or overestimate the actual amount of pesticide being applied.

Effective Pest Control. Manufacturers of pesticides spend millions of dollars researching ways to use their products, including determining the correct amount of pesticide to apply for effective control of target pests. Using less than the labeled amount of pesticide may result in inadequate control and could be a waste of time and money. Despite this, a

TABLE 11-1.

Useful Conversion Factors for Calibration.

STANDARD MEASURE	METRIC CONVERSIONS
LENGTH: 1 ft = 12 in 1 yd = 3 ft 1 mi = 5,280 ft	1 in = 25.4 mm = 2.54 cm 1 ft = 304.8 mm = 30.48 cm 1 yd = 914.4 mm = 91.44 cm 　　　 = 0.914 m 1 mi = 1,609 m = 1.61 km 1 mm = 0.03937 in 1 cm = 0.394 in = 0.0328 ft 1 m = 39.37 in = 3.281 ft 1 km = 3,281 ft = 0.6214 mi
AREA: 1 sq in = 0.007 sq ft 1 sq ft = 144 sq in = 0.000023 sq ac 1 sq yd = 1,296 sq in = 9 sq ft 1 ac = 43,560 sq ft = 4,840 sq yd	1 sq in = 6.45 sq cm 1 sq ft = 929 sq cm 1 sq yd = 8,361 sq cm = 0.8361 sq m 1 ac = 4,050 sq m = 0.405 h 1 sq cm = 0.155 sq in 1 sq m = 1,550 sq in = 10.76 sq ft 1 h = 107,600 sq ft = 2.47 ac
VOLUME: 1 tsp = 0.17 fl oz 1 tbs = 3 tsp 1 fl oz = 2 tbs = 6 tsp 1 cup = 8 fl oz = 16 tbs 1 pt = 2 cups = 16 fl oz 1 qt = 2 pt = 32 fl oz 1 gal = 4 qt = 8 pt = 128 fl oz 　　　 = 231 cu in	1 fl oz = 29.5 ml = 0.0295 l 1 pt = 437 ml = 0.437 l 1 qt = 945 ml = 0.945 l 1 gal = 3785 ml = 3.785 l 1 ml = 0.033 fl oz 1 l = 33.8 fl oz = 2.112 pt 　　 = 1.057 qt = 0.264 gal
WEIGHT: 1 oz = 0.0625 lb 1 lb = 16 oz 1 ton = 2,000 lb 1 gallon of water = 8.34 lb	1 oz = 28.35 g 1 lb = 454 g = 0.4536 kg 1 ton = 907 kg 1 gallon of water = 3.786 kg 1 g = 0.035 oz 1 kg = 35.27 oz = 2.205 lb

ac: acre
fl oz: fluid ounce
ft: foot or feet
gal: gallon
in: inch
lb: pound
mi: mile
oz: ounce
pt: pint
qt: quart
sq: square
tbs: tablespoon
tsp: teaspoon
yd: yard

cm: centimeter
g: gram
h: hectare (1 h = 10,000 sq m)
kg: kilogram
km: kilometer
l: liter
m: meter
ml: milliliter
mm: millimeter

recent study showed that a third of the applicators surveyed unknowingly underapplied pesticides by an average of 30%. One third of the applicators in this same survey overapplied pesticides, averaging 35% more pesticide than the maximum label rate. Inadequate amounts of pesticide lead to problems such as pest resistance and resurgence, while using too much pesticide has adverse effects on natural predators, target surfaces, and the environment; excessive use also wastes materials and is illegal.

Human Health Concerns. Pesticides that are applied at rates higher than label recommendations may endanger human health. Illegal residues may occur on produce when a pesticide is overapplied; the entire crop may be confiscated to protect consumers if residues are above allowable tolerances. Although reentry intervals have a relatively large safety factor, field-workers in agricultural commodities may receive unnecessary exposure to residues resulting from overapplication. Application equipment operators can be exposed to more concentrated mixtures of pesticide when the equipment is poorly calibrated or nozzles are severely worn. Amounts of pesticides used in residential, industrial, and institutional settings must likewise be carefully controlled to prevent excessive exposure to people living or working in these areas.

Environmental Concerns. Pesticides may cause environmental problems when they are not used properly. Harm to beneficial insects, such as honey bees, and to wildlife must be avoided by carefully calibrating equipment to maintain application rates within label recommendations.

Protection of Treated Surfaces. Certain pesticides are phytotoxic and can be damaging to sprayed surfaces when used at higher than recommended rates. Manufacturers evaluate these potential problems while testing their chemicals to determine safe concentrations. Applying more than the labeled amount of pesticide may increase risks of damage. Chances of building up excessive residues in the soil are also increased when too much pesticide is used, sometimes seriously limiting the types of subsequent crops that can be grown in an area.

Preventing Waste of Resources. Using the improper amount of pesticide wastes time and adds unnecessary costs to the application. Not only are pesticides expensive, but the fuel, labor, and equipment wear and tear required to make extra applications is costly, too.

Legal Aspects. Applicators who use pesticides improperly are subject to criminal and civil charges, resulting in fines, imprisonment, and lawsuits. Applicators are liable for injuries or damage to people, the environment, crops, and personal and public property caused by improper pesticide application.

METHODS OF EQUIPMENT CALIBRATION

A few simple tools are needed to calibrate pesticide application equipment (Table 11-2). Put these items in a small toolbox and use them only for calibration purposes (Figure 11-1). Keep your tools clean and in good working condition; make equipment calibration a professional operation.

TABLE 11-2.

Tools Needed for Calibrating Pesticide Application Equipment.

TOOLS NEEDED FOR CALIBRATION

1. *Stopwatch.* A stopwatch is essential for timing travel speed and flow rates. Never rely on a wristwatch unless it has a stopwatch function.

2. *Measuring tape.* A 100 foot moisture- and stretch-resistant measuring tape is used for marking off the distance to be traveled and for measuring spray swath width.

3. *Calibrated liquid container.* A container having a capacity of one or two quarts, calibrated for liquid ounces, is needed for measuring spray nozzle output.

4. *Scale.* A small scale, capable of measuring pounds and ounces, is used for weighing granules collected from a granule applicator. Most accurate weight measurements can be obtained from scales having maximum capacities between 5 and 10 pounds.

5. *Pocket calculator.* A pocket calculator is needed for making calculations in the field.

6. *Pressure gauge.* An accurate, calibrated pressure gauge, with fittings compatible with spray nozzle fittings, is helpful for checking boom pressure and for calibrating the sprayer pressure gauge.

7. *Flow meter.* A flow meter, attached to a flexible hose or filling pipe, can be used for measuring the amount of water put into a tank. This device can be used for measuring tank capacity and for determining the amount of liquid used during a calibration run. Both mechanical and electronic flow meters are available. If these are not available, a calibrated 5 gallon pail can be used instead.

8. *Flagging tape.* Colored plastic flagging tape is useful for marking off measured distances when determining applicator speed.

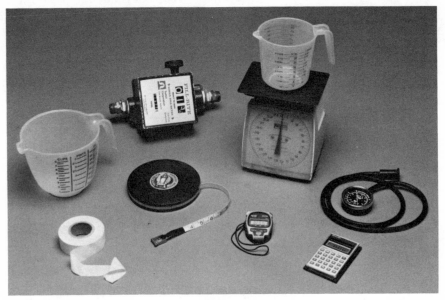

FIGURE 11-1.

A few simple tools are required for calibrating a pesticide sprayer. These include a stopwatch, measuring tape, several calibrated containers, a scale, pocket calculator, pressure gauge, flow meter, and flagging tape.

Different calibration techniques are used for liquid application equipment and dust or granular application equipment.

NOTE: Pesticide application equipment and the discharge from application equipment being calibrated may contain pesticide residue. Always wear rubber gloves and other protective equipment to prevent pesticide contamination of your eyes, hair, skin, clothing, and shoes. Read Chapter 7 for information on selecting the proper protective equipment.

Calibrating Liquid Sprayers

Equipment designed to apply pesticides dissolved or suspended in water or some other liquid needs to be frequently calibrated to monitor pump and nozzle wear. Pump wear may decrease the amount and pressure of fluid output; nozzle wear increases the volume of output, may lower the output pressure, and may produce a poor spray pattern. Abrasive pesticides, such as wettable powders, increase the rate of wear.

Before making any calibration measurements, be sure to service the sprayer; follow the servicing directions outlined in Table 11-3. Once the sprayer is serviced, begin the calibration process. The final goal is to determine how much area will be covered by each tank of spray when the sprayer is moving at a known speed and operating at a known pressure. Four factors must be measured: tank capacity, speed of travel, flow rate, and width of spray swath.

Capacity of Tank. The capacity of the spray tank, or tanks, if more than one is used, needs to be measured only one time, but it *must* be measured. You must know *exactly* how much liquid a spray tank holds. Never rely on manufacturer's ratings, because these could be approximate volumes or may not take into account fittings installed inside the tank or the capacity of spray lines, pump, and filters.

Fill the tank either with a bucket or other container of known volume, or by using a flow meter attached to a hose. Always use clean water. A 5-gallon bucket works well for smaller sprayers; the bucket must be filled

TABLE 11-3.

How to Service Liquid Application Equipment Before Beginning Calibration.

SERVICING SPRAY EQUIPMENT

1. Flush tank and pumping system with clean water to remove debris and dirt.

2. Clean and replace all filter screens.

3. Check nozzles for wear and replace if necessary or if in doubt. All nozzles must be clean.

4. Lubricate all bearings and appropriate moving parts.

5. Inspect hoses for cracks and leaks, and replace if necessary.

6. Make sure pressure gauge is working properly by testing it against another gauge known to be accurate.

FIGURE 11-2.

Flow meters, similar to the one shown, can be used to measure the volume of spray tanks.

each time to hold exactly 5 gallons and accordingly, must be calibrated and marked before use. The spray tank should be perfectly level during filling. Close all valves to prevent water from leaking out. Add water, 5 gallons at a time, until the tank is nearly filled. Use smaller-volume calibrated containers to top off the tank. Record the total volume of water required to fill the tank (paint or engrave this figure onto the outside of the tank for permanent reference). The tank's sight gauge should be calibrated while the tank is being filled by making marks on the tank or gauge as measured volumes of water are put in. If the unit is not equipped with a sight gauge, mark volume increments on a dipstick that can be kept with the tank. Use 1-gallon marks for tanks with a capacity of 10 gallons or less, and increments of 5 or 10 gallons for tanks having a total capacity of 50 gallons or less. Increments of 10 to 20 gallons are used on larger tanks. Once the sight gauge or dipstick is calibrated and labeled, it will be possible to determine how much liquid is in the tank when the tank is not entirely full. Tanks must always be returned to a level surface when reading the sight gauge or dipstick.

A flow meter attached to a filling hose may be used for measuring the volume of larger tanks (Figure 11-2); flow meters enable accurate and fairly rapid determinations of large volumes. Be sure to calibrate and label the sight gauge or dipstick as the tank is being filled.

Speed of Travel. Always measure travel speed under actual working conditions. If you are calibrating an orchard sprayer, use a filled tank in an orchard; similarly, row crop and field sprayers should be calibrated under actual conditions. Tractors travel faster on paved or smooth surfaces than on soft dirt or clods. Never rely on tractor speedometers for mile-per-hour measurements because wheel slippage and variation in tire size due to wear may cause as much as a 30% difference in actual over indicated speed. When calibrating a backpack or hand-held sprayer, walk on terrain similar to the area that will be sprayed.

Using a 100-foot tape, measure off any convenient distance. It can be more or less than 100 feet, but calibration is more accurate if longer distances (between 200 and 300 feet) are used, especially if equipment moves at several miles per hour. Sometimes multiples of 88 feet are chosen because 88 feet is the distance covered in 1 minute while traveling 1 mile per hour. In orchards or vineyards, a given number of tree or vine spaces of known length can serve as a convenient reference. Indicate the beginning and end of the measured distance with colored flagging tape.

Have someone drive (or walk, if calibrating a backpack sprayer) the sprayer through the measured distance at the speed desired for an actual application. Choose a speed within a range appropriate for the application equipment. When using a tractor, note the throttle setting, gear, and rpm of the engine. The use of a positive throttle stop is helpful so the engine can always be returned to the same speed. Be sure actual application speed is attained before crossing the first marker. Use a stopwatch to determine the time, in minutes and seconds, required to traverse the measured distance (Figure 11-3). For best results, repeat this process 2 or 3 times and take an average. Follow the procedure in Table 11-4 to calculate the actual speed of the equipment.

Flow Rate. Measure the actual output of the sprayer when nozzles are new, then periodically thereafter to accommodate for nozzle wear. Although manufacturers provide charts showing output of given nozzle

FIGURE 11-3.

Measure off a known distance when calculating the speed of travel of the application equipment. Use a stopwatch to time the travel of the sprayer through the measured distance.

sizes at specified sprayer pressures, you should check output under actual conditions of operation. Manufacturer's charts are most accurate when using new nozzles because used nozzles will be worn and may have different output rates; however, even new nozzles may have slight variations in actual output. Sprayer pressure gauges may not be accurate, adding further error to the output estimate determined from charts.

Liquid sprayer output is usually measured in *gallons per minute*. Two methods can be used depending on the type of sprayer being calibrated. The first method is designed for low-pressure sprayers and small hand-held units; it involves collecting a volume of water emitted out of individual nozzles over a measured period of time. The second method, for large airblast and high-pressure sprayers, measures the total output of the sprayer over a known period of time.

Collection Method for Low-Pressure and Small Hand-Held Sprayers. Low-pressure sprayers, including low-pressure boom sprayers, backpack sprayers, and controlled droplet applicators, can be calibrated by measuring the amount of spray emitted from nozzles. If the sprayer is equipped with more than one nozzle, collect liquid from each separately to get a comparison of their output; this operation will point out any malfunction or wear. A stopwatch and calibrated container are required for making measurements. Wear rubber gloves to avoid skin contact with the liquid, and stand upwind from the nozzles to prevent fine mist or spray from contacting your face and clothing. Wear eye protection to prevent getting spray droplets in your eyes.

For low-pressure power sprayers used in agricultural, right-of-way, and landscape application, fill the tank at least half full with water, start the sprayer, and bring the system up to normal operating pressure. Operate hydraulic agitators if they are to be used during the application, because hydraulic agitators divert some liquid from the nozzles and often lower the pressure in the system. Most power sprayers have a limited operating pressure range depending on the type of pump and type of power unit; never attempt to operate equipment beyond its normal

TABLE 11-4.

Calculating Speed of Application Equipment.

1. Convert minutes and seconds into minutes by dividing the seconds (and any fraction of a second) by 60.

> EXAMPLE: Your trip took:
> 1 min and 47.5 sec

$$\frac{47.5 \text{ sec}}{60 \text{ sec/min}} = 0.79 \text{ min}$$

> ... add these amounts together:
>
> 1 min + 0.79 min = 1.79 min

2. Add the *converted* minutes from each run and divide by the number of runs.

> EXAMPLE: Three runs were made. . .
>
> Run #1 = 1 min, 47.5 sec = 1.79 min
> Run #2 = 1 min, 39.8 sec = 1.66 min
> Run #3 = 1 min, 52.0 sec = 1.87 min
>   Total = 5.32 min

$$\frac{5.32 \text{ min}}{3 \text{ runs}} = 1.77 \text{ min/run average time}$$

3. Divide the measured distance by the average time; this will tell you how many feet were traveled in one minute.

> EXAMPLE: The measured distance for this example is 227 feet.

$$\frac{227 \text{ ft}}{1.77 \text{ min}} = 128.25 \text{ ft/min}$$

4. If you wish to determine the speed in miles per hour, divide the feet-per-minute figure by 88 (the number of feet traveled in 1 minute at 1 mile per hour).

> EXAMPLE:

$$\frac{128.25 \text{ ft/min}}{88 \text{ ft/min/mi/hr}} = 1.46 \text{ mi/hr}$$

FIGURE 11-4.

To determine the output from each nozzle, collect liquid over a measured period of time. Make sure the sprayer is operating at the pressure that would be used under actual field conditions. Wear rubber gloves and eye protection because the liquid may contain traces of pesticide.

working range because this may cause premature failure of the pump. If the sprayer is powered by a tractor PTO, be sure that the tractor engine rpm is the same as that determined in the speed calibration; otherwise the pump output pressure will be different. Adjust the pressure to the requirements of the spray situation and nozzle manufacturer's recommendations. Be sure appropriate nozzles are installed on the equipment. Check the pressure by attaching a calibrated pressure gauge at either end of the boom, replacing one of the nozzles. Open the valves to all nozzles and note the pressure, make adjustments as necessary, then remove the gauge.

While all nozzles are operating at the proper pressure, collect about 15 to 30 fluid ounces of liquid from each (Figure 11-4). Use a stopwatch to determine the time in seconds required to collect each volume.

When calibrating backpack sprayers, pump the unit as you would during an actual application. Collect spray into a calibrated container for a measured period of time. Compressed air sprayers lose pressure during operation, so they must be pumped frequently. To calibrate, fill the tank about half full with water to provide a sufficient volume of air to keep the pressure more uniform. For some types of controlled droplet applicators, it is possible to disconnect the hose and orifice from above the spinning disc or cup and collect liquid into a calibrated container over a measured period of time; the liquid must flow through the orifice.

Record the volume of liquid collected from each nozzle or orifice and the time in seconds required to collect each amount. Use a format similar to Table 11-5. Determine the fluid-ounces-per-second output for each nozzle by dividing the volume by the the seconds required to collect it (Table 11-6). Convert ounces-per-second into gallons-per-minute by

TABLE 11-5.

Recording Nozzle Output.

NOZZLE #	VOLUME	TIME
1	12.5 fl oz	23.2 sec
2	12.0	22.5
3	15.5	24.8
4	14.5	26.1
5	19.0	27.2
6	13.0	23.9

TABLE 11-6.

Calculating Gallons-per-Minute for Low-Pressure Sprayers.

1. Determine the gallons-per-minute output of each nozzle by dividing the fluid ounces collected by the time (in seconds) and multiplying the result by 0.4688.

EXAMPLE:

NOZZLE	fl oz/sec		× 0.4688	=	gpm
1	12.5/23.2	= 0.539	× 0.4688	=	0.253
2	12.0/22.5	= 0.533	× 0.4688	=	0.250
3	15.5/24.8	= 0.625	× 0.4688	=	0.293
4	14.5/26.1	= 0.556	× 0.4688	=	0.261
5	19.0/27.2	= 0.699	× 0.4688	=	0.328
6	13.0/23.9	= 0.544	× 0.4688	=	0.255

Total Output = 1.640 gpm

2. Compute the percent variation from the rated nozzle output. Divide the actual gallons-per-minute output by the rated output. Subtract 1 from this number and multiply by 100.

EXAMPLE:

NOZZLE	Actual gpm / Rated gpm	SUBTRACT 1.00	MULTIPLY BY 100	= PERCENT VARIATION
1	0.253/0.250 = 1.012	− 1.00 = 0.012	× 100	= 1.2%
2	0.250/0.250 = 1.000	− 1.00 = 0.000	× 100	= 0.0
3	0.293/0.250 = 1.172	− 1.00 = 0.172	× 100	= 17.2
4	0.261/0.250 = 1.044	− 1.00 = 0.044	× 100	= 4.4
5	0.328/0.250 = 1.312	− 1.00 = 0.312	× 100	= 31.2
6	0.255/0.250 = 1.020	− 1.00 = 0.020	× 100	= 2.0

multiplying the result by the constant 0.4688 (60 seconds per minute divided by 128 fluid ounces per gallon equals 0.4688).

Output among nozzles will usually vary. In the example in Table 11-6, part 1, the output ranges from 0.250 gallons per minute to 0.373 gallons per minute. Assume that the rated capacity (as given by the manufacturer) for these nozzles at the recommended operating pressure is 0.250 gallons per minute. The variation *among* nozzles should not be greater than 5%, and the output of any nozzle should not exceed the manufacturer's rated output by more than 10%. The percentage of variation (example in Table 11-6, part 2) can be computed by dividing the actual output by the rated output. Subtract 1.00 from this figure, then multiply by 100 to obtain the variation in percent. Nozzles 3 and 5 in this example exceed these amounts and therefore must be replaced. However, whenever any nozzles are replaced, the flow rate of all the nozzles must be rechecked because changing one nozzle may affect the pressure in the whole system; after changing nozzles, readjust the pressure regulator to maintain the desired pressure. Table 11-7, part 1 shows how to recalculate the gallons-per-minute output after replacing worn nozzles.

Spray check devices are calibration aids that provide a visual representation of the spray pattern produced by nozzles on spray booms. This portable device is placed under a boom and the output from several

TABLE 11-7.

Recalculating Output After Replacing Worn Nozzles.

1. Replace worn nozzles (numbers 3 and 5 in this example) and remeasure the output of all nozzles on the boom. Recalculate the gallons-per-minute for each nozzle. Add these rates together to determine the total output of the sprayer.

EXAMPLE:

NOZZLE	fl oz/sec			× 0.4688	= gpm
1	12.5/23.2	=	0.539	× 0.4688	= 0.253
2	12.0/22.5	=	0.533	× 0.4688	= 0.250
3	13.3/24.5	=	0.542	× 0.4688	= 0.254
4	14.5/26.1	=	0.556	× 0.4688	= 0.261
5	15.2/28.3	=	0.537	× 0.4688	= 0.252
6	13.0/23.9	=	0.544	× 0.4688	= 0.255
				Total output	= 1.525 gpm

2. Check to see that all nozzles are within 5% of the rated capacity of these nozzles.

EXAMPLE:

NOZZLE	Actual gpm / Rated gpm	SUBTRACT 1.00	MULTIPLY BY 100	= PERCENT VARIATION
1	0.253/0.250 = 1.012	− 1.00 = 0.012	× 100	= 1.2%
2	0.250/0.250 = 1.000	− 1.00 = 0.000	× 100	= 0.0
3	0.254/0.250 = 1.016	− 1.00 = 0.016	× 100	= 1.6
4	0.261/0.250 = 1.044	− 1.00 = 0.044	× 100	= 4.4
5	0.252/0.250 = 1.008	− 1.00 = 0.008	× 100	= 0.8
6	0.255/0.250 = 1.020	− 1.00 = 0.020	× 100	= 2.0

nozzles is collected into a series of evenly spaced cells. After collection, the device is rotated from a horizontal to a vertical position; collected liquid drains into glass vials corresponding to the individual cells. Floats inside these vials rise to the top of the liquid level. Variation in levels can be readily seen, pinpointing nozzle problems and/or poor nozzle height adjustment.

Measured Release Method for Airblast or High-Pressure Sprayers. Due to the air blast or high pressures of larger sprayers, it is not possible to collect ejected liquid into a container. Therefore you must measure the output of the sprayer over a period of time by determining how much water was used.

Start by moving the sprayer to a level surface and fill the tank to its maximum with clean water; liquid must be at a level that can be duplicated when refilling. A convenient technique is to fill the tank with clean water to the point where it just begins to overflow. Use low-volume, low-pressure water, such as from a garden hose, for topping off the tank. Check for leaks around tank seals and in hoses. All nozzles must be clean and operating properly or the results will be inaccurate.

Stand upwind and operate the sprayer at its normal operating speed and pressure. Open the valves to all nozzles, starting a stopwatch at the

FIGURE 11-5.

It is not possible to collect the sprayed liquid from some types of sprayers. To determine the amount of liquid expelled by these sprayers: (1) fill the tank to a known level; (2) run the sprayer under normal conditions for a timed period; and (3) refill the tank to its original level, measuring the amount of water used.

same time. Continue to run the sprayer for several minutes, then close the valves to all nozzles and record the elapsed time (Figure 11-5).

If the tank has been calibrated and marked, the amount of liquid used will be apparent or can be determined with a calibrated dipstick. Otherwise, use a flow meter attached to a low-pressure filling hose and refill the sprayer to the original level. Record the gallons of water used; this volume is the amount of liquid sprayed during the timed run. Repeat this process two more times to get an average of sprayer output. Determine the gallons-per-minute output of the sprayer by using the calculations shown in Table 11-8.

TABLE 11-8.

Calculating Gallons-per-Minute for High Pressure Sprayers.

1. Record the elapsed time during each trial run and the amount of liquid sprayed:

EXAMPLE:

Run #	Time	Volume
1	1 min 45 sec	37.5 gal
2	1 min 30 sec	33.5 gal
3	1 min 50 sec	38.0 gal

2. Convert the time from minutes and seconds to minutes by dividing the seconds by 60 and adding this decimal to the minutes.

EXAMPLE:

Run #	min	sec	sec/60	= min
1	1	45	0.75	1.75
2	1	30	0.50	1.50
3	1	50	0.83	1.83

3. Divide the collected gallons for each run by the minutes, resulting in gallons-per-minute.

EXAMPLE:

Run #	gal/min	= gpm
1	37.5/1.75	= 21.4
2	33.5/1.50	= 22.3
3	38.0/1.83	= 20.8

4. Add all the gallon-per-minute figures and divide this total by the number of runs (3 in this example) to get the average gallon-per-minute output.

EXAMPLE:

Run #	gpm
1	21.4
2	22.3
3	20.8
Total =	64.5

64.5/3 = 21.5 gpm average output

Width of Spray Swath. The final measurement needed to complete calibration is the width of spray swath being applied by the sprayer. Figure 11-6 illustrates spray swath widths for various application situations. For multiple nozzle boom-type sprayers, the swath width is the width of the boom plus the distance between one pair of nozzles; swath width also can be calculated by multiplying the number of nozzles by the nozzle spacing. (Number of Nozzles × Nozzle Spacing = Swath Width.) When making a pesticide application, position the end nozzle of each subsequent pass to leave a space equal to the nozzle spacing on the boom (Figure 11-7). Spray boom height must be adjusted so that there is approximately a 30% overlap of spray from adjacent nozzles on the boom (Figure 11-8). Position nozzles at the exact height they would be during an actual application. Check the spray boom to make sure it is level; an unlevel boom will cause uneven distribution of spray (Figure 11-9). Fan nozzles must be aligned as illustrated in Figure 11-10 to give an even spray distribution.

When spray is emitted as separate bands or strips, the swath width is equal to the combined width of each individual band, but does not include the unsprayed spaces between bands (Figure 11-11).

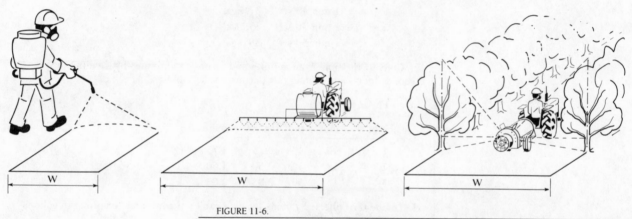

FIGURE 11-6.

A spray swath is the horizontal width being covered with spray material during a single pass. Swath width is measured differently, depending on the type of pesticide application.

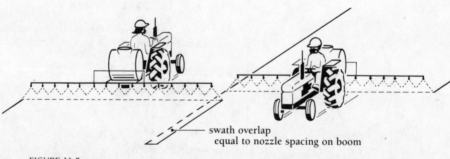

swath overlap
equal to nozzle spacing on boom

FIGURE 11-7.

Spray from adjacent swaths should overlap by the same amount as spray from nozzles on the spray boom overlaps (usually about 30% of the spray pattern of one nozzle). To do this, allow one nozzle-width spacing between swaths as illustrated here.

FIGURE 11-8.

Under normal conditions, flat fan nozzles on a spray boom must be spaced so there is a 30% overlap of the spray emitted by adjacent nozzles. This provides for a uniform distribution of spray.

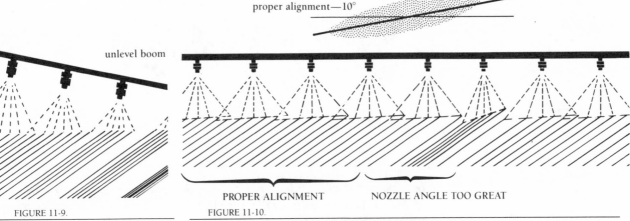

FIGURE 11-9.

An unlevel spray boom will cause an uneven pesticide application.

FIGURE 11-10.

The spray pattern will be uneven if nozzles are not aligned properly on the spray boom. Rotate nozzles about ten degrees from the axis of the boom to prevent droplets from adjacent nozzles from touching, but still allow for proper overlap of the spray pattern.

If the sprayer is used to apply pesticides to crop plants in an orchard or vineyard and plants on both sides of the sprayer are treated with one pass, the spray swath is equal to the width of the tree or vine row (Figure 11-12). But if spray is being applied only out of one side of the sprayer and the sprayer is moved along both sides of the tree or vine row, the swath is one-half the width of the tree or vine spacing (Figure 11-13). Use a tape measure to determine tree or vine row width; take several measurements within the orchard or vineyard to check if row spacing is uniform and consistent; average the results if any variation is found (Figure 11-14). To adjust an air blast orchard or vineyard sprayer to apply a given volume of water per acre, it is convenient to know the number of trees or vines per acre. After a tank of known volume is sprayed out, the number of trees or vines completely sprayed can be counted and the proportion of an acre covered is then easily figured. Travel speed can be

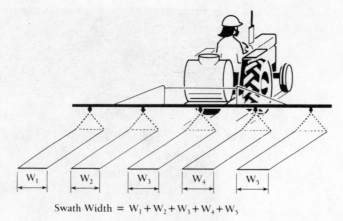

Swath Width = $W_1 + W_2 + W_3 + W_4 + W_5$

FIGURE 11-11.

Swath width from banded applications is determined by adding the widths of the individual bands.

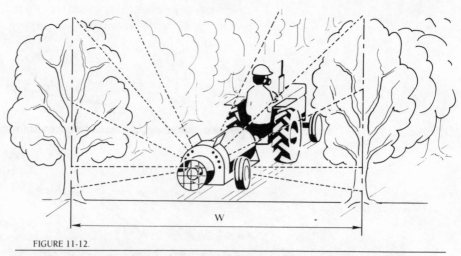

FIGURE 11-12.

In orchards or vineyards, if plants on both sides of the sprayer are being sprayed simultaneously with an air blast sprayer or high-pressure boom sprayer, the swath width is the distance between plant rows.

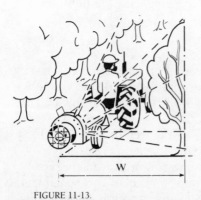

FIGURE 11-13.

When spray is emitted from only one side of of an orchard or vineyard airblast sprayer, the swath width of each pass is one-half the plant row spacing.

increased or decreased slightly to apply less or more liquid per acre, or nozzle sizes can be changed.

Swath width for herbicide strip sprays in orchards and vineyards should be measured only to the center of the tree or vine row and should not include overlap (Figure 11-15). Unless the herbicide is applied to the entire orchard or vineyard floor, the actual sprayed area will be less than the total planted area.

Sometimes nozzles are attached to an inverted "U"-shaped boom so that pesticides can be applied to the top and both sides of vines or plants in a row (Figure 11-16). Swath width for this type of equipment is equal to the distance between opposing nozzles.

Pesticides are often injected into the soil through special subsoil chisels spaced along a tractor-mounted tool bar. It is assumed that pesticides are being applied to the entire subsurface area in most soil injection applications, therefore swath width is equal to the number of chisels multiplied by the space between the chisels on the tool bar (Figure 11-17). Occasionally pesticides are injected as a band, so swath width is the sum of all the band widths, similar to surface band applications.

FIGURE 11-14.

Swath width for pesticide sprays in orchards and vineyards should be measured from the center of one tree or vine row to the center of the adjacent row. Take several measurements in different locations to check for variation in plant spacing. If variation exists, average the measurements.

FIGURE 11-15.

Swath width for herbicide strip sprays in orchards and vineyards should be measured only to the center of the tree or vine row and should not include overlap.

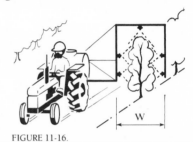

FIGURE 11-16.

Sometimes spray can be applied to both sides of a plant or vine row through a specially designed, horseshoe-shaped boom arrangement. Several plant rows can often be sprayed at the same time with these applicators. Spray swath width is the distance between opposing nozzles. If multiple rows are sprayed, the swath width is the sum of the distances.

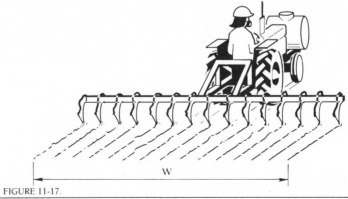

FIGURE 11-17.

Subsoil chisels, spaced along a tractor's tool bar, are used to inject pesticides into the soil. When pesticides are injected into the soil, the swath width is usually considered to be the width of the tool bar.

Measure the swath width of a backpack sprayer used for ground application of pesticides from the spray pattern produced on the ground in a test run. Keep the nozzle at the height held during an actual application; maintain this height at all times to prevent variation in swath width. Nozzles of these types of sprayers usually provide a uniform spray pattern, so swaths need to be overlapped only enough to assure a uniform application pattern. Use the same method to measure swath width of controlled droplet applicators.

Determining the Amount of Pesticide to Use. Use the tank volume, the speed of the applicator, the flow rate of the sprayer, and the spray swath width to calculate the total area that can be covered with each tank of material. Knowing this value allows you to determine how much pesticide to put into the tank. Two calculation methods may be used, one for pesticides applied by the acre (Table 11-9) and the other for applications, such as landscape treatments or sprays in confined areas, made by the square foot (Table 11-10).

Table 11-11 is an example of how calibration formulae can be combined onto a single sheet for in-field use. This example shows a calibration worksheet designed for orchard sprayers. Similar sheets can be prepared for other types of pesticide sprayers.

To prevent waste of pesticide material, accurately measure the area to be treated and mix only the amount of chemical needed.

TABLE 11-9.

How Much Pesticide to Put into the Spray Tank
(Pesticides applied on a per-acre basis.)

1. First, determine the area that can be treated in one minute. Divide the spray swath width by 43560 (the number of square feet in one acre) and multiply the result by the travel speed, in feet-per-minute. The result will be the acres treated per minute. In the example in Table 11-4, page 331, travel speed was calculated to be 128.25 feet per minute. Assuming the swath width is 12 feet, the calculation would be:

EXAMPLE:

$$\frac{12 \text{ ft}}{43560 \text{ sq ft/ac}} \times 128.25 \text{ ft/min} = 0.0353 \text{ ac/min}$$

In this example, when a swath 12 feet wide is being sprayed, 0.0353 acres are covered in one minute.

2. Next, determine the gallons of liquid being applied per acre. Divide the gallons-per-minute figure by the acres-per-minute:

EXAMPLE:

$$\frac{1.525 \text{ gal/min}}{0.0353 \text{ ac/min}} = 43.2 \text{ gal/ac}$$

3. Then, determine the number of acres that can be treated with a full tank. Divide the actual measured volume of the spray tank (or tanks) by the gallon-per-acre figure. Assume the tank holds 252.5 gallons when filled:

EXAMPLE:

$$\frac{252.5 \text{ gal/tank}}{43.2 \text{ gal/ac}} = 5.84 \text{ ac/tank}$$

4. Finally, determine how much pesticide to put in the tank. Multiply acres-per-tank by the recommended rate per acre of pesticide; check the pesticide label for this information. (If the label calls for "active ingredient" see the "Active Ingredient Calculations" section on page 351.)

EXAMPLE:

Pesticide Label Says		Acres per tank		Amount of Pesticide to Put in tank
1.5 lb/ac	×	5.84	=	8.76 lb
3 qt/ac	×	5.84	=	17.52 qt
2 gal/ac	×	5.84	=	11.68 gal
1 pt/ac	×	5.84	=	5.84 pt

Changing Sprayer Output

Once a sprayer has been calibrated, its output rate is determined for a specific speed. There may be times when this output rate needs to be changed to accommodate for variations in foliage, plant spacing, other aspects of the treatment area, or requirements to travel at a faster or slower speed. Also, the spray output will change as nozzles or pumps begin to wear. Several adjustments can be made, either alone or in combination, to effectively increase or decrease sprayer output within a limited range.

Changing Speed. The simplest way to alter the volume of spray being applied to an area is by changing the travel speed of the sprayer. A slower speed results in more liquid being applied, while a faster speed reduces the application rate. Such adjustments may be needed when swath width changes slightly, such as in orchards or vineyards where plant spacing may differ from block to block (Figure 11-18). Changing the travel speed

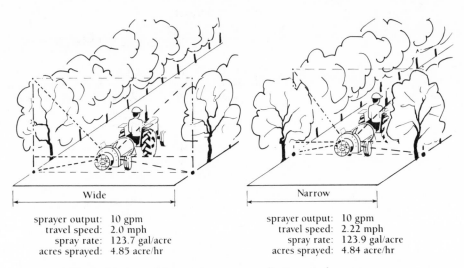

	Wide	Narrow
sprayer output:	10 gpm	10 gpm
travel speed:	2.0 mph	2.22 mph
spray rate:	123.7 gal/acre	123.9 gal/acre
acres sprayed:	4.85 acre/hr	4.84 acre/hr

Sprayer must travel faster to keep spray application rate the same.

FIGURE 11-18.

Changes in row spacing in an orchard or vineyard affect the amount of spray being applied per acre. Increasing or decreasing ground speed can accommodate for the difference in spacing so the correct amount of pesticide per acre will be applied. Variations in the size of trees or vines may also influence the rate of application.

eliminates the need for altering the concentration of chemical in the spray tank, although there are limits to the amount of speed change that can be made. Operating application equipment too fast is a common error and will result in poor coverage. Operating it too slow results in runoff, waste, and an increase in application time and cost. To determine how much to increase or decrease your speed, rework the calculations shown in Table 11-9 or 11-10, inserting the new swath width.

TABLE 11-10.

How Much Pesticide to Put into the Spray Tank
(For pesticides applied by the square foot).

1. Determine how many square feet can be treated in one minute. Multiply the speed as determined by the procedures in Table 11-4, page 331, by the swath width. In this example, assume a single nozzle hand-operated sprayer is being used to apply a swath width of 2.5 feet at a speed of 128.25 feet per minute.

EXAMPLE:

128.25 ft/min × 2.5 ft = 320.63 sq ft/min

2. Next, determine the volume of spray, in gallons, that will be applied to one square foot. Divide the gallon-per-minute output (see Table 11-5 for procedures) of the sprayer by the square-feet-per-minute. For this example, assume that the backpack unit sprays 0.05 gallons per minute.

EXAMPLE:

$$\frac{0.05 \text{ gal/min}}{320.63 \text{ sq ft/hr}} = 0.000156 \text{ gal/sq ft}$$

3. Then, find out how many square feet can be sprayed with one tank. Divide the gallons-per-square-foot into the measured tank capacity. For this example assume that the tank holds 3 gallons.

EXAMPLE:

$$\frac{3 \text{ gal/tank}}{0.000156 \text{ gal/sq ft}} = 19{,}230 \text{ sq ft/tank}$$

4. Finally, determine how much pesticide to put in the tank. First, read the pesticide label; it will tell you the amount of pesticide to apply. Normally, the label will tell you how much to apply per square foot (or 100 or 1000 square feet) or per acre. (If the label calls for "active ingredient" see the "Active Ingredient Calculations" section on page 351.)

EXAMPLE A: If the label gives the dosage rate per 1, 100, or 1000 square feet, multiply that rate by the square-feet-per-tank as determined in step 3:

Pesticide Label Says	×	Square Feet per Tank	=	Amount of Pesticide to Put in Tank
$\dfrac{3 \text{ fl oz}}{1000 \text{ sq ft}}$	×	19,230	=	57.69 fl oz
$\dfrac{3/4^* \text{ fl oz}}{1000 \text{ sq ft}}$	×	19,230	=	14.42 fl oz
$\dfrac{1 \text{ oz}}{100 \text{ sq ft}}$	×	19,230	=	192.3 oz

*The fraction 3/4 is converted to its decimal equivalent 0.75 to complete this calculation.

EXAMPLE B: If the pesticide label gives the dosage rate in units of pesticide per acre, convert square-feet-per-tank (from step 3) to acres-per-tank by dividing it by 43560 (there are 43560 square feet in one acre):

$$\frac{19230 \text{ sq ft/tank}}{43560 \text{ sq ft/ac}} = 0.441 \text{ ac/tank}$$

Then, multiply the labeled rate per acre by the acres-per-tank figure:

Pesticide Label Says	×	Acres per Tank	=	Amount of Pesticide to Put in Tank
1.5 lb/ac	×	0.441	=	0.661 lb (10.6 oz)
3 qt/ac	×	0.441	=	1.32 qt (42.2 fl oz)
2 gal/ac	×	0.441	=	0.882 gal (7.1 pt)
1 pt/ac	×	0.441	=	0.441 pt (7.1 fl oz)

Changing Output Pressure. As nozzles begin to wear, the spray volume will increase. When a pump becomes worn it becomes less efficient and therefore the nozzle output drops off. Adjusting the pressure regulator to increase or decrease output pressure will change the spray volume slightly. Increasing pressure increases the output, while decreasing pressure lowers it. However, in order to double the output volume it is necessary to increase the pressure by a factor of four; this is usually beyond the capabilities of the spraying system because the amount of adjustment that can be achieved is limited by the working pressure range of the sprayer pump. Whenever pressure in the system is changed, the nozzle output must be remeasured (see Table 11-5, page 332) and the calibration calculations reworked. Increasing pressure breaks spray up into finer droplets, while lowering pressure too much may reduce the effectiveness of nozzles by altering the spray pattern.

Changing Nozzle Size. The most effective way to change the output volume of a sprayer is to install different sized nozzles. Larger nozzles increase volume, while smaller ones reduce spray output. Changing nozzles usually alters the pressure of the system and requires an adjustment of the pressure regulator. The volume output of disc-core nozzles may be adjusted by changing either the disc or the core, or by replacing both. Be aware that changes in either the core or disc will also change the droplet size and spray pattern. Use tables included in nozzle manufacturers' catalogs as a guide for estimating output of different combinations. Whenever any nozzles are changed, recalibrate the sprayer and refigure its new total output.

TABLE 11-11.

Orchard Sprayer Calibration Worksheet. A worksheet such as this can be helpful in recording and computing the figures necessary for calibration. Similar worksheets can be developed for other types of sprayers. (In this example, notice the difference between the rated output of the nozzles and the actual output. Nozzles are worn).

ORCHARD SPRAYER CALIBRATION

Grower: **D. BROWN** ____ Date: **1-29-88** _____ Sprayer Type: **AIRBLAST** _____

CHECK: ☑ 1. Filter screens and strainers clean?
      ☑ 2. Tank clean and free of scale and sediment?
      ☑ 3. Pressure gauge operating?
      ☑ 4. Nozzles working properly?

Sprayer operating pressure: **100** psi

I-A. GALLONS/HOUR:
    (Method I – using nozzle chart from manufacturer's catalog)

Nozzle Size	Number (N)	Rated Output (gallons/minute)	Minutes Per Hour	Gallons Per Hour
D2-25	8	× 0.25	× 60	= 120
D4-25	8	× 0.45	× 60	= 216
____	____	× ____	× 60	= ____

                    TOTAL GALLONS PER HOUR = **336**

I-B. GALLONS/HOUR: (Method 2 – measurement)
    1. Fill sprayer to verifiable level.
    2. Run sprayer for a measured period of time (T), spraying under the same conditions as in the orchard.   T = **3.53**
    3. Refill sprayer, measuring the amount of water used (GAL) in gallons.   GAL = **20.4**
    4. Calculate: gallons/hour = (GAL × 60)/T
                    TOTAL GALLONS/HOUR = **346.7**

II. MILES/HOUR:
    1. Establish distance (D) in feet.   D = **253**
    2. Measure elapsed time for sprayer to travel the distance. Make 3 runs and average results.
      a. First run: Time = **1.05** minutes.
      b. Second run: Time = **1.15** minutes.
      c. Third run: Time = **1.13** minutes.
    3. Average of three runs (T) = **1.11** minutes.
    4. Calculate miles per hour:
        MPH = (D/T)/88   MPH = **2.59**

III. ACRES/HOUR:
    1. Measure width of tree row (W) in feet.   W = **22**
    2. Calculate miles per acre:
      miles/acre = (43560/W)/5280   MILES/ACRE = **0.375**
    3. Calculate acres per hour:
      acres/hour = MPH/(miles/acre)   ACRES/HOUR = **6.91**

IV. GALLONS/ACRE:
    (gallons/hour)/(acres/hour) = gallons/acre   GALLONS/ACRE = **50.17**

V. ACRES/TANK:
    Tank size = **500** gallons/tank
    (gallons/tank)/(gallons/acre) = acres/tank   ACRES/TANK = **9.97**

VI. AMOUNT OF PESTICIDE/TANK:
Recommended amount of pesticide/acre = **2.5 lb.**
(pesticide/acre) × (acres/tank) = pesticide/tank

PESTICIDE/TANK = **24.9 lb.**

VII. CALIBRATION CHECK:
1. Tree spacing (S) = **22** × **22** feet   S = **484**
2. Trees per acre (T) = 43560/S   T = **90**
3. Count the actual number of trees sprayed (N) with one tank: N = **918**
4. Actual acres sprayed = N/T   ACTUAL ACRES = **10.2**
5. Calculated acres per tank (from "V" above)   CALCULATED ACRES/TANK = **9.97**
6. Percent accuracy = calculated acres/actual acres × 100

ACCURACY = **97.7%**

Calibrating Dry Applicators

The techniques for calibrating dry applicators are similar in many ways to those used for liquids. However, granule applicators must be calibrated for each type of granular pesticide being applied and for each change in weather or field conditions. Granules vary in size and shape from one pesticide to the next, influencing their flow rate from the applicator hopper. Temperature and humidity, as well as field conditions, also influence granule flow.

Before beginning to calibrate a dry applicator, be sure that it is clean and all parts are working properly; most equipment requires periodic lubrication. Chapter 10 has instructions on cleaning and servicing application equipment. Always wear rubber gloves to prevent contact with residues on the equipment. Calibration of granule applicators involves using actual pesticides, so special precautions must be taken. Some formulations are dusty, and may require respiratory protection.

Three factors need to be measured when calibrating a dry applicator: the travel speed, rate of output, and swath width.

Travel Speed. Determine travel speed in feet-per-minute in the same manner as you would for liquid applicators, following the instructions given on page 329. Applicator hoppers should be filled so that speed can be measured under actual operating conditions.

Rate of Output. To determine the rate of output, fill the hopper or hoppers with the granular formulation to be used. Most granule applicator hoppers have ports with adjustable openings for granules to pass through; refer to charts supplied by the manufacturer to determine the approximate opening for the rate and speed you will be using. Once the approximate opening is set, use one of the following three methods to determine the actual output rate:

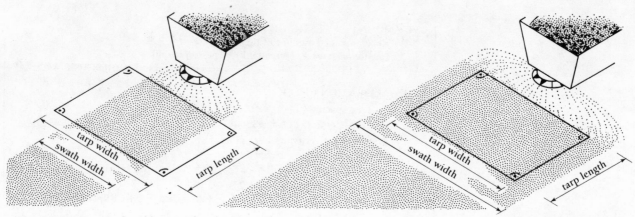

Area measurement: swath width × tarp length

Area measurement: tarp width × tarp length

FIGURE 11-19.

To determine the area of granules being applied, measure the swath width across a plastic tarp and multiply this by the length of the tarp. If the swath is wider than the tarp, the area is computed by multiplying the length by the width of the tarp.

1. *Measure the Quantity of Granules Applied to a Known Area.* Collecting and weighing the granules actually applied to a known area is often the easiest way to calibrate a granule applicator and should be used when working with broadcast applicators. Spread out a plastic tarp of known size on the ground, then operate the broadcast applicator at a known speed across the tarp (Figure 11-19). Place the granules collected by the tarp into a container and weigh them. Use the calculations in Table 11-12 to figure the amount of granular pesticide being applied per acre or other unit of area.

2. *Collect a Measured Amount of Granules Over a Known Period of Time.* Collecting and weighing quantities of granules over a known period of time is similar to calibrating a liquid boom sprayer with multiple nozzles; this method is used for granule applicators with multiple ports. While operating the applicator at a normal speed, collect granules from each port into a container; record the time required to collect each sample. Weigh samples separately, then use the calculations shown in Table 11-13 to find the output rate.

3. *Refill the Hopper After a Measured Period of Time.* This method may be used with hand-operated equipment or when small quantities are being applied, but is most useful when multiple applicators are used together on a boom. Fill the hopper or hoppers to a known level and operate the equipment for a measured period of time. When finished, weigh the quantity of granules required to refill the hoppers to their original levels. Use the calculations shown in Table 11-14 to compute the output rate.

Swath Width. To measure the swath width of granules dispersed by the applicator, operate the equipment under actual field conditions. Whenever possible, place cans, trays, or other containers at even intervals across the width of the application swath to collect granules. Weigh the granules collected in each container separately to determine the distribution pattern. Some spreaders can be operated over a strip of black cloth or plastic to provide a rapid visual assessment of granule distribution and swath width.

Granule applicators that apply bands or inject granules into the soil do not have devices to disperse granules from side to side. Swath width is determined by adding the widths of individual bands.

TABLE 11-12.

Calculating Granule Output Rate by Measuring the Quantity Applied to a Known Area.

---

1. Spread a 10 foot by 10 foot (or larger) plastic tarp on the ground and measure its length and width. Multiply the length by the width to determine the area of the tarp.

    EXAMPLE:
    Tarp size = 10 feet by 12 feet.
    Tarp area is:
    $10 \times 12 = 120$ square feet.

2. Fill the hopper or hoppers of the granule applicator, adjust the output ports to the recommended opening, and travel across the tarp at a known speed while granules are being broadcast.

3. Measure the swath width of the granules that were applied (see Figure 11-19) and compute the area of the swath. If the swath is wider than the tarp, the area figure to be used is equal to the area of the tarp. If the swath is narrower than the tarp, multiply the swath width by the length of the tarp.

4. Transfer all the granules on the tarp into a container and weigh them.

5. Multiply the weight of the granules collected (in pounds) by the area (acre, 1000 square feet, or 100 square feet) given on the label. (An acre is 43,560 square feet.) Divide the result by the area of the swath.

    EXAMPLE:
    Assume the swath width of granules being applied equals 15 feet. Therefore use the tarp area of 120 square feet in the calculations. (If the swath width was less than the tarp width, for example 8 feet, the area would then be 8 feet $\times$ 12 feet = 96 square feet.) Multiply the weight (in pounds) by the labeled area and divide the result by the tarp or swath area:

    Weight = 8 oz $\times$ 16 oz/lb = 0.5 lb

$$\frac{0.5 \text{ lb} \times 43560 \text{ sq ft/ac}}{120 \text{ sq ft}} = 181.5 \text{ lb/ac}$$

In this illustration, the granule applicator is broadcasting 181.5 pounds of material per acre. If the label calls for a greater amount, open the port more or slow the speed of travel of the applicator. If the label calls for a lesser amount, close the port some or speed up the rate of travel. Once an adjustment has been made, repeat this calibration procedure.

TABLE 11-13.

Calculating Granule Output Rate by Collecting a Measured Amount over a Known Period of Time.

1. Adjust the hopper opening according to manufacturer's instructions suggested for your required application rate. If no information is available, begin with an intermediate setting.

2. Operate the equipment at the speed of an actual application. Collect granules into a clean container, such as a pan or bag, before they drop to the ground. Use a stopwatch to determine the time required to collect each volume. If granules are dispersed through more than one opening, collect and time the output from each. Because some units drop granules onto a spinning disc for dispersal, it may be necessary to disable the disc by disconnecting the drive chain or belt to prevent granule loss during collection. For smaller units, collect the discharge into a bag placed over the outlet. Be sure granules are moved away from the port quickly enough to prevent clogging.

3. Weigh the output from each port separately to detect any variability; if necessary, adjust ports to equalize flow rates. Collections should be weighed in ounces.

4. Determine the output in pounds-per-hour. Divide each weight by the collection time and multiply by 0.0625 (0.0625 is the number obtained by dividing 1 minute by 16 ounces per pound; this number will convert ounces-per-minute into pounds-per-minute).

EXAMPLE: The following is an example of an output collected from a granule applicator with six ports, although the same calculations would apply if only one port were used. Hopper openings were adjusted following manufacturer's recommendations for an application of 200 pounds per acre:

Port #	Ounces	Time
1	29.5	0.25 min
2	33.0	0.28
3	31.5	0.26
4	29.0	0.25
5	33.0	0.27
6	30.0	0.26

Port #	$\dfrac{oz}{min}$		=	oz/min	×	0.0625	=	lb/min
1	29.5/0.25		=	118.0	×	0.0625	=	7.375
2	33.0/0.28		=	117.9	×	0.0625	=	7.369
3	31.5/0.26		=	121.2	×	0.0625	=	7.575
4	29.0/0.25		=	116.0	×	0.0625	=	7.250
5	33.0/0.27		=	122.2	×	0.0625	=	7.638
6	30.0/0.26		=	115.4	×	0.0625	=	7.213
								44.42

5. Determine total pounds-per-minute output by adding the individual outputs of each port. In this example the total output is 44.42 pounds per minute.

6. Use the technique shown in Table 11-15, page 350, to calculate the rate per acre or other unit of area.

TABLE 11-14.

Calculate the Rate of Output by Refilling the Hopper After a Measured Period of Time.

1. Fill the hopper or hoppers to a known level with granules.

2. Operate the equipment for a measured period of time at a known speed.

3. Weigh the amount of granules required to refill the hopper or hoppers to their original level. If multiple hoppers are being used, be sure each is applying approximately the same amount of granules. If a significant variation exists, adjust the ports and repeat steps 1 through 3.

EXAMPLE: In this example, six applicators are used together on a boom. They have been adjusted so that they all apply approximately the same amount of granules:

Hopper Number	Operating Time	Weight of Granules
1	2.5 minutes	6.2 lbs
2	2.5	6.1
3	2.5	6.1
4	2.5	6.3
5	2.5	6.1
6	2.5	5.9
		36.7

4. Convert the output to pounds-per-minute by dividing the total weight from all hoppers by the time they were operated.

EXAMPLE:

$$\frac{36.7 \text{ lbs}}{2.5 \text{ min}} = 14.68 \text{ lbs/min}$$

5. Use the technique shown in Table 11-15, page 350, to calculate the rate-per-acre or other unit of area.

Application Rate. Use Table 11-15 to calculate the actual rate of granules being applied per acre or other unit of area. If your calculations do not correspond to the labeled rate, adjust the equipment and repeat the calibration procedure.

Motorized and hand-operated applicators apply granules at a fixed output, independent of ground speed. If ground speed increases, the effect will be to reduce the amount of granules applied per unit of area; conversely, when ground speed decreases, more material is applied. Application rate with this type of equipment can therefore be adjusted not only by the size of the port opening, but also by the speed of travel. The output of ground-wheel-driven granule applicators, however, varies according to the ground speed. If ground speed increases, the applicator runs faster and the output rate is greater. When the ground speed is slowed down, output decreases because the applicator runs slower. The result of this automatic change in output is that the equipment will apply the same amount of material per acre or other unit of area, no matter what speed it is driven (the equipment will have minimum and maximum operating speeds determined by the manufacturer, however). The application rate also can be adjusted by increasing or decreasing the size of the port openings or by changing drive gears or sprockets to change the speed ratio of the metering mechanism to the ground wheel.

TABLE 11-15.

Calculating Rate-per-Acre or Other Unit of Area.

---

1. Determine the acres-per-minute being treated by dividing the swath width by 43560 (the number of square feet in an acre) and multiplying the result by the speed of travel. In this example the swath width is 30 feet and the application speed is 352 feet per minute (4 miles per hour).

EXAMPLE:

$$\frac{30 \text{ ft (swath)}}{43560 \text{ sq ft/ac}} \times 352 \text{ ft/min} = 0.242 \text{ ac/min}$$

2. Determine the pounds of formulated pesticide being applied per acre by dividing the output rate of the granule applicator (as computed from the calculations performed in Tables 11-13 or 11-14) by the acres-per-minute calculated in step 1. This example uses 44.42 pounds per minute as the output rate.

EXAMPLE:

$$\frac{44.42 \text{ lb/min}}{0.242 \text{ ac/min}} = 183.6 \text{ lb/ac}$$

CALCULATION FOR ACTIVE INGREDIENT, PERCENT SOLUTION, AND PARTS-PER-MILLION SOLUTIONS

Not all pesticide recommendations call for dry or liquid *formulated* amounts of pesticide per unit of area. Sometimes recommendations require the pesticide to be applied in pounds of active ingredient (a.i.) per unit of area, be mixed as a percent solution, or be diluted to parts per million (ppm). Before adding pesticide to the spray tank, read and understand the dilution instructions on the label.

Active Ingredient (a.i.) Calculations

Pesticides are seldom available in their pure state; they are normally formulated into a ready-to-use product by combining them with adjuvants and inert ingredients such as carriers and solvents. Therefore, only a percentage of any formulated product, whether dry or liquid, is pure pesticide, known as *active ingredient* (a.i.). Some University of California and other pesticide use guidelines call for a.i. when there are several formulations available, often from different manufacturers. Using a.i calculations rather than formulation calculations enables the same amount of actual pesticide to be applied to a unit of area, no matter what formulation is used.

The a.i. of any pesticide will be listed on its label. Labels of liquid pesticides give the percentage by weight of active ingredient and also tell how many pounds of active ingredient are in 1 gallon of formulation (Figure 11-20); labels of dry formulations list the percentage by weight of active ingredient. Use Table 11-16 to make active ingredient calculations with liquid formulations, Table 11-17 for dry formulations, and Table 11-18 for granular formulations.

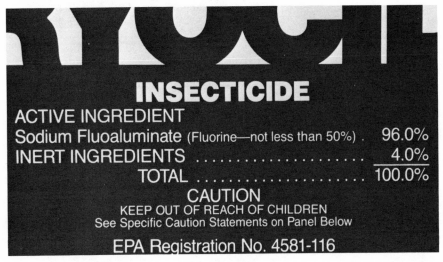

FIGURE 11-20.

To determine the percentage of active ingredient in a pesticide formulation, check the pesticide label. Liquid formulations list active ingredient as the number of pounds per gallon of formulation. Dry formulations list active ingredient as the total percent of the weight.

TABLE 11-16.

Calculating Active Ingredient with Liquid Formulations.

Assume that a sprayer has been calibrated and found to spray 7.5 acres per tank. You have a recommendation to apply 1.5 pounds a.i. of chlorothalonil per acre to control rust on snap beans, and have been supplied with a liquid formulation containing 4.17 pounds (a.i.) of chlorothalonil per gallon.

1. Determine the number of acres that can be treated with 1 gallon of formulation by dividing the pounds of a.i. per gallon by the recommended pounds of a.i. per acre.

   EXAMPLE:
   $$\frac{4.17 \text{ lb a.i per gal}}{1.5 \text{ lb a.i. per ac}} = 2.78 \text{ ac/gal}$$

2. Divide the known acre capacity of your tank by the acres per gallon:

   EXAMPLE:
   $$\frac{7.5 \text{ ac/tank}}{2.78 \text{ ac/gal}} = 2.7 \text{ gal/tank}$$

This is the number of gallons of formulated chlorothalonil that should be put into the tank for spraying 7.5 acres of crop.

TABLE 11-17.

Calculating Active Ingredient with Powder Formulations.

The calibrated sprayer you are using covers 7.5 acres per tank, and you have a recommendation to apply 1.5 pounds a.i. of chlorothalonil per acre for control of rust on snap beans. You are provided with a wettable powder formulation that, according to the label, contains 75% chlorothalonil.

1. Convert the percent a.i. to a decimal by dividing by 100 (or simply move the decimal point two places to the left).

EXAMPLE:
75% = 0.75 lb a.i./lb form

2. Divide the recommended amount of a.i. by the amount of a.i. in the formulation.

EXAMPLE:
$$\frac{1.5 \text{ lb a.i. per ac}}{0.75 \text{ lb a.i per lb form}} = 2 \text{ lb form/ac}$$

3. Multiply the pounds of formulation per acre by the number of acres-per-tank to find out how much material to put into the tank.

EXAMPLE:
2 lb form/acre × 7.5 ac/tank = 15 lb/tank

TABLE 11-18.

Calculating Active Ingredient with Granular Formulations.

You are given a recommendation for application of 0.50 lb a.i. of ethoprop per 1000 square feet of turf for control of nematodes. You are provided with a granular formulation containing 10% active ingredient (0.1 pound of a.i. per pound of formulation).

1. Convert the percent a.i. into a decimal and divide this into the recommended application rate:

EXAMPLE:
$$\frac{0.5 \text{ lb a.i. per 1000 sq ft}}{0.1 \text{ lb a.i. per lb form}} = 5 \text{ lb form}$$

2. Calibrate the granule applicator so that it applies 5 pounds of formulated ethoprop per 1000 square feet.

Percent Solutions

Sometimes label recommendations require that the pesticide be mixed as a *percent solution*. The active ingredient is mixed to get a known concentration, regardless of the volume per unit area of spray put out by the sprayer. Percent solutions are mixed on a weight/weight basis (w/w), meaning pounds of a.i. per pound of water. Table 11-19 provides an example of calculating a percent solution with liquid formulations and Table 11-20 shows those for dry formulations.

TABLE 11-19.

Calculating a Percent Solution with a Liquid Formulation.

To prepare a percent solution using liquid formulations, you need to know the volume of the spray tank, the weight of active ingredient per gallon of formulation, and the weight of a gallon of water. The weight of water is a constant, being approximately 8.34 pounds. Assume you have measured the volume of the spray tank and find that it holds 264.5 gallons of water. You are given a recommendation to apply a 1% solution of glyphosate for control of aquatic weeds, using a high-pressure sprayer with a hand-held spray nozzle. The formulation of glyphosate that you are to use contains 5.4 pounds of active ingredient per gallon.

1. Find the total weight of the liquid in the filled tank by multiplying 264.5 gallons by 8.34 pounds per gallon:

EXAMPLE:
264.5 gal × 8.34 lb/gal = 2205.93 lb

2. Multiply this weight by 0.01 (1%) to determine the weight of a.i. required to mix a 1% solution:

EXAMPLE:
2205.93 × 0.01 = 22.06 lb

3. Divide the required weight of a.i. by the weight of of a.i. in the formulation. The result is the number of gallons of liquid formulation that should be added to 264.5 gallons of water to achieve a 1% solution:

EXAMPLE:
$$\frac{22.06 \text{ lb a.i.}}{5.4 \text{ lb a.i./gal}} = 4.1 \text{ gal formulation}$$

In this example, one tank of liquid should contain 4.1 gallons of glyphosate formulation. The total volume of water combined with the glyphosate formulation should equal 264.5 gallons, the capacity of the tank. You would therefore use 260.4 gallons of water and 4.1 gallons of formulated glyphosate.

NOTE: These calculations give a close approximation of the amount of liquid formulation to add to the tank to achieve a known percent solution. The mathematics for a more exact figure are more complex and unnecessary for this type of work.

TABLE 11-20.

Calculating a Percent Solution with a Dry Formulation.

Dry formulations require similar calculations for percent solutions. First, from the label, determine the percent of a.i. in the dry formulation. Assume for this example that it is 75% a.i.; 1 pound of dry formulation would contain 0.75 pound of pesticide active ingredient. You need to mix a 1% spray solution of this formulation in the 264.5 gallon tank.

1. Find the total weight of the liquid in the filled tank by multiplying 264.5 gallons by 8.34 pounds per gallon:

EXAMPLE:
264.5 gal × 8.34 lb/gal = 2205.93 lb

2. Multiply this weight by 0.01 (1%) to determine the weight of a.i. required to mix a 1% solution:

EXAMPLE:
2205.93 × 0.01 = 22.06 lb

3. Divide the weight of a.i. by the decimal equivalent of the percent of a.i. in the formulation. The result is the number of pounds of formulation that should be added to 264.5 gallons of water to achieve a 1% solution:

EXAMPLE:
$$\frac{22.06 \text{ lb a.i.}}{0.75} = 29.41 \text{ lb formulation}$$

Add 29.41 pounds of wettable powder to 264.5 gallons of water to achieve a 1% solution.

Parts-Per-Million (ppm) Solutions

Certain pesticides need to be mixed in parts-per-million (ppm) concentrations, which are essentially the same as percent solutions. For example, a 100 ppm solution is equal to a 0.01% solution (Table 11-21). The ppm designation represents the parts of active ingredient of pesticide

TABLE 11-21.

Parts Per Million (ppm).

ppm	DECIMAL	PERCENT SOLUTION
1 ppm	0.000001	0.0001%
10 ppm	0.00001	0.001%
100 ppm	0.0001	0.01%
1,000 ppm	0.001	0.1%
10,000 ppm	0.01	1.0%
100,000 ppm	0.1	10.0%
1,000,000 ppm	1.0	100.0%

per million parts of water; ppm dilutions are a common way of measuring very diluted concentrations of pesticides. When calculating parts per million, use the formulae in Table 11-22 if you are mixing dry formulations with water and those in Table 11-23 for liquid formulations.

TABLE 11-22.

Calculating a Parts-Per-Million Dilution for Dry Formulations.

Assume you are given a recommendation requiring a 100 ppm concentration of oxytetracycline to be mixed in a 500 gallon tank. Oxytetracycline is used for control of fire blight on pear trees. The formulation you have is a wettable powder, containing 17% a.i.

1. Find the total weight of the liquid in the filled tank by multiplying 500 gallons by 8.34 pounds per gallon:

EXAMPLE:
500 gal × 8.34 lb/gal = 4170 lb/tank

2. Determine how many pounds of a.i. are required for a pound of spray solution:

EXAMPLE:

$$100 \text{ ppm} = \frac{100 \text{ parts a.i}}{1,000,000 \text{ parts sol}} = 0.0001$$

It will require 0.0001 pounds of a.i. for each pound of solution to achieve a 100 ppm mixture.

3. Determine how many pounds of a.i. are required for a tank of solution, using the weight of the liquid in the tank:

EXAMPLE:
4170 lb/tank × 0.0001 lb a.i. = 0.417 lb a.i.

4. Divide the weight of a.i. by the decimal equivalent of the percent of a.i. in the formulation. The result is the number of pounds of formulation that should be added to 500 gallons of water to achieve a 100 ppm solution:

EXAMPLE:

$$\frac{0.417 \text{ lb a.i.}}{0.17 \text{ lb a.i./lb form}} = 2.45 \text{ lb formulation}$$

TABLE 11-23.

Calculating a Parts-Per-Million Dilution for Liquid Formulations.

Assume a pesticide contains 5.4 pounds of a.i. in one gallon of formulation. You are required to prepare a 100 ppm concentration in a 500 gallon tank.

1. Find the total weight of the liquid in the filled tank by multiplying 500 gallons by 8.34 pounds per gallon:

EXAMPLE:
500 gal/tank × 8.34 lb/gal = 4170 lb/tank

2. Determine how many pounds of a.i. are required for a pound of spray solution:

EXAMPLE:

$$100 \text{ ppm} = \frac{100 \text{ parts a.i}}{1,000,000 \text{ parts sol}} = 0.0001$$

It will require 0.0001 pounds of a.i. for each pound of solution to achieve a 100 ppm mixture.

3. Determine how many pounds of a.i. are required for a tank of solution, using the weight of the liquid in the tank:

EXAMPLE:
4170 lb/tank × 0.0001 lb a.i. = 0.417 lb a.i./tank

4. Divide the required weight of a.i. by the pounds of a.i. per gallon to determine how many gallons of formulation are required. Since this will probably be a small number, multiply by 128 ounces per gallon to convert to ounces.

EXAMPLE:

$$\frac{0.417 \text{ lb a.i./tank}}{5.4 \text{ lb a.i./gal}} = 0.0772 \text{ gal/tank}$$

0.0722 gal/tank × 128 fl oz/gal = 9.88 fl oz/tank

Adding 9.88 fluid ounces of this formulated pesticide to 500 gallons of water will result in a 100 ppm solution.

USE OF PERSONAL COMPUTERS IN CALIBRATION

Personal computers are useful to perform the calculations needed for calibrating pesticide application equipment. Changes such as application speed, swath width, and output volume can be reentered and quickly recalculated. The results can be printed out so they can be carried with the application equipment. Commercially available software programs are available for use with many types of personal computers.

References

Chapter 1: General Study Guides

Bohmont, B. L. 1983. *The New Pesticide User's Guide*. Reston Publishing Company, Fort Collins, CO

Frishman, A. M. 1974. *Preparation for Pesticide Certification Examinations*. Arco Publishing Company, New York, NY

Weekman, G. T. 1975. *Apply Pesticides Correctly—A Guide for Commercial Applicators*. U.S. Environmental Protection Agency, Washington, D.C.

Chapter 2: Pest Identification

Applied Biochemists, Inc. 1979. *How to Identify and Control Water Weeds and Algae*, 2nd Edition. Applied Biochemists, Inc., Mequon, WI

Arnett, R. H. Jr. 1985. *American Insects*. Van Nostrand, New York, NY

Committee on Common Names of Insects 1982. *Common Names of Insects and Related Organisms*. Entomological Society of America, College Park, MD

Commonwealth Mycological Institute 1983. *Plant Pathologist's Pocketbook*, 2nd Edition. Commonwealth Agricultural Bureau, Slough, England

Davidson, R. H., and W. F. Lyon 1979. *Insect Pests of Farm, Garden, and Orchard*. John Wiley and Sons, New York, NY

Dixon, G. R. 1981. *Vegetable Crop Diseases*. AVI Publishing Company, Inc., Westport, CT

Ebeling, Walter 1975. *Urban Entomology*. University of California Publication 4057, Berkeley, CA

Fletcher, W. W., Ed. 1983. *Recent Advances in Weed Research*. Commonwealth Agricultural Bureau, Slough, England

Frazier, N. W., et al., Eds. 1970. *Virus Diseases of Small Fruits and Grapevines*. University of California, Berkeley, CA

Johnson, W. T., and H. H. Lyon 1976. *Insects that Feed on Trees and Shrubs: An Illustrated Practical Guide*. Comstock Publishing Associates, Ithaca, New York, NY

Jones, D. G., and B. C. Clifford 1978. *Cereal Diseases: Their Pathology and Control*. John Wiley and Sons, New York, NY

King, R., et al. 1985. *Farmer's Weed Control Handbook*. Doane Publishing Company, St. Louis, MO

Kofoid, C. A., Ed. 1934. *Termites and Termite Control*. University of California Press, Berkeley, CA

Kono, T., and C. S. Papp 1977. *Handbook of Agricultural Pests*. CDFA, Sacramento, CA

Mallis, A. 1982. *Handbook of Pest Control*, 6th Edition. Franzak and Foster Company, Cleveland, OH

Peterson, A. 1973. *Larvae of Insects*. (2 volumes). Edwards Brothers, Inc., Ann Arbor, MI

Pirone, Pascal P. 1970. *Diseases and Pests of Ornamental Plants*. The Ronald Press Company, New York, NY

Powell, J. A., and C. L. Hogue 1979. *California Insects*. University of California Press, Berkeley, CA

Robbins, W. W., M. K. Bellue, and W. S. Ball 1970. *Weeds of California*. CDFA, Sacramento, CA

Simmons, S. E. 1985. *Parklands Pest Management*. CDFA, Sacramento, CA

Swan, L. A., and C. S. Papp 1972. *The Common Insects of North America*. Harper and Row, New York, NY

Tattar, T. A. 1978. *Diseases of Shade Trees*. Academic Press, New York, NY

Truman, L.C., G. W. Bennett, and W. L. Butts 1982. *Scientific Guide to Pest Control Operations*, 3rd Edition. Harcourt Brace Jovanovich, Duluth, MN

Weed Science 1984. *Composite List of Weeds*. Weed Science Society of America, Champaign, IL

Wilson, M. C., A. C. York, and A. V. Provonsha 1982. *Insects of Vegetables and Fruit*, 2nd Edition. Waveland Press, Inc., Prospect Heights, IL

University of California Publications:
A Glossary of Insects, Mites, and Spiders. Publication 3314.
A Revisionary Study of the Leaf-mining Flies (Agromyzidae) of California. Publication 3273.
A Study of Insects. Publication 2949.
Almond Disease Guide. Publication 2609.
Almond Orchard Management. Publication 4092.
American Foulbrood Disease (Afb) of Honey Bees. Publication 2757.
An Illustrated Guide to the Genera of the Staphylinidae of America North of Mexico. Publication 4093.
Answers to Questions About Leptospirosis in Cattle. Publication 2271.
Ants and Their Control. Publication 2526.
Apple Scab Management. Publication 21412.
Bacterial Canker and Blast of Deciduous Fruits. Publication 2155.
Bark Beetles in California Forest Trees. Publication 21034.
Bitter Pit of Apples. Publication 2712.
Black Flies, Horse, and Deer Flies. Publication 21357.
Blackmold of Ripe Tomato Fruit. Publication 21154.
Borers in Landscape Trees and Shrubs. Publication 21316.
Broad Mite: A Pest of Coastal Lemons. Publication 21374.
Brown Rot of Stone Fruits. Publication 2206.
Canine Heartworm Disease. Publication 21359.
Carpet Beetles and Clothes Moths. Publication 2524.
Citrus Growing in the Sacramento Valley. Publication 2443.
Citrus Industry (The), Volume IV. Publication 4088.
Citrus Thrips: A Major Pest of California Citrus. Publication 21224.
Codling Moth Management. Publication 1918.
Common Flies Associated with Livestock and Poultry. Publication 21142.
Common Pantry Pests and Their Control. Publication 2711.
Common Parasites of Horses (The). Publication 4006.
Common Poultry Lice Control. Publication 2254.
Control Guide for Olive Pests and Diseases. Publication 21370.
Control of External Parasites of Chickens and Pigeons. Publication 2267.
Controlling Household Cockroaches. Publication 21035.
Controlling Olive Scale with Parasites. Publication 2507.
Cuban Laurel Thrips. Publication 2536.
Diseases and Insects of Modesto Ash. Publication 2538.
Diseases of Alfalfa in California. Publication 2594.
Diseases of Camellias in California. Publication 2151.
Dutch Elm Disease in California. Publication 21189.
Ear Tick (The). Publication 2295.

Elm Leaf Beetle. Publication 2238.
Eutypa Dieback of Apricot and Grape in California. Publication 21182.
Fall Webworm: A Tentmaking Caterpillar. Publication 21060.
Foliage and Branch Diseases of Landscape Trees. Publication 2616.
Fruittree Leafroller on Ornamentals and Fruit Trees. Publication 21053.
Fusarium Blight: A Major Disease of Kentucky Bluegrass in California. Publication 21269.
General Recommendations for Nematode Sampling. Publication 21234.
Grape Pest Management. Publication 4105.
Grape Pests in the Southern San Joaquin Valley. Publication ANRP003.
Green Fruit Beetle: A Common Fruit Pest. Publication 21191.
Grower's Weed Identification Handbook. Publication 4030.
Gypsy Moth in California. Publication 21387.
Horn Fly (The). Publication 2296.
Horse Bots and Their Control. Publication 2337.
Horsehair Worms. Publication 21238.
Identification and Biology of the Face Fly. Publication 2207.
Insect Identification Handbook. Publication 4099.
Insect, Mite, and Disease Guide for Christmas Trees. Publication 2994.
Integrated Management of Pest Flies on Horse Ranches. Publication 2335.
Integrated Management of Pest Flies on the Dairy. Publication 2329.
Integrated Pest Management for Alfalfa Hay. Publication 3312.
Integrated Pest Management for Almonds. Publication 3308.
Integrated Pest Management for Citrus. Publication 3303.
Integrated Pest Management for Cole Crops and Lettuce. Publication 3307.
Integrated Pest Management for Cotton in the Western Region of the United States. Publication 3305.
Integrated Pest Management for Potatoes in the Western United States. Publication 3316.
Integrated Pest Management for Rice. Publication 3280.
Integrated Pest Management for Tomatoes. Publication 3274.
Integrated Pest Management for Walnuts. Publication 3270.
Leaf Curl Control in Peaches and Nectarines. Publication 2613.
Lice on Livestock and Horses. Publication 2298.
Micronutrient Deficiencies of Citrus. Publication 2115.
Mistletoe Control in Shade Trees. Publication 2571.
Monterey Pine Tip Moth. Publication 2809.
Mosquitoes of California, Third Edition. Publication 4084.
Nantucket Pine Tip Moth: Biology and Control. Publication 21423.
Nematode Diseases of Food and Fiber Crops in the Southwestern United States. Publication 4083.
Oak Worm (Oak Moth) and Its Control. Publication 2542.
Omnivorous Looper on Avocados in California. Publication 21101.
Pajaroello Tick. Publication 2503.
Pest Management Guide for Insects and Nematodes of Cotton in California. Publication 4089.
Pit Scales on Oak. Publication 2543.
Poria Wood Rot of Deciduous Fruit and Nut Trees. Publication 21033.
Postharvest Diseases of Citrus Fruits in California. Publication 21407.
Postharvest Pathology of Fruits and Vegetables: Postharvest Losses in Perishable Crops. Publication 1914.
Powderpost Beetles and Their Control. Publication 21017.
Predaceous and Parasitic Arthropods in California Cotton Fields. Publication 1820.
Psoroptic Cattle Scabies. Publication 21236.
Red Turpentine Beetle: A Pest of Pines. Publication 21055.
Redhumped Caterpillar: A Pest of Many Trees. Publication 21064.
Reducing Root Rot in Plants. Publication 4004.
Root-knot Nematode on Cotton. Publication 2819.
Rose Diseases. Publication 2607.
Scale Insects and Their Control. Publication 2237.
Scaly-Leg Mites. Publication 21237.
Sequoia Pitch Moth on Monterey Pine. Publication 2544.
Sheep Keds and Nasal Bots. Publication 21358.
Shot Hole of Stone Fruits. Publication 21363.
Silverfish and Firebrats: How to Control Them. Publication 21001.
Sowbugs and Pillbugs. Publication 21015.
Spider Mite Pests of Cotton. Publication 2888.
Spiders. Publication 2531.

Strawberry Production in California. Publication 2959.
Studies on the Population Dynamics of the Western Pine Beetle. Publication 4042.
Sugarbeet Pest Management Series:
    Aphid-borne Diseases. Publication 3277.
    Leaf Diseases. Publication 3278.
    Nematodes. Publication 3272.
Termites and Other Wood-destroying Insects. Publication 2532.
Turfgrass Pests. Publication 4053.
Virus Diseases of Small Fruits and Grapevines. Publication 4056.
Walnut Orchard Management. Publication 21410.
Weed Management in Sugarbeets. Publication 21375.
Western Grapeleaf Skeletonizer in California (The). Publication 21395.
Winter Tick (The). Publication 2301.
Yellow Bud Mosaic. Publication 2862.
Yellow Leaf Roll of Peaches. Publication 21092.

Chapter 3: Pest Management

Applied Biochemists, Inc. 1979. *How to Identify and Control Water Weeds and Algae,* 2nd Edition. Applied Biochemists, Inc., Mequon, WI

Bond, E. J. 1984. *Manual of Fumigation for Insect Control.* Food and Agriculture Organization of the United Nations, Rome, Italy

California Weed Conference 1985. *Principles of Weed Control in California.* Thompson Publications, Fresno, CA

Coulson, R. N., and J. A. Witter 1984. *Forest Entomology—Ecology and Management.* John Wiley and Sons, New York, NY

Davidson, R. H., and W. F. Lyon 1979. *Insect Pests of Farm, Garden, and Orchard.* John Wiley and Sons, New York, NY

Dixon, G. R. 1981. *Vegetable Crop Diseases.* AVI Publishing Co., Inc., Westport, CT

Ebeling, Walter 1975. *Urban Entomology.* University of California Publication 4057, Berkeley, CA

Edwards, S. R., B. M. Bell, and M. E. King 1981. *Pest Control in Museums: A Status Report (1980).* Association of Systematics Collections, Lawrence, KS

Fletcher, W. W., Ed. 1983. *Recent Advances in Weed Research.* Commonwealth Agricultural Bureau, Slough, England

Flint, M. L., and R. van den Bosch 1981. *Introduction to Integrated Pest Management.* Plenum Press, New York, NY

Frazier, N. W., et al., Eds. 1970. *Virus Diseases of Small Fruits and Grapevines.* University of California, Berkeley, CA

Huffaker, C. B. 1980. *New Technology of Pest Control.* John Wiley and Sons, New York, NY

Jones, D. G., and B. C. Clifford 1978. *Cereal Diseases: Their Pathology and Control.* John Wiley and Sons, New York, NY

King, R., et al. 1985. *Farmer's Weed Control Handbook.* Doane Publishing Company, St. Louis, MO

Kofoid, C. A., Ed. 1934. *Termites and Termite Control.* University of California Press, Berkeley, CA

Mallis, A. 1982. *Handbook of Pest Control,* 6th Edition. Franzak and Foster Company, Cleveland, OH

Matthews, G. A. 1984. *Pest Management.* Longman Publishing Company, London, England

Palti, J. 1981. *Cultural Practices and Infectious Crop Diseases.* Springer-Verlag, New York, NY

Ross, M. A., and C. A. Lembi 1985. *Applied Weed Science.* Burgess Publishing Company, Minneapolis, MN

Simmons, S. E. 1985. *Parklands Pest Management.* CDFA, Sacramento, CA

Smith, E. H., and D. Pimentel 1978. *Pest Control Strategies*. Academic Press, New York, NY

Truman, L.C., G. W. Bennett, and W. L. Butts 1982. *Scientific Guide to Pest Control Operations*, 3rd Edition. Harcourt Brace Jovanovich, Duluth, MN

Ware, G. W. 1980. *Complete Guide to Pest Control*. Thompson Publications, Fresno, CA

University of California Publications:
A Guide to Controlling Almond Pests, Diseases, and Micronutrient Deficiencies. Publication 21343.
A Slide Rule for Cotton Crop and Insect Management. Publication 21361.
Almond Disease Guide. Publication 2609.
Almond Orchard Management. Publication 4092.
Ants and Their Control. Publication 2526.
Apple Scab Management. Publication 21412.
Bacterial Canker and Blast of Deciduous Fruits. Publication 2155.
Bark Beetles in California Forest Trees. Publication 21034.
Biological Control and Insect Pest Management. Publication 1911.
Biological Control of Pest Mites. Publication 3304.
Bitter Pit of Apples. Publication 2712.
Black Flies, Horse, and Deer Flies. Publication 21357.
Borers in Landscape Trees and Shrubs. Publication 21316.
Branched Broomrape. Publication 2182.
Broad Mite: A Pest of Coastal Lemons. Publication 21374.
Broadleaf Weed Control in Wheat and Barley. Publication 21012.
Brown Rot of Stone Fruits. Publication 2206.
Carpet Beetles and Clothes Moths. Publication 2524.
Chemical Control Guide for Walnuts. Publication 21261.
Chemical Weed Control in Vineyards. Publication 2216.
Cherry Crinkle and Deep Suture Disease. Publication 2454.
Citrus Growing in the Sacramento Valley. Publication 2443.
Citrus Industry (The), Volume IV. Publication 4088.
Citrus Thrips: A Major Pest of California Citrus. Publication 21224.
Codling Moth Management. Publication 1918.
Common Pantry Pests and Their Control. Publication 2711.
Common Parasites of Horses (The). Publication 4006.
Common Poultry Lice Control. Publication 2254.
Control Guide for Olive Pests and Diseases. Publication 21370.
Control Guide for Prune Pests, Diseases, and Micronutrient Deficiencies. Publication 21394.
Control of External Parasites of Chickens and Pigeons. Publication 2267.
Controlling Ceratocystis Canker of Stone Fruit Trees. Publication 2205.
Controlling Household Cockroaches. Publication 21035.
Controlling Olive Scale with Parasites. Publication 2507.
Cuban Laurel Thrips. Publication 2536.
Degree-days: The Calculation and Use of Heat Units in Pest Management. Publication 21373.
Diseases of Alfalfa in California. Publication 2594.
Ear Tick (The). Publication 2295.
Elm Leaf Beetle. Publication 2238.
European Canker of Apple in California. Publication 2612.
Eutypa Dieback of Apricot and Grape in California. Publication 21182.
Fall Webworm: A Tentmaking Caterpillar. Publication 21060.
Foliage and Branch Diseases of Landscape Trees. Publication 2616.
Fruittree Leafroller on Ornamentals and Fruit Trees. Publication 21053.
Fusarium Blight: A Major Disease of Kentucky Bluegrass in California. Publication 21269.
Gossyplure-baited Traps as Pink Bollworm Survey Detection, Research, and Management Tools in Southwestern Desert Cotton-growing Areas. Publication 1915
Grape Pest Management. Publication 4105.
Green Fruit Beetle: A Common Fruit Pest. Publication 21191.
Guide to Turfgrass Pest Control. Publication 2209.
Gypsy Moth in California. Publication 21387.
Horn Fly (The). Publication 2296.
Horse Bots and Their Control. Publication 2337.
Horsehair Worms. Publication 21238.
Host List of Powdery Mildews of California. Publication 2217.
Identification and Biology of the Face Fly. Publication 2207.

Insect and Disease Control Recommendations for Rice. Publication 2748.

Insect and Mite Control on Lawns. Publication 2540.

Insect and Mite Control Program for Grapes. Publication 21102.

Insect and Nematode Control Recommendations for Asparagus, Eggplant, Okra, Peppers, and Sweet Corn. Publication 21140.

Insect and Nematode Control Recommendations for Celery, Cole Crops, Head Lettuce, and Spinach. Publication 21141.

Insect and Nematode Control Recommendations for Field Corn and Sorghum. Publication 2746.

Insect and Nematode Control Recommendations for Sugarbeets. Publication 21139.

Insect and Nematode Control Recommendations for Tomatoes. Publication 21138.

Insect and Nematode Recommendations for Cotton. Publication 2083.

Insect and Rodent Control in Stored Grains. Publication 2378.

Insect and Spider Mite Control Program for Beans. Publication 21386.

Insect Control Guide for Alfalfa Hay. Publication 2763.

Insect Control Guide for Barley, Wheat, and Oats. Publication 2268.

Insect, Mite, and Disease Guide for Christmas Trees. Publication 2994.

Integrated Management of Pest Flies on Horse Ranches. Publication 2335.

Integrated Management of Pest Flies on Poultry Ranches. Publication 2505.

Integrated Management of Pest Flies on the Dairy. Publication 2329.

Integrated Pest Management for Alfalfa Hay. Publication 3312.

Integrated Pest Management for Almonds. Publication 3308.

Integrated Pest Management for Citrus. Publication 3303.

Integrated Pest Management for Cole Crops and Lettuce. Publication 3307.

Integrated Pest Management for Cotton in the Western Region of the United States. Publication 3305.

Integrated Pest Management for Potatoes in the Western United States. Publication 3316.

Integrated Pest Management for Rice. Publication 3280.

Integrated Pest Management for Tomatoes. Publication 3274.

Integrated Pest Management for Walnuts. Publication 3270.

Leaf Curl Control in Peaches and Nectarines. Publication 2613.

Lice on Livestock and Horses. Publication 2298.

Microbial/Biorational Pesticide Registration. Publication 3318.

Micronutrient Deficiencies of Citrus. Publication 2115.

Mistletoe Control in Shade Trees. Publication 2571.

Monterey Pine Tip Moth. Publication 2809.

Mosquito Control on the Farm. Publication 2850.

Nantucket Pine Tip Moth: Biology and Control. Publication 21423.

Nematode Diseases of Food and Fiber Crops in the Southwestern United States. Publication 4083.

Nontillage and Strip Weed Control in Almond Orchards. Publication 2770.

Oak Worm (Oak Moth) and Its Control. Publication 2542.

Omnivorous Looper on Avocados in California. Publication 21101.

Pajaroello Tick. Publication 2503.

Pest Control and Water Management in Rice. Publication 21298.

Pest Management Guide for Insects and Nematodes of Cotton in California. Publication 4089.

Pit Scales on Oak. Publication 2543.

Planning Dairy Wastewater Systems for Mosquito Control. Publication 21398.

Plants in California Susceptible to Phytophthora cinnamomi. Publication 21178.

Plants Resistant or Susceptible to Verticillium Wilt. Publication 2703.

Poisonous Larkspurs: Identification and Control. Publication 2129.

Poria Wood Rot of Deciduous Fruit and Nut Trees. Publication 21033.

Postharvest Diseases of Citrus Fruits in California. Publication 21407.

Postharvest Pathology of Fruits and Vegetables: Postharvest Losses in Perishable Crops. Publication 1914.

Postharvest Treatment of Pear Trees for Control of Pear Decline. Publication 2614.

Powderpost Beetles and Their Control. Publication 21017.

Psoroptic Cattle Scabies. Publication 21236.

Red Turpentine Beetle: A Pest of Pines. Publication 21055.

Redhumped Caterpillar: A Pest of Many Trees. Publication 21064.

Reducing Loss from Crown Gall Disease. Publication 1845.

Reducing Root Rot in Plants. Publication 4004.

Resistance or Susceptibility of Certain Plants to Armillaria Root Rot. Publication 2591.

Root-knot Nematode on Cotton. Publication 2819.

Rose Diseases. Publication 2607.

Scale Insects and Their Control. Publication 2237.
Scaly-Leg Mites. Publication 21237.
Selective Chemical Weed Control. Publication 1919.
Sequoia Pitch Moth on Monterey Pine. Publication 2544.
Sheep Keds and Nasal Bots. Publication 21358.
Shot Hole of Stone Fruits. Publication 21363.
Silverfish and Firebrats: How to Control Them. Publication 21001.
Soil Solarization: A Nonchemical Method for Controlling Diseases and Pests. Publication 21377.
Sowbugs and Pillbugs. Publication 21015.
Spider Mite Pests of Cotton. Publication 2888.
Spiders. Publication 2531.
Strawberry Production in California. Publication 2959.
Sugarbeet Pest Management Series:
    Aphid-borne Diseases. Publication 3277.
    Leaf Diseases. Publication 3278.
    Nematodes. Publication 3272.
Termites and Other Wood-destroying Insects. Publication 2532.
Treatment Guide for California Citrus Thrips, 1984–86. Publication 2903.
Turfgrass Disease Control Guide. Publication 2619.
Turfgrass Pests. Publication 4053.
Virus Diseases of Small Fruits and Grapevines. Publication 4056.
Walnut Orchard Management. Publication 21410.
Weed Control in Cucurbits. Publication 21326.
Weed Control in Dichondra. Publication 2204.
Weed Control in Grain Sorghum. Publication 21030.
Weed Control in Lettuce. Publication 2987.
Weed Control in Red and Ladino Clover. Publication 21263.
Weed Control in Seedling Alfalfa. Publication 2917.
Weed Management Guide for Citrus. Publication 2979.
Weed Management in Sugarbeets. Publication 21375.
Western Grapeleaf Skeletonizer in California (The). Publication 21395.
Winter Tick (The). Publication 2301.
Yellow Bud Mosaic. Publication 2862.
Yellow Leaf Roll of Peaches. Publication 21092.
Yellow Starthistle Control. Publication 2741.

Chapter 4: Pesticides

Aizawa, Hiroyasu 1982. *Metabolic Maps of Pesticides.* Academic Press, New York, NY

Bohmont, B. L. 1983. *The New Pesticide User's Guide.* Reston Publishing Company, Fort Collins, CO

Bond, E. J. 1984. *Manual of Fumigation for Insect Control.* Food and Agriculture Organization of the United Nations, Rome, Italy

Brown, V. K. 1980. *Acute Toxicity in Theory and Practice: with Special Reference to the Toxicology of Pesticides.* John Wiley and Sons, New York, NY

Chambers, J.E. and J. D. Yarbrough, Eds. 1982. *Effect of Chronic Exposures to Pesticides on Animal Systems.* Raven Press, New York, NY

Coats, J. R. 1982. *Insecticide Mode of Action.* Academic Press, New York, NY

Corbett, J. R., K. Wright, and A. C. Baillie 1984. *The Biochemical Mode of Action of Pesticides,* 2nd Edition. Academic Press, New York, NY

Hudson, R. H., R. K. Tucker, and M. A. Haegele 1984. *Handbook of Toxicity of Pesticides to Wildlife,* 2nd Edition. U.S. Department of the Interior, Fish and Wildlife Service, Resource Publication 153

Jacobson, M., and D.G. Crosby, Eds. 1971. *Naturally Occurring Insecticides.* Marcel Dekker, New York, NY

Kohn, G. K., Ed. 1974. *Mechanism of Pesticide Action.* American Chemical Society, Washington, D.C.

Lal, R., Ed. 1984. *Insecticide Microbiology.* Springer-Verlag, New York, NY

Magee, P. S., G. K. Gustave, and J.J. Menn, Eds. 1984. *Pesticide Synthesis Through Rational Approaches*. American Chemical Society, Washington, D.C.

Matsumura, F. 1976. *Toxicology of Insecticides*. Plenum Press, New York, NY

McFarlane, N.R. 1977. *Crop Protection Agents—Their Biological Evaluation*. Academic Press, New York, NY

O'Brien, R. D. 1967. *Insecticides: Action and Metabolism*. Academic Press, New York, NY

Street, J.C., Ed. 1975. *Pesticide Selectivity*. Marcel Dekker, New York, NY

Wagner, S. L. 1983. *Clinical Toxicology of Agricultural Chemicals*. Noyes Data Corp., Park Ridge, NJ

Ware, G. W. 1983. *Pesticides, Theory and Application*. W. H. Freeman Co., San Francisco, CA

Wilkinson, C. F. 1976. *Insecticide Biochemistry and Physiology*. Plenum Press, New York, NY

Chapter 5: Pesticide Laws and Regulations

Bohmont, B. L. 1983. *The New Pesticide User's Guide*. Reston Publishing Company, Fort Collins, CO

California Department of Food and Agriculture 1980. *Laws and Regulations Study Guide for Agricultural Pest Control Adviser, Agricultural Pest Control Operator, Pesticide Dealer, and Pest Control Aircraft Pilot Examinations*. CDFA, Sacramento, CA

California Department of Food and Agriculture 1983. *Extracts from the Food and Agricultural Code and Title 3 Administrative Code Pertaining to Pesticides and Pest Control Operations*. CDFA, Sacramento, CA

Keller, J. J., and Associates, Inc. 1985. *Pesticides Guide: Registration, Classification, and Applications*. J. J. Keller & Associates, Inc., Neenah, WI

State of California Resources Agency 1982. *California's New Pesticide Regulations and You*. Sacramento, CA

Ware, G. W. 1983. *Pesticides, Theory and Application*. W. H. Freeman Co., San Francisco, CA

University of California Publications:
Microbial/Biorational Pesticide Registration. Publication 3318.
Pesticide Registration Procedures and Requirements. Publication 3313.

Chapter 6: Hazards Associated With Pesticide Use

Biggar, J. W., and J. N. Seiber, Eds. 1987. *Fate of Pesticides in the Environment*. University of California Publication 3320, Berkeley, CA

Bond, E. J. 1984. *Manual of Fumigation for Insect Control*. Food and Agriculture Organization of the United Nations, Rome, Italy

Brown, V. K. 1980. *Acute Toxicity in Theory and Practice: with Special Reference to the Toxicology of Pesticides*. John Wiley and Sons, New York, NY

California Assembly Office of Research 1985. *The Leaching Fields: A Nonpoint Threat to Groundwater*. Joint Publications Office, Sacramento, CA

Cardozo, C. L., S. Nicosia, and J. Troiano 1985. *Agricultural Pesticide Residues in California Well Water: Development and Summary of a Well Inventory Data Base for Non-Point Sources*. Environmental Assessment Program, CDFA, Sacramento, CA

Chambers, J.E. and J. D. Yarbrough, Eds. 1982. *Effect of Chronic Exposures to Pesticides on Animal Systems*. Raven Press, New York, NY

Holden, P. 1986. *Pesticides and Groundwater Quality: Issues and Problems in Four States*. National Academy Press, Washington, D.C.

Hudson, R. H., R. K. Tucker, and M. A. Haegele 1984. *Handbook of Toxicity of Pesticides to Wildlife*, 2nd Edition. U.S. Department of the Interior, Fish and Wildlife Service, Resource Publication 153, Washington, D.C.

Kennedy, M.V., Ed. 1978. *Disposal and Decontamination of Pesticides.* American Chemical Society Symposium Series 73, Washington, D.C.

Matsumura, F. 1976. *Toxicology of Insecticides.* Plenum Press, New York, NY

McEwen, F. L., and G. R. Stephenson 1979. *The Use and Significance of Pesticides in the Environment.* John Wiley and Sons, New York, NY

McFarlane, N.R. 1977. *Crop Protection Agents—Their Biological Evaluation.* Academic Press, New York, NY

Morgan, Donald P. 1982. *Recognition and Management of Pesticide Poisonings,* 3rd Edition. U.S. Environmental Protection Agency, Office of Pesticide Programs, Washington, D.C.

Siewierski, M., Ed. 1984. *Determination and Assessment of Pesticide Exposure.* Elsevier, New York, NY

Swann, R. L., and A. Eschenroeder, Eds. 1983. *Fate of Chemicals in the Environment.* American Chemical Society, Washington, D.C.

Tordoir, W. F., and E. A. H. van Heemstra 1980. *Field Worker Exposure During Pesticide Application.* Elsevier Scientific Publishing Co., New York, NY

United States—Task Group on Occupational Exposure to Pesticides 1974. *Occupational Exposure to Pesticides.* Federal Working Group on Pest Management, Washington, D.C.

Wagner, S. L. 1983. *Clinical Toxicology of Agricultural Chemicals.* Noyes Data Corp., Park Ridge, NJ

Watson, D. L., and A. W. A. Brown 1977. *Pesticide Management and Insecticide Resistance.* Academic Press, New York, NY

White-Stevens, R., Ed. 1977. *Pesticides in the Environment,* Volume 3. Marcel Dekker, Inc., New York, NY

University of California Publications:
Pesticides in Soil and Groundwater. Publication 3300.
Pesticide Toxicities. Publication 21062.
Toxicology: The Science of Poisons. Publication 21221.

Chapter 7: Protecting People and the Environment

Barker, R. L., and G. C. Coletta, Eds. 1986. *Performance of Protective Clothing.* ASTM Special Technical Publication 900, Philadelphia, PA

Biggar, J. W., and J. N. Seiber, Eds. 1987. *Fate of Pesticides in the Environment.* University of California Publication 3320, Berkeley, CA

Bohmont, B. L. 1983. *The New Pesticide User's Guide.* Reston Publishing Company, Fort Collins, CO

Flint, M. L., and R. van den Bosch 1981. *Introduction to Integrated Pest Management.* Plenum Press, New York, NY

Haskell, P. T. 1985. *Pesticide Application: Principles and Practice.* Clarendon Press, Oxford

Huffaker, C. B. 1980. *New Technology of Pest Control.* John Wiley and Sons, New York, NY

Magee, P. S., G. K. Gustave, and J.J. Menn, Eds. 1984. *Pesticide Synthesis Through Rational Approaches.* American Chemical Society, Washington, D.C.

Matthews, G. A. 1984. *Pest Management.* Longman Publishing Company, London, England

Matthews, G. A. 1982. *Pesticide Application Methods.* Longman Publishing Company, London, England

Siewierski, M., Ed. 1984. *Determination and Assessment of Pesticide Exposure.* Elsevier, New York, NY

Smith, E. H., and D. Pimentel 1978. *Pest Control Strategies.* Academic Press, New York, NY

Street, J.C., Ed. 1975. *Pesticide Selectivity.* Marcel Dekker, New York, NY

United States—Task Group on Occupational Exposure to Pesticides 1974. *Occupational Exposure to Pesticides*. Federal Working Group on Pest Management, Washington, D.C.

Walker, J.O. 1980. *Spraying Systems for the 1980s*. Proc. Symp. Royal Holloway College, Monograph 24, Croydon, England

Watson, D. L., and A. W. A. Brown 1977. *Pesticide Management and Insecticide Resistance*. Academic Press, New York, NY

Weekman, G. T. 1975. *Apply Pesticides Correctly—A Guide for Commercial Applicators*. U.S. Environmental Protection Agency, Washington, D.C.

University of California Publications:
Reducing Pesticide Hazards to Honey Bees with Integrated Management Strategies. Publication 2883.
Safe Handling of Agricultural Pesticides. Publication 2768.
Turfgrass Pests. Publication 4053.
Using Pesticides Safely in the Home and Yard. Publication 21095.

Chapter 8: Pesticide Emergencies

Bohmont, B. L. 1983. *The New Pesticide User's Guide*. Reston Publishing Company

Chevron Chemical Company *Pre-Emergency Planning Guide for the Independent Dealer*. Chevron Chemical Company, San Francisco, CA

Kennedy, M.V., Ed. 1978. *Disposal and Decontamination of Pesticides*. American Chemical Society Symposium Series 73, Washington, D.C.

Morgan, Donald P. 1982. *Recognition and Management of Pesticide Poisonings*, 3rd Edition. U.S. Environmental Protection Agency, Office of Pesticide Programs, Washington, D.C.

Shell Chemical Company *Agricultural Chemicals Safety Manual*. Shell Chemical Company, Houston, TX

Weekman, G. T. 1975. *Apply Pesticides Correctly—A Guide for Commercial Applicators*. U.S. Environmental Protection Agency, Washington, D.C.

Chapter 9: Effective Use of Pesticides

Akesson, N. B., and W. E. Yates 1979. *Pesticide Application Equipment and Techniques*. Food and Agricultural Organization of the United Nations, Rome, Italy

Bohmont, B. L. 1983. *The New Pesticide User's Guide*. Reston Publishing Company, Fort Collins, CO

Bond, E. J. 1984. *Manual of Fumigation for Insect Control*. Food and Agriculture Organization of the United Nations, Rome, Italy

California Weed Conference 1985. *Principles of Weed Control in California*. Thompson Publications, Fresno, CA

Flint, M. L., and R. van den Bosch 1981. *Introduction to Integrated Pest Management*. Plenum Press, New York, NY

Haskell, P. T. 1985. *Pesticide Application: Principles and Practice*. Clarendon Press, Oxford

Huffaker, C. B. 1980. *New Technology of Pest Control*. John Wiley and Sons, New York, NY

Magee, P. S., G. K. Gustave, and J.J. Menn, Eds. 1984. *Pesticide Synthesis Through Rational Approaches*. American Chemical Society, Washington, D.C.

Matthews, G. A. 1984. *Pest Management*. Longman Publishing Company, London, England

Matthews, G. A. 1982. *Pesticide Application Methods*. Longman Publishing Company, London, England

Palti, J. 1981. *Cultural Practices and Infectious Crop Diseases*. Springer-Verlag, New York, NY

Simmons, S. E. 1985. *Parklands Pest Management*. CDFA, Sacramento, CA

Smith, E. H., and D. Pimentel 1978. *Pest Control Strategies*. Academic Press, New York, NY

Street, J.C., Ed. 1975. *Pesticide Selectivity*. Marcel Dekker, New York, NY

Walker, J.O. 1980. *Spraying Systems for the 1980s*. Proc. Symp. Royal Holloway College, Monograph 24, Croydon, England

Ware, G. W. 1980. *Complete Guide to Pest Control*. Thompson Publications, Fresno, CA

Ware, G. W. 1983. *Pesticides, Theory and Application*. W. H. Freeman Co., San Francisco, CA

Weekman, G. T. 1975. *Apply Pesticides Correctly. A Guide for Commercial Applicators*. U.S. Environmental Protection Agency, Washington, D.C.

University of California Publications:
Codling Moth Management. Publication 1918.
Degree-days: The Calculation and Use of Heat Units in Pest Management. Publication 21373.
General Recommendations for Nematode Sampling. Publication 21234.
Gossyplure-baited Traps as Pink Bollworm Survey Detection, Research, and Management Tools in Southwestern Desert Cotton-growing Areas. Publication 1915.
Grape Pest Management. Publication 4105.
Integrated Management of Pest Flies on Horse Ranches. Publication 2335.
Integrated Pest Management for Alfalfa Hay. Publication 3312.
Integrated Pest Management for Almonds. Publication 3308.
Integrated Pest Management for Citrus. Publication 3303.
Integrated Pest Management for Cole Crops and Lettuce. Publication 3307.
Integrated Pest Management for Cotton in the Western Region of the United States. Publication 3305.
Integrated Pest Management for Potatoes in the Western United States. Publication 3316.
Integrated Pest Management for Rice. Publication 3280.
Integrated Pest Management for Tomatoes. Publication 3274.
Integrated Pest Management for Walnuts. Publication 3270.
Selective Chemical Weed Control. Publication 1919.
Turfgrass Pests. Publication 4053.
Walnut Orchard Management. Publication 21410.

Chapter 10: Pesticide Application Equipment

Akesson, N. B., and W. E. Yates 1979. *Pesticide Application Equipment and Techniques*. Food and Agricultural Organization of the United Nations, Rome, Italy

Bohmont, B. L. 1983. *The New Pesticide User's Guide*. Reston Publishing Company, Fort Collins, CO

Bond, E. J. 1984. *Manual of Fumigation for Insect Control*. Food and Agriculture Organization of the United Nations, Rome, Italy

Ebeling, Walter 1975. *Urban Entomology*. University of California Publication 4057, Berkeley, CA

Haskell, P. T. 1985. *Pesticide Application: Principles and Practice*. Clarendon Press, Oxford

Huffaker, C. B. 1980. *New Technology of Pest Control*. John Wiley and Sons, New York, NY

Mallis, A. 1982. *Handbook of Pest Control*, 6th Edition. Franzak and Foster Company, Cleveland, OH

Matthews, G. A. 1982. *Pesticide Application Methods*. Longman Publishing Company, London, England

Truman, L.C., G. W. Bennett, and W. L. Butts 1982. *Scientific Guide to Pest Control Operations*, 3rd Edition. Harcourt Brace Jovanovich, Duluth, MN

Walker, J.O. 1980. *Spraying Systems for the 1980s*. Proc. Symp. Royal Holloway College, Monograph 24, Croydon, England

Weekman, G. T. 1975. *Apply Pesticides Correctly—A Guide for Commercial Applicators*. U.S. Environmental Protection Agency, Washington, D.C.

Chapter 11: Calibration of Pesticide Application Equipment

Akesson, N. B., and W. E. Yates 1979. *Pesticide Application Equipment and Techniques.* Food and Agricultural Organization of the United Nations, Rome, Italy

Bohmont, B. L. 1983. *The New Pesticide User's Guide.* Reston Publishing Company, Fort Collins, CO

Bond, E. J. 1984. *Manual of Fumigation for Insect Control.* Food and Agriculture Organization of the United Nations, Rome, Italy

Kroon, Cornelius W. 1978. *Liquid Calibration Handbook.* Thompson Publications, Fresno, CA

Matthews, G. A. 1982. *Pesticide Application Methods.* Longman Publishing Company, London, England

Walker, J.O. 1980. *Spraying Systems for the 1980s.* Proc. Symp. Royal Holloway College, Monograph 24, Croydon, England

Weekman, G. T. 1975. *Apply Pesticides Correctly—A Guide for Commercial Applicators.* U.S. Environmental Protection Agency, Washington, D.C.

University of California Publication:
How Much Chemical Do You Put in the Tank?. Publication 2718.

Glossary

abiotic. nonliving factors, such as wind, water, temperature, or soil type or texture.

absorb. to soak up or take in a liquid or powder.

acaricide. a pesticide used to control mites.

accumulate. to increase in quantity within an area, such as the soil or tissues of a plant or animal.

acetylcholine. an enzyme that transmits nerve signals between nerves and muscles, sensory organs, or other nerves.

acetylcholinesterase. an enzyme that deactivates acetylcholine to allow nerve signals to stop. Organophosphate and carbamate pesticides block the action of acetylcholinesterase, resulting in loss of control of nerve function.

acidifier. an adjuvant used to lower the pH (or acidify) the water being mixed with a pesticide. Pesticides often break down more slowly if the spray water is slightly acid. Acidifiers are also referred to as acidulators.

acidulator. see acidifier.

activator. an adjuvant that increases the activity of a pesticide by reducing surface tension or speeding up penetration through insect or plant cuticle.

active ingredient (a.i.). the material in the pesticide formulation that actually destroys the target pest or performs the desired function.

additive effect. an increase in toxicity brought about by combining one pesticide with another. The increased toxicity is no greater than if an equal volume of either pesticide is used alone, however.

adjuvant. a material added to a pesticide mixture to improve or alter the deposition, toxic effects, mixing ability, persistence, or other qualities of the active ingredient.

adsorb. to take up and hold on surface.

aerosol. very fine liquid droplets or dust particles often emitted from a pressurized can or aerosol generating device.

aestivation. dormancy during summer or periods of high temperature or a dry season.

agitator. a mechanical or hydraulic device that stirs the liquid in a spray tank to prevent the mixture from separating or settling.

alga. an aquatic, nonvascular plant. (plural: algae).

all-terrain cycle. a three- or four-wheeled motorcyclelike vehicle used for applying low volumes of pesticides in agricultural areas and open lands.

amphibian. a cold blooded organism such as a frog, toad, or salamander.

anemometer. an instrument used for measuring wind speed.

anionic. negatively charged; a characteristic of some types of surfactants which help prevent pesticides from being washed off treated surfaces.

annual. a type of plant that passes through its entire life cycle in one year or less.

antagonistic effect. reduced toxicity or effectiveness as a result of combining one pesticide with another.

antibiotic. a substance produced by a living organism, such as a fungus, that is toxic to other types of living organisms. Sometimes used as a pesticide.

anticoagulant. a type of rodenticide that causes death by preventing normal blood clotting.

apiary. a place where bees are kept, such as a bee hive.

aquatic. pertaining to water, such as aquatic weeds or aquatic pest control.

aquifer. an underground formation of sand, gravel, or porous rock that contains water. The place where groundwater is found.

arsenical pesticide. a type of pesticide that contains some form of arsenic.

arthropod. an animal having jointed appendages and an external skeleton, such as an insect, a spider, a mite, a crab, or a centipede.

attractant. a substance that attracts a specific species of animal to it. When manufactured to attract pests to traps or poisoned bait, attractants are considered to be pesticides.

auger. a spiral-shaped shaft used for moving pesticide dusts or granules from a hopper to a moving belt or disk for application.

avicide. a pesticide used to control pest birds.

back siphoning. the process that permits pesticide-contaminated water to be sucked from a spray tank back into a well or other water source. Back siphoning is prevented by providing an air gap or check valve in the pipe or hose used to fill a spray tank.

bacterium. a unicellular microscopic plantlike organism that lives in soil, water, organic matter, or the bodies of plants and animals. Some bacteria cause plant or animal diseases. (plural: bacteria).

bait. a food or foodlike substance that is used to attract and often poison pest animals.

barrier cream. a preparation that can be applied to the skin to help reduce pesticide exposure. Barrier creams provide limited protection and only for short periods of time. They are normally used on areas of the face and head which cannot be protected satisfactorily by other means.

beneficial. pertaining to being helpful in some way to people, such as a beneficial plant or insect.

biennial. a plant that completes part of its life cycle in one year and the remainder of its life cycle the following year.

bifluid nozzle. a special nozzle used for producing extremely fine droplets. Fluid is broken up into small droplets by passing through a high-velocity airstream.

bioaccumulation. the gradual buildup of certain pesticides within the tissues of living organisms after feeding on lower organisms containing smaller amounts of these pesticides. Animals higher up on the food chain accumulate greater amounts of these pesticides in their tissues.

biochemical. pertaining to a chemical reaction that takes place within the cells or tissues of living organisms.

biological control. the action of parasites, predators, pathogens, or competitors in maintaining another organism's density at a lower average than would occur in their absence. Biological control may occur naturally in the field or be the result of manipulation or introduction of biological control agents by people.

biotic. pertaining to living organisms, such as the influences living organisms have on pest populations.

boom. a structure attached to a truck, tractor, or other vehicle, or held by hand, to which spray nozzles are attached.

botanical. derived from plants or plant parts.

broad-spectrum pesticide. a pesticide that is capable of controlling many different species or types of pests.

broadcast application. a method of applying granular pesticides by dispersing them over a wide area using a spinning disc or other mechanical device.

broadleaf. one of the major plant groups, known as dicots, with netveined leaves usually broader than grasses. Seedlings have two seed leaves (cotyledons); broadleaves include many herbaceous plants, shrubs, and trees.

buffer. an adjuvant that lowers the pH of a spray solution and, depending on its concentration, can maintain the pH within a narrow range even if acidic or alkaline materials are added to the solution.

buffer area. a part of a pest infested area that is not treated with a pesticide to protect adjoining areas from pesticide hazards.

calibration. the process used to measure the output of pesticide application equipment so that the proper amount of pesticide can be applied to a given area.

California Department of Food and Agriculture (CDFA). the state agency responsible for regulating the use of pesticides in California.

carbamate. a class of pesticides commonly used for control of insects, mites, fungi, and weeds.

carcinogenic. having the ability to produce cancer.

carrier. the liquid or powdered inert substance that is combined with the active ingredient in a pesticide formulation. May also apply to the water or oil that a pesticide is mixed with prior to application.

cationic. pertaining to materials that contain positively charged ions. Some surfactants include cationic materials to improve mixing and absorption by the target pest.

caution. the signal word used on labels of pesticides in toxicity Category III or IV; these pesticides have an oral LD_{50} greater than 500 and a dermal LD_{50} greater than 2000.

chemigation. the application of pesticides to target areas through an irrigation system.

chlorotic. a yellowing or bleaching of normally green leaves due to a nutrient deficiency, disease, pest damage, or other disorder.

chronic. pertaining to long duration or frequent recurrence.

chronic onset. symptoms of pesticide poisoning that occur days, weeks, or months after the actual exposure.

closed mixing system. a device used for measuring and transferring liquid pesticides from their original container to the spray tank. Closed mixing systems reduce chances of exposure to concentrated pesticides. Closed mixing systems are usually required when mixing Category I materials.

coalescent effect. the unique mode of action observed when two or more pesticides having different modes of action are combined.

common name. the recognized, nonscientific name given to plants or animals. The Weed Science Society of America and the Entomological Society of America publish lists of recognized common names. Many pesticides also have common names.

compatible. the condition in which two or more pesticides mix without unsatisfactory chemical or physical changes.

compatibility agent. an adjuvant that improves the ability of two or more pesticides to combine.

competition. the struggle between pests and nonpests for the same resources, such as water, light, nutrients, or space.

confined area. enclosed spaces such as attics, crawl spaces, closed rooms, warehouses, greenhouses, holds of ships, and other areas that may be treated with pesticides.

contact poison. a pesticide that provides control when target pests come in physical contact with it.

controlled droplet applicator (CDA). an application device that produces more uniformly sized liquid droplets by passing liquid over a notched, spinning disk.

cotyledon. the first leaf or pair of leaves of a sprouted seed. Grasses (monocots) have a single cotyledon while broadleaved plants (dicots) have a pair of cotyledons.

coverage. the degree to which a pesticide is distributed over a target surface.

cumulative effect. poisoning symptoms that only appear after several repeated doses over a period of time, indicating that the toxic material is building up in the system of the poisoned individual.

cuticle. the outer protective covering of plants and arthropods that aids in preventing moisture loss.

danger. the signal word used on labels of pesticides in toxicity Category I—those pesticides with an oral LD_{50} less than 50 or a dermal LD_{50} less than 200 or those having specific, serious health or environmental hazards.

deactivation. the process by which the toxic action of a pesticide is reduced or eliminated by impurities in the spray tank, by water being used for mixing, or by biotic or abiotic factors in the environment.

defoaming agent. an adjuvant that eliminates foaming of a pesticide mixture in a spray tank.

defoliant. a pesticide used to remove leaves from target plants, often as an aid in harvesting the plant.

degradation. the breakdown of a pesticide into an inactive or less active form. Environmental conditions, impurities, or microorganisms can contribute to the degradation of pesticides.

delayed mixture. an incompatibility or adverse effect between two pesticides that were applied to the same target but at different times.

dehydration. the process of a plant or animal losing water or drying up.

deposition. the placement of pesticides on target surfaces.

deposition aid. an adjuvant that improves the ability of a pesticide spray to reach the target.

dermal. pertaining to the skin. One of the major ways pesticides can enter the body to possibly cause poisoning.

desiccant. a pesticide that destroys target pests by causing them to lose body moisture.

diluent. the inert liquid or powdered material that is combined with the active ingredient during manufacture of a pesticide formulation.

disease. a condition, caused by biotic or abiotic factors, that impairs some or all of the normal functions of a living organism.

dissolve. to pass into solution.

dormant. to become inactive during winter or periods of cold weather.

dose. the measured quantity of a pesticide. Often the size of the dose determines the degree of effectiveness, or, in the case of poisoning of nontarget organisms, the degree of injury.

drift. the movement of pesticide dust, spray, or vapor away from the application site.

dry flowable. a dry, granular pesticide formulation intended to be mixed with water for application. When combined with water, a dry flowable will be similar to a wettable powder. Dry flowable formulations are measured by volume rather than weight.

dust. finely ground pesticide particles, sometimes combined with inert materials. Dusts are applied without mixing with water or other liquid.

ecological. an approach that considers the interrelationship between living organisms and the environment.

economic damage. damage caused by pests to plants, animals, or other items which results in loss of income or a reduction of value.

economic threshold. the point at which the value of the damage caused by a pest exceeds the cost of controlling the pest, therefore it becomes practical to use the control method.

efficacy. the ability of a pesticide to produce a desired effect on a target organism.

electrostatic. an electrical charge that causes a pesticide liquid or dust to be attracted to the target surface.

emergence. the appearance of a plant through the surface of the soil.

emulsifier. an adjuvant added to a pesticide formulation to permit petroleum-based pesticides to mix with water.

emulsion. droplets of petroleum-based liquids (oils) suspended in water.

emulsifiable concentrate. a pesticide formulation consisting of a petroleum-based liquid and emulsifiers that enable it to be mixed with water for application.

encapsulation. a process by which tiny liquid droplets or dry particles are contained in polymer plastic capsules to slow their release into the environment and prolong their effectiveness. Sometimes encapsulation lowers hazards to people mixing or applying pesticides.

endangered species. rare or unusual living organisms whose existence is threatened by people's activities, including the use of some types of pesticides.

environment. all of the living organisms and nonliving features of a defined area.

Environmental Protection Agency (EPA). the federal agency responsible for regulating pesticide use in the United States.

enzyme. a complex chemical compound produced and used by a living organism to induce or speed up chemical reactions without being itself permanently altered.

epidermis. the outer layer of skin of vertebrates or the cellular layer of tissue beneath the cuticle of invertebrates.

eradication. the pest management strategy that attempts to eliminate all members of a pest species from a defined area.

evaporate. the process of a liquid turning into a gas or vapor.

exclusion. a pest management technique that uses physical or chemical barriers to prevent certain pests from getting into a defined area.

exotic. a pest from another country, one that is not native to the local area.

exposure. coming in contact with a pesticide.

extender. an adjuvant that enhances the effectiveness or effective life of a pesticide by some means such as screening ultraviolet light, slowing down volatilization, or improving sticking qualities.

fallow. cultivated land that is allowed to lie dormant during a growing season.

flowable. a pesticide formulation of finely ground particles of insoluble active ingredient suspended in a petroleum-based liquid combined with emulsifiers; flowables are mixed with water for final application.

fog. a spray of very small pesticide-laden droplets that remain suspended in the air.

foliage. the leaves of plants.

formulation. a mixture of active ingredient combined during manufacture with inert materials. Inert materials are added to improve the mixing and handling qualities of a pesticide.

frass. solid fecal material produced by insect larvae.

fumigant. vapor or gas form of a pesticide used to penetrate porous surfaces for control of soil dwelling pests or pests in enclosed areas or storage.

fungicide. a pesticide used for control of fungi.

fungus. multicellular lower plant lacking chlorophyll, such as a mold, mildew, rust, or smut. The fungus body normally consists of filamentous strands called the mycelium and reproduces through dispersal of spores (plural: *fungi*).

general-use pesticide. pesticides that have been designated for use by the general public as well as by licensed or certified applicators. General-use pesticides usually have minimal hazards.

generic pesticide. a pesticide not protected by a patent; one that may be manufactured by many different companies.

granule. a dry formulation of pesticide active ingredient and inert materials compressed into small, pebblelike shapes.

groundwater. fresh water trapped in aquifers beneath the surface of the soil; one of the primary sources of water for drinking, irrigation, and manufacturing.

ground-wheel-driven. a trailer-mounted dry or liquid pesticide applicator that gets the power to drive a pump, auger, or spinning disc from one of the trailer wheels as the unit is towed.

habitat. the place where plants or animals live and grow.

half life. the period of time that must elapse for a pesticide to lose half of its original toxicity or effectiveness.

herbaceous. a plant that is herblike, usually having little or no woody tissue.

herbicide. a pesticide used for the control of weeds.

hibernation. the process of passing the winter in a resting or nonactive state.

hormone. a chemical produced in the cells of a plant or animal that produces changes in cells in another part of the organism's structure.

host. a plant or animal species that provides sustenance for another organism.

host resistance. the ability of a host plant or animal to ward off or resist attack by pests or to be able to tolerate damage caused by pests.

hydrolysis. a chemical process that involves incorporating a water molecule into another molecule.

hyphae. the threadlike fibers that make up the mycellium of a fungus.

impregnate. an item, such as a flea collar, that has been manufactured with a certain pesticide in it; impregnates usually emit small, localized quantities of pesticide over an extended period of time.

incompatibility. a condition in which two or more pesticides are unable to mix properly or one of the materials chemically alters the other to reduce its effectiveness or produce undesirable effects on the target.

incorporate. to move a pesticide below the surface of the soil by discing, tilling, or irrigation. To combine one pesticide with another.

inert ingredients. materials in the pesticide formulation that are not the active ingredient. Some inert ingredients may be toxic or hazardous to people.

infection. the establishment of a microorganism within the tissues of a host plant or animal.

infestation. a troublesome invasion of pests within an area such as a building, greenhouse, agricultural crop, or landscaped location.

inhalation. the method of entry of pesticides through the nose or mouth into the lungs.

inhibit. to prevent something from happening, such as a biochemical reaction within the tissues of a plant or animal.

insect growth regulator (IGR). a type of pesticide used for control of certain insects. Insect growth regulators disrupt the normal process of development from immature to mature life stages.

insecticide. a pesticide used for the control of insects. Some insecticides are also labeled for control of ticks, mites, spiders, and other arthropods.

instar. the period between molts in larvae of insects. Most larvae pass through several instars; these are usually given numbers such as 1st instar, 2nd instar, etc.

integrated pest management (IPM). a pest management program that uses life history information and extensive monitoring to understand a pest and its potential for causing economic damage. Control is achieved through multiple approaches including prevention, cultural practices, pesticide applications, exclusion, natural enemies, and host resistance. The goal is to achieve long-term suppression of target pests with minimal impact on nontarget organisms and the environment.

interactive effect. interaction when two or more pesticides are mixed, producing greater or lesser toxicity to the target pests or changing the mode of action.

interval. the legal period of time between when a pesticide is applied and workers are allowed to enter the treated area or produce can be harvested. See *preharvest interval* and *reentry interval*.

invert emulsion. an emulsion where water droplets are suspended in an oil rather than the oil droplets being suspended in water.

invertebrate. any animal having an external skeleton or shell, such a insects, spiders, mites, worms, nematodes, and snails and slugs.

ion. an atom or molecule that carries a positive or negative electrical charge due to losing or gaining electrons through a chemical reaction.

ionize. the process in which a chemical converts into ions when it dissolves in water or other liquid.

knapsack sprayer. a small portable sprayer carried on the back of the person making the pesticide application. Some knapsack sprayers are hand-operated and others are powered by small gasoline engines.

larva. the immature form of insects that undergo metamorphosis (plural: *larvae*).

LC$_{50}$. the lethal concentration of a pesticide in the air or in a body of water that will kill half of a test animal population. LC$_{50}$ values are given in micrograms per milliliter of air or water (μg/ml).

LD$_{50}$. the lethal dose of a pesticide that will kill half of a test animal population. LD$_{50}$ values are given in milligrams per kilogram of test animal body weight (mg/kg).

leaching. the process by which some pesticides move down through the soil, usually by being dissolved in water, with the possibility of reaching groundwater.

lethal. capable of causing death.

material safety data sheet. an information sheet provided by a pesticide manufacturer describing chemical qualities, hazards, safety precautions, and emergency procedures to be followed in case of a spill, fire, or other emergency.

mesh. the term used to describe the number of wires per inch in a screen, such as one used to filter foreign particles out of spray solutions to keep nozzles from becoming clogged. Mesh is also the term used to describe the size of pesticide granules, pellets, and dusts.

metabolism. the total chemical process that takes place in a living organism to utilize food and manage wastes, provide for growth and reproduction, and accomplish all other life functions.

metal organic. a type of pesticide made up of organic molecules that include metal ions such as zinc, copper, iron, arsenic, and mercury.

microbial pesticide. pertaining to pesticides that consist of bacteria, fungi, or viruses used for control of weeds, invertebrates, or (rarely) vertebrates.

microencapsulated. a pesticide formulation in which particles of the active ingredient are encased in plastic capsules; pesticide is released after application when the capsules break down.

micron. a very small unit of measure: 1/1,000,000th of a meter.

microorganism. an organism of microscopic size, such as a bacterium, virus, fungus, viroid, or mycoplasma.

mimic. to copy or appear to be like something else.

mitigating. the process of making a problem, such as a pest infestation, less severe.

mode of action. the way a pesticide reacts with a pest organism to destroy it.

monitoring. the process of carefully watching the activities, growth, and development of pest organisms over a period of time, often utilizing very specific procedures.

monocot. a member of a group of plants whose seedlings have a single cotyledon; monocots are often known as grasses.

mutagenic. a chemical that is capable of causing deformities in living organisms.

mycelium. the vegetative body of a fungus, consisting of a mass of slender filaments called *hyphae* (plural: *mycelia*).

mycoplasma. a microorganism intermediate between viruses and bacteria, capable of causing diseases in plants.

narcotic. the mode of action of some insecticides resulting in a prolonged sleeplike state from which the target insects may not recover.

National Institute for Occupation Safety and Health (NIOSH). the federal agency that tests and certifies respiratory equipment for pesticide application.

natural enemy. an organism that causes premature death of a pest organism; includes predators, pathogens, parasites, and competitors.

necrosis. localized death of living tissue.

nematicide. a pesticide used to control nematodes.

nematode. elongated, cylindrical, nonsegmented worms. Nematodes are commonly microscopic; some are parasites of plants or animals.

neoprene. a synthetic rubber material used to make gloves, boots, and clothing for protection against pesticide exposure.

NIOSH. National Institute of Occupational Safety and Health.

NOEL. no observable effect level. The NOEL is the maximum dose or exposure level of a pesticide that produces no noticeable toxic effect on test animals.

nonionic. pertaining to an adjuvant that dissolves in the spray solution to produce no positive or negative ions.

nonorganic. pesticides that do not contain organic molecules. ·

nonpoint pollution source. pollution from pesticides or other materials that arises from their normal or accepted use over a large general area and extended period of time.

nonselective. a pesticide that has an action against many species of pests rather than just a few.

nontarget organism. animals or plants within a pesticide treated area that are not intended to be controlled by the pesticide application.

noxious. something that is harmful to living organisms, such as a noxious weed.

ocular. pertaining to the eye—this is one of the routes of entry of pesticides into the body.

oral. through the mouth—this is one of the routes of entry of pesticides into the body.

organic. a pesticide whose molecules contain carbon and hydrogen atoms. Also may refer to plants or animals which are grown without the addition of synthetic fertilizers or pesticides.

organism. any living thing.

organochlorine. a class of pesticides, commonly used as insecticides, that contain a chlorine atom incorporated into an organic molecule. Organochlorines are often highly persistent. Many organochlorine compounds are no longer used as pesticides.

organophosphate. a commonly used class of pesticides; organophosphates are organic molecules containing phosphorous. Some organophosphates are highly toxic to people. Most break down in the environment very rapidly.

ornamental. cultivated plants that are grown for other purposes than food or fiber.

parasite. a plant or animal that derives all its nutrients from another organism. Parasites often attach themselves to their host or invade the host's tissues. Parasitism may result in injury or death of the host.

pathogen. a microorganism that causes a disease.

pellet. a pesticide formulation consisting of the dry active ingredient and inert materials pressed into uniform sized granules.

penetrate. to pass through a surface such as skin, protective clothing, plant cuticle, or insect cuticle. Also refers to the ability of an applied spray to pass through dense foliage.

percolation. the process by which water flows downward through permeable soil. During percolation, water may dissolve or leach out pesticides and other chemicals in the soil and carry them downward.

perennial. a plant that lives longer than two years—some may live indefinitely. Some perennial plants lose their leaves and become dormant during winter; others may die back and resprout from underground root structures each year. The evergreens are perennial plants that do not die back or become dormant.

persistent pesticide. a pesticide that remains active in the environment for long periods of time because it is not easily broken down by microorganisms or environmental factors.

pesticide. any substance or mixture of substances intended for preventing, destroying, repelling, or mitigating any insects, rodents, nematodes, fungi, or weeds, or any other forms of life declared to be pests; and any other substance or mixture of substances intended for use as a plant regulator, defoliant, or desiccant.

pesticide formulation. the pesticide as it comes from its original container, consisting of the active ingredient blended with inert materials.

pesticide resistance. genetic qualities of a pest population that enable individuals to resist the effects of certain types of pesticides that are toxic to other members of that species.

pH. a measure of the concentration of hydrogen ions in a solution—as the number of hydrogen ions increase, the solution becomes more acid.

pheromone. a chemical produced by an animal to attract other animals of the same species.

photosynthesis. the process by which plants convert sunlight into energy.

physiological. pertaining to the functions and activities of living tissues.

phytotoxic. injurious to plants.

plantback restriction. a restriction which limits the type of commodity that can be grown in an area for a designated period of time after a certain pesticide has been used.

plant growth regulator (PGR). a pesticide used to regulate or alter the normal growth of plants or development of plant parts.

point pollution source. pollution of the soil or groundwater caused by spilling or dumping quantities of a toxic material in one location.

postemergent. an herbicide applied after emergence of a specified weed or crop.

posting. the placing of signs around an area to inform workers and the public that the area has been treated with a pesticide.

potency. pertaining to the toxicity of a pesticide.

potentiation. an increase in the toxicity of a pesticide brought about by mixing it with another pesticide or chemical.

pour-on. a ready-to-use formulation or diluted mixture of pesticide for control of external parasites on livestock. The liquid is usually poured along the back of the animal.

powder. a finely ground dust containing active ingredient and inert materials. This powder is mixed with water before application as a liquid spray.

ppb. parts per billion.

ppm. parts per million.

precipitation. the process by which solid particles settle out of a solution, such as a formulated pesticide in a spray tank.

predaceous. the habit of hunting and eating other animals.

predacide. a pesticide used for control of predaceous mammals such as coyotes.

preemergent. the action of an herbicide that controls specified weeds as they sprout from seeds before they push through the soil surface.

preharvest interval. a period of time as set by law that must elapse after a pesticide has been applied to an edible crop before the crop can be harvested legally. Pesticide labels provide information on preharvest intervals.

preplant. an herbicide that has been incorporated into the soil to control weeds prior to planting crop seeds.

propellant. a material, such as compressed air or gas, used to propel spray liquids or dusts to target surfaces.

protectant. a pesticide that provides a chemical barrier against pest attack.

protozoan. minute, single-celled organisms belonging to the phylum Protozoa. Protozoans are common in the soil and water; some are parasitic on animals.

psi. pounds per square inch.

pupa. in insects having complete metamorphosis, the resting life stage between larval and adult forms.

pyrethroid. a synthetic pesticide that mimics pyrethrin, a botanical pesticide derived from certain species of chrysanthemum flowers.

quarantine. a condition in which the movement of certain items (such as produce) within a designated area are restricted to prevent the spread of pests.

rate. the quantity or volume of liquid spray, dust, or granules that is applied to an area over a specified period of time.

recombination. an occurrence in which a pesticide breaks down and then combines with other chemicals in the environment to produce a different compound than what was originally applied.

recommendation. a written document prepared by a licensed Pest Control Adviser that prescribes the use of a specific pesticide or other pest control method.

reentry interval. the period of time specified by law that must elapse after a pesticide is applied before people can resume work in the treated area.

regulatory control. management of pests by the passage of laws and regulations that restrict activities which would promote pest buildup.

repellent. a pesticide used to keep target pests away from a treated area by saturating the area with an odor that is disagreeable to the pest.

reservoir. a population of pests within a local area. Also an organism harboring plant or animal pathogens.

residue. traces of pesticide that remain on treated surfaces after a period of time.

resistance. see *pesticide resistance* or *host resistance*.

respiration. the metabolic process in plants and animals in which, among other things, oxygen is exchanged for carbon dioxide or carbon dioxide is exchanged for oxygen.

restricted-use pesticide. a pesticide, usually in toxicity Category I, that can only be used by commercial applicators who have a valid Qualified Pesticide Applicator license or certificate or private applicators who have demonstrated to the local agricultural commissioner that they understand the proper methods of handling, using, and disposing of these materials.

restrictive statement. a statement on a pesticide label that restricts the use of that pesticide to specific areas or by designated individuals.

rhizome. an underground rooting structure of certain types of plants.

rinsate. the liquid derived from rinsing pesticide containers or spray equipment.

rodenticide. a pesticide used for control of rats, mice, gophers, squirrels, and other rodents.

rope wick applicator. a device used to apply contact herbicides onto target weed foliage with a saturated rope or cloth pad.

rpm. revolutions per minute.

ruffling. the condition in which plant foliage is separated by air during a pesticide application to allow pesticide droplets to come in contact with all surfaces.

runoff. the liquid spray material that drips from the foliage of treated plants or from other treated surfaces. Also the rainwater or irrigation water that leaves an area—this water may contain trace amounts of pesticide.

safety cab. an enclosed cab installed on a tractor to protect the operator from pesticide exposure. The cab includes an air filtering system.

saprophytes. an organism that lives on dead or decaying organic matter.

secondary pest. an organism that becomes a pest only after a natural enemy, competitor, or primary pest has been eliminated through some type of pest control method.

selective pesticide. a pesticide that has a mode of action against only a single or small number of pest species.

service container. any container designed to hold concentrate or diluted pesticide mixtures, including the sprayer tank, but not the original pesticide container.

shelf life. the maximum period of time that a pesticide can remain in storage before losing some of its effectiveness.

shingling. the clumping or sticking together of plant foliage caused by the force of a liquid spray. Shingling prevents spray droplets from reaching all surfaces of the foliage, and may result in poor pest control.

signal word. the word "Danger," "Warning," or "Caution" that appears on a pesticide label that signifies how toxic the pesticide is and what toxicity category it belongs to.

site of action. the location within the tissues of the target organism where a pesticide acts.

SLN. special local needs registration.

slurry. a watery mixture containing pesticide powder; slurries leave a thick coating of pesticide residue on treated surfaces.

soil mobility. a variable characteristic of a pesticide, based on its chemical nature. Highly mobile pesticides leach rapidly through the soil and may contaminate groundwater. Immobile pesticides or those with low soil mobility remain tightly attached to soil particles and are resistant to leaching.

soluble. a material that dissolves completely in a liquid.

soluble powder. a pesticide formulation where the active ingredient and all inert ingredients completely dissolve in water to form a true solution.

solution. a liquid that contains dissolved substances, such as a soluble pesticide.

solvent. a liquid capable of dissolving certain chemicals.

sorptive dust. a fine powder used to destroy arthropods by removing the protective wax coating that prevents water loss.

source reduction. sanitation practices.

special local needs registration (SLN). the registration of a pesticide for treatment of a local or specific pest problem where no registered pesticide is available.

spore. a reproductive structure produced by some plants and microorganisms that is resistant to environmental influences.

spot treatment. a method of applying pesticides only in small, localized areas where pests congregate rather than treating a larger, general area.

spreader. an adjuvant that lowers the surface tension of treated surfaces to enable the pesticide to be absorbed.

sterilant. a pesticide used for control of rodents by preventing their reproduction.

sticker. an adjuvant used to prevent pesticides from being washed or abraded off treated surfaces.

stolon. an aboveground runner or rooting structure found in some species of plants.

stomach poison. a pesticide that kills target animals who ingest it.

structural pest. a pest such as a termite or wood rot fungus that destroys structural wood in buildings.

sublethal dose. a pesticide dose insufficient to cause death in the exposed organism.

suppress. to lower the level of a pest population.

surface active ingredient. see *surfactant*.

surface tension. forces on the surfaces of liquid droplets that keep them from spreading out over treated surfaces.

surface water. water found in ponds, lakes, reservoirs, streams, and rivers.

surfactant. an adjuvant used to improve the ability of the pesticide to stick to and be absorbed by the target surface.

suspension. fine particles of solid material distributed evenly throughout a liquid such as water or oil.

swath. the area covered by one pass of the pesticide application equipment.

symptom. a sign which indicates the presence of a disease or disorder.

synergism. a reaction in which a chemical that has no pesticidal qualities can enhance the toxicity of a pesticide it is mixed with.

synthesized. applying to pesticides that are manufactured through chemical processes rather than occurring naturally.

systemic pesticide. a pesticide that is taken up into the tissues of the organism and transported to other locations where it will affect pests.

tag-along. a liquid or dry pesticide applicator mounted on a wheeled trailer and pulled behind a tractor or other powered vehicle.

tank mix. a mixture of pesticides or pesticides and fertilizers applied at the same time.

target. either the pest that is being controlled or surfaces within an area that the pest will contact.

temperature inversion. a condition in which air above an area is warmer than air near the ground. This warm air forms a cap that may cause pesticide vapor or droplets to collect and concentrate.

teratogenic. a chemical that is capable of causing birth deformities.

thickener. an adjuvant that increases the viscosity of the spray solution so that larger droplets are formed by the nozzles; thickeners are used to control drift.

threshold limit value (TLV). the airborne concentration of a pesticide in parts per million (ppm) that produces no adverse effects over a period of time.

triple rinse. a process used to remove most traces of liquid pesticide from a pesticide container. After draining the container into a spray tank for 30 seconds, the container is partially filled with water, capped, agitated, and drained into the tank. Rinsing and draining is repeated three times.

TLV. threshold limit value

tolerance. the ability to endure the effects of a pesticide or pest without exhibiting adverse effects.

toxicity. the potential a pesticide has for causing harm.

toxicity testing. a process in which known doses of a pesticide are given to groups of test animals and the results observed.

tracking powder. a fine powder that is dusted over a surface to detect or control certain pests such as cockroaches or rodents. For control, the inert powder is combined with a pesticide; the animal ingests this powder and becomes poisoned when it cleans itself.

translocate. the movement of pesticides from one location to another within the tissues of a plant.

tuber. an underground reproductive structure of some types of plants. Tubers are characterized by small scalelike leaves.

ultra-low-volume (ULV). a pesticide application technique in which very small amounts of liquid spray are applied over a unit of area; usually ½ gallon or less of spray per acre in row crops to about 5 gallons of spray per acre in orchards and vineyards.

unloader. a sensitive valvelike mechanism used on high-pressure applicators that diverts the liquid back into the tank when nozzles are shut off to prevent a rapid buildup of pressure in the system and possibly damage the pump. When the flow to the nozzles is turned back on the unloader quickly restores pressure to them.

unsulfonated residue (UR). a measure of purity of petroleum oils used as pesticides. Oils used for insecticides and acaricides must have a minimum UR rating based on the grade or type of oil. Oils with higher UR ratings are safer for use on plants.

vaporize. to transform from a spray of droplets to a foglike vapor or gas.

vector. an organism, such as an insect, that can transmit a pathogen to plants or animals.

vertebrate. the group of animals that have an internal skeleton and segmented spine, such as fish, birds, reptiles, and mammals.

viroid. a microorganism which is much smaller than a virus and not enclosed in a protein coat. Some viroids produce disease symptoms in certain plants.

virus. a very small organism that multiplies in living cells and is capable of producing disease symptoms in some plants and animals.

viscosity. a physical property of a fluid that affects its flowability; more viscous fluids flow less easily and produce larger spray droplets.

volatile. able to pass from liquid into a gaseous stage readily at low temperatures.

volute. a metal ductlike structure used to direct the air flow from a sprayer fan. Spray nozzles are often positioned near the outlet of the air flow. Volutes enable pesticide laden air to be directed to tree tops or other hard to reach areas.

warning. the signal word used on labels of pesticides in toxicity Category II, having an oral LD_{50} between 50 and 500 and a dermal LD_{50} between 200 to 2000.

watershed. an area of land that drains its surface water into a defined watercourse or body of water.

water-soluble concentrate. a liquid pesticide formulation that dissolves in water to form a true solution.

wettable powder. a type of pesticide formulation consisting of an active ingredient that will not dissolve in water combined with a mineral clay and other inert ingredients and ground into a fine powder.

wetting agent. an adjuvant used in pesticide mixtures to lower the surface tension of spray droplets, enabling them to come in close contact and spread out over target surfaces, especially those containing fine hairs or waxy layers.

Index